SPRINGER PROCEEDINGS IN PHYSICS 109

SPRINGER PROCEEDINGS IN PHYSICS

91 **The Dense Interstellar Medium in Galaxies**
Editors: S. Pfalzner, C. Kramer, C. Straubmeier, and A. Heithausen

92 **Beyond the Standard Model 2003**
Editor: H.V. Klapdor-Kleingrothaus

93 **ISSMGE**
Experimental Studies
Editor: T. Schanz

94 **ISSMGE**
Numerical and Theoretical Approaches
Editor: T. Schanz

95 **Computer Simulation Studies in Condensed-Matter Physics XVI**
Editors: D.P. Landau, S.P. Lewis, and H.-B. Schüttler

96 **Electromagnetics in a Complex World**
Editors: I.M. Pinto, V. Galdi, and L.B. Felsen

97 **Fields, Networks, Computational Methods and Systems in Modern Electrodynamics**
A Tribute to Leopold B. Felsen
Editors: P. Russer and M. Mongiardo

98 **Particle Physics and the Universe**
Proceedings of the 9th Adriatic Meeting, Sept. 2003, Dubrovnik
Editors: J. Trampetić and J. Wess

99 **Cosmic Explosions**
On the 10th Anniversary of SN1993J (IAU Colloquium 192)
Editors: J. M. Marcaide and K. W. Weiler

100 **Lasers in the Conservation of Artworks**
LACONA V Proceedings, Osnabrück, Germany, Sept. 15–18, 2003
Editors: K. Dickmann, C. Fotakis, and J.F. Asmus

101 **Progress in Turbulence**
Editors: J. Peinke, A. Kittel, S. Barth, and M. Oberlack

102 **Adaptive Optics for Industry and Medicine**
Proceedings of the 4th International Workshop
Editor: U. Wittrock

103 **Computer Simulation Studies in Condensed-Matter Physics XVII**
Editors: D.P. Landau, S.P. Lewis, and H.-B. Schüttler

104 **Complex Computing-Networks**
Brain-like and Wave-oriented Electrodynamic Algorithms
Editors: I.C. Göknar and L. Sevgi

105 **Computer Simulation Studies in Condensed-Matter Physics XVIII**
Editors: D.P. Landau, S.P. Lewis, and H.-B. Schüttler

106 **Modern Trends in Geomechanics**
Editors: W. Wu and H.S. Yu

107 **Microscopy of Semiconducting Materials**
Proceedings of the 14th Conference, April 11–14, 2005, Oxford, UK
Editors: A.G. Cullis and J.L. Hutchison

108 **Hadron Collider Physics 2005**
Proceedings of the 1st Hadron Collider Physics Symposium, Les Diablerets, Switzerland, July 4–9, 2005
Editors: M. Campanelli, A. Clark, and X. Wu

109 **Progress in Turbulence II**
Proceedings of the iTi Conference in Turbulence 2005
Editors: M. Oberlack et al.

Volumes 66–90 are listed at the end of the book.

Martin Oberlack · George Khujadze ·
Silke Günther · Tanja Weller ·
Michael Frewer · Joachim Peinke ·
Stephan Barth (Eds.)

Progress in Turbulence II

Proceedings of the iTi Conference in Turbulence 2005

With 179 Figures and 5 Tables

Springer

Martin Oberlack
oberlack@fdy.tu-darmstadt.de

George Khujadze
khujadze@fdy.tu-darmstadt.de

Silke Günther
guenther@fdy.tu-darmstadt.de

Tanja Weller
weller@fdy.tu-darmstadt.de

Michael Frewer
frewer@fdy.tu-darmstadt.de

Fachgebiet für Strömungsdynamik
Fachbereich Maschinenbau
TU Darmstadt University
Gebäude L5 / 01
Petersenstraße 13
64287 Darmstadt
Germany

Joachim Peinke
peinke@uni-oldenburg.de

Stephan Barth
barth@uni-oldenburg.de

Hydrodynamics Group
Faculty of Physics
Carl v. Ossietzky University
26111 Oldenburg
Germany

Library of Congress Control Number: 2007923194

ISSN 0930-8989
ISBN 978-3-540-32602-1 Springer Berlin Heidelberg New York

Springer is a part of Springer Science+Business Media
springer.com

Cover design: WMXDesign, Heidelberg
Typesetting: Integra Software Services Pvt. Ltd., Pondicherry, India

Printed on acid-free paper SPIN: 11589525 45/3100/Integra 5 4 3 2 1 0

Preface 2005

Overlooking the 20th century research in the engineering and physical sciences we observe that in particular one topic is rather notorious difficult and essentially unsolved both from a fundamental and from a more engineering point of few: high Reynolds number fluid flow and turbulence. Though a rather old field of research it appears that its importance may not be underestimated at all and in fact its weight is still increasing as has just been identified by the *US national committee on Theoretical and Applied Mechanics*. Areas such as environment, transportation, manufacturing, medicine, biology, energy, defense and even homeland security have been named or, of specific technical and scientific significance, areas such as engines, aerodynamics, process engineering, cosmic flows to reference only a few.

Our ability to numerically calculate high Reynolds number turbulent flows from Navier-Stokes equations at high precision, say the drag coefficient of a complex flow way below one percent say is rather limited, not to speak of our lack of knowledge to compute this analytically from first principles. This is rather remarkable since the fundamental equations of fluid flow, the Navier-Stokes equations, have been known for more than 150. Still, the ability of Navier-Stokes equation to precisely model turbulence, though at moderate Reynolds numbers and for canonical flows, has been proven in numerous high fidelity direct numerical simulations.

These difficulties go alongside with the mathematical difficulties to prove existence and uniqueness of the Navier-Stokes equations in three dimensions, which has been defined as one of the seven Millennium problems in mathematics by the Clay Institute in Cambridge, Massachusetts. Interesting enough the corresponding problem in 2D has already been solved in the 1960th. Still, even if we would be able to prove the full 3D problem it remains unclear how to construct concrete solutions for a specific technical or scientific problem.

For this reason for the time being further research will be conducted in rather diverse fields of turbulence. This spans from pure mathematical analysis over turbulence physics to applied turbulence research. In the last decades this has led to a broad diversification of turbulence research where contact between different sub-communities sometimes has been lost. This was the stimulation of the interdisciplinary turbulence initiative (ITI) which was founded 1999 as a cooperation between physicists and engineers working in turbulence and funded by the German science foundation DFG.

It established itself as a platform both for common research as well as for interdisciplinary communication across communities. This covers many different areas in turbulence research e.g. local isotropy and its controversial theories, the rather broad area of turbulence modeling, more recent approaches such as Lattice Boltzmann methods or wavelets and the wide range of applications. Since turbulence progress was in the past very often method based in the last ten years the three main techniques i.e. analytical methods, mostly symmetries, asymptotic techniques and numerical solutions moved closer to each other and fertilized each other.

Since its establishment in 1999 the ITI conference 2005 was the second after the one in 2003. Both were held in Bad Zwischenahn/Germany. The ITI conference series is planned to continue with a third conference in 2008 and from then on proceeding in a three-year rhythm.

The structure of the present book is as such that contributions have been bundled according to covering topics i.e. I Fundamentals, II Experiments, III Coherent Structures, IV Turbulent Boundary Layers, V Convection, VI Scalar Mixing and Particles, VII Polymer and Scalars and VIII Large Eddy Simulation.

At this point we would like to thank all authors for their contributions to this proceedings and the referees giving critical comments to the contributions and therewith considerably raising the scientific quality. We would like to thank Thomas Ditzinger from Springer for his patience during the production of the book and DFG, particularly Dr. Lachenmeier, for the financial support of the initiative and the aid for the conference. Finally we gratefully acknowledge Stephan Barth handling the author and referee communication and integrating the book to its final form.

M. Oberlack, G. Khujadze, S. Guenther, T. Weller, M. Frewer, J. Peinke, S. Barth (Darmstadt and Oldenburg, Fall 2006)

Contents

Part VI Scalar Mixing and Particles

Part I

Fundamentals

What Rayleigh-Bénard, Taylor-Couette and Pipe Flows have in Common

Bruno Eckhardt[1], Siegfried Grossmann[1] **, and Detlef Lohse[2]

[1] Fachbereich Physik, Philipps-Universität, Renthof 6, D-35032 Marburg, Germany, bruno.eckhardt@physik.uni-marburg.de grossmann@physik.uni-marburg.de

[2] Department of Applied Physics, University of Twente, 7500 AE Enschede, The Netherlands, d.lohse@utwente.nl

The very close correspondence between the three types of thermally or shear driven fluid flows is elucidated. Expressions for the relevant currents of temperature, angular momentum, and axial velocity by transverse convective flow in the profile direction are derived from the Navier-Stokes equations. Also the dissipation rates of the advective flows are calculated. Exact relations between the respective currents, the dissipation rates and the external control parameters are presented.

1 Introduction

Although Rayleigh-Bénard, Taylor-Couette, and pipe flow have rather different driving mechanisms – externally controlled temperature differences, concentric cylinders rotating at different speeds, or an externally controlled pressure drop or mean flow – the also different physical quantities of interest – the heat flow, the torque, or the skin friction – have a remarkable feature in common. Instead of the expected scaling behavior of these quantities of interest in terms of the driving forces, the corresponding scaling exponents turn out to be valid only locally. In fact they depend on the driving force and change for increasing forcing. This striking correspondence prompts the idea, that the corresponding mechanism might be very similar. In this paper we report on this. Section 2 deals with the main ideas to calculate the Nusselt number in Rayleigh-Bénard flow. In the next Sect. 3 we show that the corresponding ideas can be developed also for flow in the gap between independently rotating cylinders. In the last step, Sect. 4, we derive the very corresponding relations

** Evening talk, delivered by SGn, reprinted from Nonlinear Phenomena and Complex Systems, Vol 9, (2006)

also for pipe flow. We close (Sect. 5) by indicating the answer to the question of a forcing dependent scaling: It is the varying weight of the boundary layer relative to the bulk contributions that changes the relevant scaling exponent with increasing external forcing!

2 Rayleigh-Bénard Convetion

Thermal heat convection in fluid layers heated from below, known as Rayleigh-Bénard (RB) flow, has been intensively studied in the recent years, see [1, 2, 3] and others. Considerable, impressive experimental progress has been obtained. The standards of precision today are very high, with a scatter $\leq 0.1\%$ and a temperature stabilization of $\approx 0.001^o$ C or better.

The quantities of main interest are the heat flux Q and the amplitude U of the convection field. The corresponding dimensionless quantities are the Nusselt number $Nu = Q/\Lambda\Delta L^{-1} = J^\theta/\kappa\Delta L^{-1}$ and the Reynolds number $Re = U/\nu L^{-1}$. Here Λ and κ are the heat and thermal conductivity, Δ is the temperature surplus of the hotter bottom plate relative to the colder top plate of the container, L is the height of the fluid layer (or container) and ν the fluid's kinematic viscosity. The flow is assumed to be incompressible, the density of the liquid ρ_{fluid} is constant, $\kappa = \Lambda/c_p\rho_{fluid}$, with c_p the isobaric heat capacity per mass.

The external flow parameters are the Rayleigh number $Ra = g\alpha_p\Delta L^3/(\nu\kappa)$ and the material property $Pr = \nu/\kappa$ is the Prandtl number. Ra measures the buoyancy due to the gravitational acceleration g, with α_p the isobaric thermal expansion coefficient. The geometry of the container is parametrized by the aspect ratio $\Gamma = D/L$, lateral extension D in multiples of the height L. In experiment Γ is in the range of about $1/2, \ldots, 20$, in theory we consider the ideal limit $\Gamma \to \infty$.

The physical understanding of Nu, Re as functions of the varying control parameters Ra and Pr has been developed in a series of papers [4, 5, 6, 7]. It basically consists of two arguments. First, there are close and exact relations between the thermal current J^θ and the dissipation rates ε_u and ε_θ of the velocity and the temperature fields $\mathbf{u}$ and θ. Second, the dissipation rates can be modelled in terms of their dimensional amplitudes.

The exact relations, which can be derived from the Oberbeck-Boussinesq equations of motion, are

$$\varepsilon_u/(\nu^3 L^{-4}) \equiv \tilde{\varepsilon}_u = Pr^{-2} Ra(Nu - 1) \tag{1}$$

$$\text{and} \qquad \varepsilon_\theta/(\kappa\Delta^2 L^{-2}) \equiv \tilde{\varepsilon}_\theta = Nu \;. \tag{2}$$

They are valid in the case $\Gamma \to \infty$ or with laterally periodic boundary conditions.

In order to derive them, a well defined expression for the thermal current J^θ is needed,

$$J^{\theta} = -\kappa \partial_z \langle \theta \rangle_{A,t} + \langle u_z \theta \rangle_{A,t} = Nu \cdot \kappa \Delta L^{-1} \ . \tag{3}$$

This definition is a unique consequence of the θ-equation, if averaged over planes A parallel to the bottom and top plates at any height z, with $0 < z < L$ from bottom to top. Also the time average is taken (or stationarity assumed). The defining property to identify J^{θ} is that it is **independent** of z. This expresses the conservation of heat, the current must be the same at each height z.

While J^{θ} consists of area averages, the dissipation rates are volume averages, $\varepsilon_u = \nu \langle u^2_{i|j} \rangle_{V,t}$ and $\varepsilon_{\theta} = \kappa \langle \mathrm{grad}^2 \theta \rangle_{V,t}$. The average convective dissipation rate $\tilde{\varepsilon}_u$ is decomposed now into its isotropic bulk contribution and the anisotropic boundary layer part. The thermal dissipation, being equal to the **area** averaged current, is decomposed according to the dimensional amplitudes of the two J^{θ}-terms.

$$\tilde{\varepsilon}_u = \ \sim U^2/(LU^{-1}) \ + \ \sim \nu (U^2/\lambda_u^2) \cdot (\lambda_u/L) \ , \tag{4}$$

$$Nu = \ \sim L/\lambda_{\theta} \ + \ \sim U\Delta \ . \tag{5}$$

The Ansatz for the width λ_u of the kinematic boundary layer is the Prandtl scaling

$$\lambda_u/L = a/\sqrt{Re} \ , \tag{6}$$

the one for the thermal boundary layer width λ_{θ} is either obtained from (3) taken at the plates or from the Pohlhausen theory

$$\lambda_{\theta}/L = 1/(2Nu) \approx \ \sim \sqrt{Re\ Pr} \ . \tag{7}$$

Depending on the size of Pr, the velocity U which is relevant in the expression for Nu either is the full amplitude U itself, namely if $\lambda_u \ll \lambda_{\theta}$, or it is only the fraction $U\lambda_{\theta}/\lambda_u$ at the edge of the thermal boundary layer, if $\lambda_u \gg \lambda_{\theta}$. These two limiting cases, known as "lower" and "upper" ranges, are connected by a switch function $f(\lambda_{\theta}/\lambda_u) = \left(\frac{(\lambda_{\theta}/\lambda_u)^n}{1+(\lambda_{\theta}/\lambda_u)^n} \right)^{1/n}$; that $n = 4$ is chosen, is not important.

These ideas lead to the following set of two equations,

$$Pr^{-2} Ra (Nu - 1) = c_1 Re^{5/2} + c_2 Re^3 \ , \tag{8}$$

$$Nu = c_3 \sqrt{Re Pr f\left(\frac{\sqrt{Re}}{Nu} \right)} + c_4 Re Pr f\left(\frac{\sqrt{Re}}{Nu} \right) \ . \tag{9}$$

Its solution leads to the quantities of interest, $Nu(Ra, Pr)$ and $Re(Ra, Pr)$ (after fitting the constants c_i to one available data set), and allows to calculate the heat current as well as the convection amplitude for any value of the external control parameters Ra and Pr. The agreement with an increasing amount of data (experimental as well as numerical) is up to now very pleasing; details are given in [4, 5, 6, 7].

The most remarkable observation is that there are **no global** scaling exponents

$$Nu \sim Ra^{\beta} Pr^{\beta'} , \tag{10}$$

but that these β's are local quantities only, changing as functions of Ra and Pr in a ball park (for β) between about $\frac{1}{5}$, $\frac{1}{4}$ and $\frac{1}{3}$ or even $\frac{1}{2}$. More precisely, Nu and Re are varying superpositions of boundary layer and bulk contributions to the dissipation rate and the current.

3 Taylor-Couette Flow

We now show that there is a very close correspondence between thermal RB flow and Taylor-Couette (TC) flow in the gap between concentrically rotating cylinders and (in the next section) also with pipe flow. The details will be published elsewhere, cf. [8, 9, 10]. Here only the main ideas of this interesting correspondence are discussed. We even restrict ourselves to the exact relations; the modelling can be (and has been) also worked out, of course, see the detailed publications.

In TC flow two cylinders with radii r_1, r_2 rotate independently with angular velocities ω_1, ω_2. The gap width between the cylinders is $d = r_2 - r_1$, the aspect ratio Γ is ℓ_{cyl}/d and in practice of the order of 10. The nondimensional control parameters are $Re_1 = r_1\omega_1 d/\nu$ and Re_2 correspondingly. Another important geometrical parameter is the radius ratio $\eta = r_1/r_2$, varying between 0 (large gap) and 1 (small gap). With decreasing η the effects of curvature become more relevant, while for $\eta \to 1$ the TC system approaches the flat narrow channel. Another interesting degree of freedom is counter rotation, $\omega_2 < 0$ and $\omega_1 > 0$. Then a neutral surface develops at some intermediate radius r_n, defined by $u_\varphi(r_n) = 0$. The results presented in the following hold in all these various cases. They are exact consequences of the Navier-Stokes equations for incompressible fluid flow.

$$\partial_t \mathbf{u} = -(\mathbf{u}\cdot\boldsymbol{\nabla})\mathbf{u} - \boldsymbol{\nabla} p + \nu \Delta \mathbf{u} , \qquad \mathrm{div}\mathbf{u} = 0 . \tag{11}$$

These have to be used, of course, in cylindrical coordinates, as found e.g. in [11].

Consider at first laminar TC flow, which apparently corresponds to the RB system at rest. As there is a profile in the RB case, namely the temperature profile, also in laminar TC flow one has a profile, this time of angular velocity, $\omega_1 < \omega(r) < \omega_2$. It is $\omega(r) = A + B/r^2$. The corresponding velocity is purely azimuthal, $u_\varphi = r\omega(r)$ and the angular momentum field is $L = r u_\varphi = r^2\omega(r)$. Due to the r-gradient of the profile there is a transverse molecular ω-current $\propto -\nu\partial_r\omega$, similar to $\propto -\kappa\partial_z\theta$ in RB. Note that it is not the state of resting cylinder which corresponds to the RB state with $\mathbf{u} = 0$; this latter still has a θ-profile. The resting TC system has no profile anymore.

If the rotation rates are sufficiently large, convection transverse to the cylinder surfaces develops, $\mathbf{u}_\perp$, having velocity components u_r and u_z. These of course couple to u_φ via (11). This transverse convection enhances the radial ω-transport.

The ω-transport leads to forces $F_{1,2}$ and thus to torques $T_1 \equiv T$ and $T_2 = -T$ on the cylinders, which can be measured [12, 13, 14, 15, 16] and more. One expects $T \propto Re_1^\alpha$, similar as in RB we wrote $J^\theta \propto Ra^\beta$. It again turns out that also this exponent α depends on Re_1, i.e. is only local but not universal.

How can one define the ω-current? In perfect analogy to the argument in RB convection we perform area and time averages the relevant equation of motion for TC flow. In RB this is the θ-equation because the driving mechanism is the θ-profile. Since in TC the driving profile is in ω or $u_\varphi = r\omega$, we start from the u_φ-equation, i.e. the φ-component of the Navier-Stokes (11). The relevant area here are cylinders $A(r) = 2\pi r \cdot \ell_{cyl}$ concentric to the TC cylinders, for any $r_1 \leq r \leq r_2$. After some algebra one finds that the following quantity has to be the same for all r,

$$r^3 \left(-\nu \partial_r \langle \omega \rangle_{A(r),t} + \langle u_r \omega \rangle_{A(r),t} \right) \equiv J^\omega \ . \tag{12}$$

Because of its r-independence we consider this J^ω as the angular momentum current and as the very analog of J^θ. Of course, an proportionality factor is still free. To prove the physical meaning of just this expression J^ω, we consider its relation to the torque. It is $T = r_1 F_1 = r_1 \cdot \sigma_{r\varphi}(r_1) \cdot 2\pi r_1 \ell_{cyl}$, with the stress tensor in cylindrical coordinates $\sigma_{r\varphi}(r_1) = -\rho_{fluid}\nu(\partial_r u_\varphi - \frac{u_\varphi}{r})_{r_1}$. Thus $T = 2\pi \ell_{cyl} \rho_{fluid} J^\omega$.

J^ω, as J^θ in RB, consists of a molecular contribution to the transport, $\propto -\nu \partial_r \langle \omega \rangle_{A(r),t}$ and a convective part $\propto \langle u_r \omega \rangle_{A(r),t}$, proportional to the transverse amplitude u_r. The latter term is missing in the laminar case. Thus the convective enhancement of the angular momentum transport is

$$N^\omega = \frac{J^\omega}{J^\omega_{lam}} = \frac{J^\omega}{2\nu B} \ , \tag{13}$$

which we call the ω-Nusselt number. This leads to

$$T \propto J^\omega_{lam} \cdot N^\omega \propto Re_1 \cdot N^\omega \ . \tag{14}$$

There is a significant influence of the curvature on J^ω, namely the explicit factor of r^3 in (12). It is only *with* this r^3 that J^ω is r-constant. It means that the molecular gradient contribution as well as the convective correlation function are much weaker near the outer than near the inner cylinder. In particular, the differences between the ω-slopes near the inner and outer cylinder increase dramatically with η,

$$\partial_{r_2} \langle \omega \rangle_2 = \eta^3 \partial_{r_1} \langle \omega \rangle_1 \ . \tag{15}$$

For decreasing radius ratio η or increasing gap width $1-\eta$ the boundary layer at the inner cylinder is (much) thinner than at the outer one.

Having identified the corrrespondence of the relevant currents, we now consider the energy dissipation rate in TC flow. It reads $\varepsilon = \frac{\nu}{2}\langle(\partial_i u_j + \partial_j u_i)^2\rangle_V$ and is calculated from the Navier-Stokes (11) by multiplying with $\mathbf{u}$ and volume averaging. After some algebra (see [9]) one finds

$$\varepsilon = \frac{2}{r_2^2 - r_1^2}(\omega_1 - \omega_2) J^\omega \ . \tag{16}$$

This total dissipation rate clearly does **not** correspond to ε_u in RB flow, cf. (1); namely, we have $\varepsilon_u = 0$ in the case of no convective transport, but ε in (16), instead, reduces to $\varepsilon_{lam} \propto J^\omega_{lam} \neq 0$ in case of missing convective transport. Therefore, the convective dissipation rate in TC flow corresponding to ε_u in RB is

$$\tilde{\varepsilon}_w = \frac{\varepsilon - \varepsilon_{lam}}{\nu^3 d^{-4}} = \frac{\varepsilon_{lam}}{\nu^3 d^{-4}}\left(N^\omega - 1\right) \ . \tag{17}$$

One can introduce a purely geometrical analog of the material property $Pr = \nu/\kappa$ by defining a quasi-Prandtl number

$$\sigma(\eta) = \left(\frac{r_a}{r_g}\right)^4 = \left(\frac{\frac{1}{2}(1+\eta)}{\sqrt{\eta}}\right)^4 , \tag{18}$$

$r_{a,g}$ denoting the arithmetic and geometric mean radii of $r_{1,2}$. A quasi-thermal conductivity then is $\kappa = \nu/\sigma$. One can also introduce a Taylor number

$$Ta = \sigma \frac{d^2 r_a^2 (\omega_1 - \omega_2)^2}{\nu^2} = \frac{d^2 r_a^2 (\omega_1 - \omega_2)^2}{\nu\kappa} \ . \tag{19}$$

In terms of these control parameters the convective dissipation rate (17) becomes

$$\tilde{\varepsilon}_w = \sigma^{-2} Ta (N^\omega - 1) \ , \tag{20}$$

the perfect analogon of $\tilde{\varepsilon}_u$ from (1). These two exact relations, (20) for the convective dissipation rate, and (12),(13) for the ω-current between the cylinders, establish the complete correspondence between RB and TC flows. It is, incidentally, precisely with the prefactors included in the definition (19) of the Taylor number, that this analogy is perfect. This suggests which of the various different, though dimensionally all correct Taylor numbers should be preferred, namely the definition (19). The current or ω-Nusselt number (13) can be written as

$$N^\omega = \left(\frac{r}{r_a}\right)^3 \frac{\langle u_r \omega \rangle_{A(r),t} - \nu \partial_r \langle \omega \rangle_{A(r),t}}{\kappa(\omega_1 - \omega_2)/d} \ . \tag{21}$$

The correspondence to (3) is striking. In a simplified version we had shown in [17] that the torque data can be rather well described along this RB-TC-correspondence. For more details in the present frame of complete correspondence, see [9].

4 Pipe Flow

We finally analyse the correspondence not only of RB with TC but now also with pipe flow, being rather brief only and giving reference to [8, 10] for more details.

The counter part to the RB state of rest but with a temperature θ profile, corresponding to laminar TC flow with an ω proffile, in the case of pipe flow is the laminar Hagen-Poiseuille flow. It also has a profile, namely of the axial component $u_z(r)$, leading to a molecular current $\propto -\nu\partial_r u_z(r)$, which in turn gives rise to a skin friction at the pipe's wall that has to be overcome by the applied external pressure drop $\Delta p/\ell_{pipe}$. To define the proper current in pipe flow we perform area and time averages of the u_z-component of the Navier-Stokes (11). Here the appropriate surfaces are virtual circular surfaces concentric to the pipe, of any radius $r \leq a$, the pipe's radius: $A^{(=)} = 2\pi r \cdot \ell_{pipe} = A(r)^{(=)}$. After some arguments one arrives at the following definition for the u_z-current, which is independent of r, constant for all radii smaller than a,

$$J^{u_z} = \frac{2}{r}\left(\langle u_r u_z\rangle_{A^{(=)},t} - \nu\partial_r\langle u_z\rangle_{A^{(=)},t}\right) \ . \tag{22}$$

It is only including the factor r^{-1} that this expression is constant. The u_z-Nusselt number is $N^u = J^{u_z}/J^{u_z}_{lam}$, where $J^{u_z}_{lam} = 8\nu a^{-2}U$. Here U is the cross-sectional mean flow velocity and $a = d/2$ the radius and diameter of the pipe.

Convective enhancement of the u_z-current, which flows from the interior to the pipe wall is provided by the radial component u_r (and, of course, u_φ), which develop in the turbulent state beyond laminarity. This gives rise to a convective dissipation rate. The full dissipation rate in the pipe is $\varepsilon = UJ^{u_z}$. The contribution due to the transverse u_r-convection is obtained by subtracting the laminar contribution. This leads to the following definition of the convective dissipation rate

$$\tilde{\varepsilon}_w = \frac{\varepsilon - \varepsilon_{lam}}{\nu^3 d^{-4}} \propto Re^2(N^u - 1) \ . \tag{23}$$

This corresponds perfectly to (1) in RB and (17),(20) in TC. We would like to draw the readers' attention to the fact, that there is one possibly severe difference between pipe and the other two flows. While in RB and TC the driving profiles are externally controlled as the bottom and top plates temperatures or as the cylinders rotation frequencies, in pipe flow the profile underlies dynamical fluctuations. The center velocity which determines the u_z-profile is a **dynamical** response of the flow already, either to the externally applied pressure drop $\Delta p/\ell_{pipe}$ or to the controlled *mean* flow velocity U.

5 Summarizing Conclusions

Having calculated Nu, N^ω, or N^u, all along the same lines, one finds the local exponents of the quantities of physical interest. In RB convection this is the Nusselt number exponent β itself. In TC flow between independently rotating cylinders it is the torque exponent α from $T \propto Re_1^\alpha$; one then has with (14) $T \propto Re_1 N^\omega \propto Re_1^{1+\beta}$, i.e. $\alpha(Re_1) = 1 + \beta(Re_1)$. In pipe flow the skin friction coefficient is of interest, $c_f = (16/Re)N^u \propto Re^{-1+\beta}$, so that the relation $\alpha(Re) = -1 + \beta(Re)$ holds. This establishes the close correspondence between RB, TC, and pipe flow. In all three cases exact relations connecting the currents and the convective dissipation rates hold.They are consequences of the Navier-Stokes equations and are valid in any situation. The further steps, namely the modelling of both the currents N and the dissipation rates $\tilde{\varepsilon}_w$ in terms of the dimensional amplitudes, need more details about the respective flows and will be discussed in [8, 9, 10].

References

1. Ahlers G, Grossmann S, Lohse D (2002) Physik Journal 2:31-37.
2. Funfschilling D, Brown E, Nikolaenko A, Ahlers G (2005) J Fluid Mech 536: 145-154.
3. Sun C, Ren L-Y, Song H, Xia K-Q (2005) J Fluid Mech 542:165-174.
4. Grossmann S, Lohse D (2000) J Fluid Mech 407:27-56.
5. Grossmann S, Lohse D (2001) Phys Rev Lett 86:3316-3319.
6. Grossmann S, Lohse D (2002) Phys Rev E 66:016305.
7. Grossmann S, Lohse D (2004) Phys of Fluids 16:4462-4472.
8. Eckhardt B, Grossmann S, Lohse D (2005) Analogy of turbulent transport in thermal convection and in shear flows, to be submitted.
9. Eckhardt B, Grossmann S, Lohse D (2005) Torque scaling in turbulent Taylor-Couette flow between independently rotating cylinders, to be submitted.
10. Eckhardt B, Grossmann S, Lohse D (2005) Scaling of skin friction in turbulent pipe flow, to be submitted.
11. Landau L D, Lifshitz E M (1987) Fluid Mechanics. Pergamon Press, Oxford.
12. Wendt F (1933) Ingenieurs-Archiv 4:577-595.
13. Taylor G I (1936) Proc R Soc London Ser A157:546-564 and 565-578.
14. Lathrop D P, Fineberg J and Swinney H S (1992) Phys Rev Lett 68:1515-1518.
15. Lathrop D P, Fineberg J and Swinney H S (1992) Phys Rev A 46:6390-6405.
16. Lewis G S and Swinney H S (1999) Phys Rev E 59:5457-5467.
17. Eckhardt B, Grossmann S and Lohse D (2000) Eur Phys J B 18:541-545.

Small-Scale Statistics in High-Resolution Direct Numerical Simulation of Turbulence

Yukio Kaneda[1] and Takashi Ishihara[2]

Department of Computational Science and Engineering, Nagoya University, Nagoya, 464-8603, Japan
[1]kaneda@cse.nagoya-u.ac.jp
[2]ishihara@cse.nagoya-u.ac.jp

We performed high-resolution direct numerical simulations (DNSs) of incompressible turbulence in a periodic box with the number of grid points up to 4096^3. The DNS data shed some light on the nature, especially on the possible universality, of small-scale statistics of turbulence at high Reynolds number Re. We report some results of the analysis of the DNS data with an emphasis on the possible dependence of the statistics on Re.

1 Introduction

The number of dynamical degrees of freedom (DOF) of turbulence increases rapidly with the Reynolds number Re. It is therefore necessary to treat a huge number of DOF for direct numerical simulation (DNS) of turbulence at high Re. Although the number of DOF that can be treated in DNS is obviously limited by available computer memory and power, it is increasing rapidly with the recent development of computational hardware and software.

Recently, we performed high-resolution DNSs of incompressible homogeneous turbulence obeying the Navier-Stokes equations in a periodic box with the number N^3 of grid points up to 4096^3 on the Earth Simulator. The number of DOF in the DNSs with $N^3 = 4096^3$ is about $4096^3 \times 4 > 2 \times 10^{11}$, where $4 = 3 + 1$ is the number of DOF (3 velocity components and pressure) on each grid point. The DNSs are based on a spectral method free from alias error, and sustained performance up to 16.4 TFlops was achieved. [1]

An essence of fully developed turbulence at high Reynolds number lies in the existence of a wide separation between the scale where most of the turbulence energy resides and the scale where it must be dissipated by viscosity. Such a separation in DNSs with $N^3 = 4096^3$ or 2048^3 is in fact much wider than that in DNSs with lower resolution. It is expected that the data of such high-resolution DNSs may shed some light on the nature, especially on the possible universality, of fully developed turbulence at high Re.

2 Small Scale Statistics of Turbulence

It is a basic premise of many turbulence studies that in fully developed turbulence at sufficiently high Re and small scales, there is a certain kind of universality that is insensitive to the large-scale flow conditions. This idea is in accordance with Kolmogorov's hypotheses. [2] However Re as well as scale range in any experiment, measurement, or DNS of turbulence cannot be infinite but only finite. We don't know a priori how large the Re should be for the appearance of the possible universality. It is therefore of interest to make use of available high Re turbulence data and to make quantitative observation on the Re-dependence of the turbulence statistics.

Our high-resolution DNS is expected to provide us with valuable data in this respect. The DNSs consist of two series: one is with $k_{\max}\eta \sim 1$ (Series 1) and the other is with $k_{\max}\eta \sim 2$ (Series 2). Here $k_{\max}$ is the highest wavenumber in each simulation, and η is the Kolmogorov length scale. The total kinetic energy E is kept almost time-independent by introducing external forcing in the form of negative viscosity at a low wavenumber range.

The Taylor scale Reynolds number R_λ and the ratio of L/η of the integral length scale L to η are about 1200 and 2000 respectively, in the DNS with $N^3 = 4096^3$ in Series 1. The DNS data support the well known scaling,

$$L/\eta \propto Re^{3/4}, \quad \lambda/\eta \propto Re^{1/4},$$

and $R_\lambda \sim C Re^{1/2}$. [3] Here λ is the Taylor micro scale, $C = (15/D)^{1/2} \sim 6.0$, $D = \langle \epsilon \rangle L/U^3$ the normalized dissipation rate, $\langle \epsilon \rangle$ the mean rate of energy dissipation per unit mass, $(3/2)U^2 = E$, and $D \sim 0.4$ (see below).

The DNS data suggest that a class of statistics approach to constants as $Re \to \infty$. Among them is the normalized energy dissipation rate $D = \langle \epsilon \rangle L/U^3$. [4] The DNS data strongly suggest that D tends to a nonzero constant (~ 0.4) as $R_\lambda \to \infty$. They also suggest that the normalized energy flux $\Pi(k)/\langle \epsilon \rangle$ across wavenumber k in the inertial subrange also tends to a constant ($= 1$), as $Re \to \infty$. The assumption of the existence of a wide wavenumber range where $\Pi(k)$ is constant independent of k with $\Pi(k) \sim \langle \epsilon \rangle$ plays key roles in turbulence theories. However, in DNS with low resolution such as $N^3 = 512^3$ or so, the range is not so wide. [5] The range in DNSs with $N^3 = 2048^3$ or 4096^3 is much wider than it is in such DNSs.

As regards the energy spectrum $E(k)$, the DNS data fits quite well with the Kolmogorov spectrum $E(k) \propto k^{-5/3}$ in the inertial subrange. But a close inspection shows that the energy spectrum in the DNSs with $N^3 = 2048^3$ or 4096^3 fits with $E(k) \propto k^{-5/3-\mu}$ where $\mu \sim 0.1$. [4] Here we plot one dimensional energy spectrum $E_{11}(k_1)$ computed from our DNS data of$E(k)$ as

$$E_{11}(k_1) = \frac{1}{2}\int_{k_1}^{\infty}\left(1 - \frac{k_1^2}{k^2}\right)\frac{E(k)}{k}dk.$$

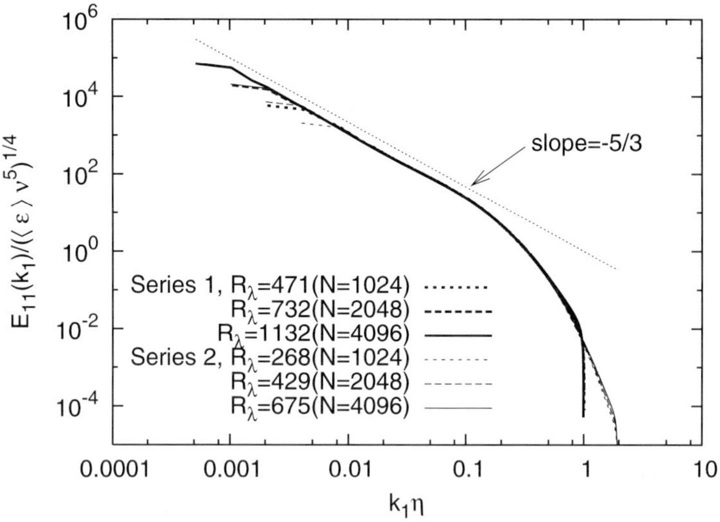

Fig. 1. Normalized one-dimensional energy spectrum $E_{11}(k_1)/(\langle\epsilon\rangle\,\nu^5)^{1/4}$ for Series 1 and 2 with N =1024, 2048, and 4096. The straight line shows the slope $\propto k^{-5/3}$

The one dimensional spectrum is easier to measure than $E(k)$ in an experiment. It is seen in Fig.1 that the width of the inertial subrange increases with R_λ.

The DNS energy spectrum in the near dissipation range ($k \sim 1/\eta$) fits well to the form

$$E(k) = C(k\eta)^\alpha \exp(-\beta k\eta),$$

where C, α and β are constants independent of k. The DNS data is consistent with the conjecture that α, β and C converge to constants, say $\alpha_\infty, \beta_\infty$ and C_∞, independent of R_λ, as $R_\lambda \to \infty$. But the convergence is very slow. A data fitting suggests that even at R_λ as large as $R_\lambda = 10,000$, the differences $\alpha - \alpha_\infty$, $\beta - \beta_\infty$, $C - C_\infty$ are still as large as approximately 3 %, 261% and 105% of $|\alpha_\infty|$, β_∞ and C_∞, respectively. [6]

As is well known, high Re turbulence exhibits strong intermittency at small scales. This can be manifested by the probability distribution functions (pdf's) of velocity gradients. One may then ask "Do the pdf's tend to universal ones independent of Re, as $Re \to \infty$?" According to our DNS, the answer is negative; the pdf's keep Re-dependence at least in the Re-range studied by our DNS. The moments, including skewness and flatness factors, of velocity gradients give simple measures characterizing the possible Re-dependence of the pdf's. The DNS data analysis so far made suggests that they have certain kind of power low dependences on Re or R_λ, but the dependences at high R_λ, say at $R_\lambda > 400$ or so are different from those at lower R_λ. The details of the analysis will be presented elsewhere.

3 Inertial Subrange Statistics

The visualization of the flow field obtained by the DNS suggests that the intense vorticity field consists of clusters of small eddies, and that the structure of the clusters at the scale of inertial subrange is distinctively different from the structure of individual eddies at smaller scale. These observations motivate us to study the inertial subrange statistics, the understanding of which may contribute to the development of proper models of turbulence such as used in large eddy simulations.

The statistics of the energy transfer T from large scale to small scale is one of the most basic statistics characterizing the inertial subrange dynamics. We analyzed the statistics of energy transfer T defined by

$$T = -\tau_{ij}\overline{S}_{ij},$$

where $\tau_{ij} = (\overline{u_i u_j} - \overline{u_i}\,\overline{u_j}) - \frac{2}{3}\delta_{ij}q$, $q = (1/2)\sum_k (\overline{u_k u_k} - \overline{u_k}\,\overline{u_k})$, $\overline{S}_{ij} = (1/2)(\partial\overline{u}_i/\partial x_j + \partial\overline{u}_j/\partial x_i)$, u_i is the i-th Cartesian velocity component, $\overline{f}$ the grid scale component of f, and we use the summation convention for repeated indices. We use here the so-called spectral cut-off filter to define $\overline{f}$, by which all the Fourier modes of f with wavenumber k larger than the cut-off wavenumber, say k_c, are removed. The reader may refer to Aoyama et al. [7] for the details of the analysis of the statistics including the variance and the skewness and flatness factors of T as well as those of the average ϵ_r of the local energy dissipation rate ϵ over a sphere of radius r.

We consider here the 2nd, 3rd, and 4th order moments of T and ϵ_r. Figure 2 shows the DNS data of these moments. It suggests that they have simple power law dependences on the wavenumber $k = k_c$, or $k = \pi/(2r)$ in the inertial subrange, at large R_λ. Various theories have been proposed for the statistics of ϵ_r. Among them, the Log-normal theory is one of the most representative ones, and gives $\langle(\epsilon_r)^n\rangle \propto (r/L)^{\mu n(n-1)/2}$. [8] Together with the DNS data, Fig. 2 also shows the slopes by this theory with $\mu = 0.25$ suggested by experiments (see [9]). Not only the DNS slopes of $\langle(\epsilon_r)^n\rangle$ but also those of $\langle T^n\rangle$ are seen in good agreement with the slopes predicted by the theory for $\langle(\epsilon_r)^n\rangle$.

4 Linear Response Theory

In the statistical mechanics of thermal equilibrium system, it is known that there are two kinds of macroscopic relations that are insensitive to the detail of the differences of microscopic state: (i) relations characterizing the equilibrium itself such as the Boyle-Charles law, and (ii) relations characterizing the response to disturbances added to the equilibrium such as the relation $J = CX$, between the generalized force X and flux J. Here (X, J) may be (scalar gradients, density flux), (temperature gradient, heat flux), (electric field, electric current), etc. The coefficient C reflects the macroscopic sate of the equilibrium system. The analogy between statistical equilibrium system

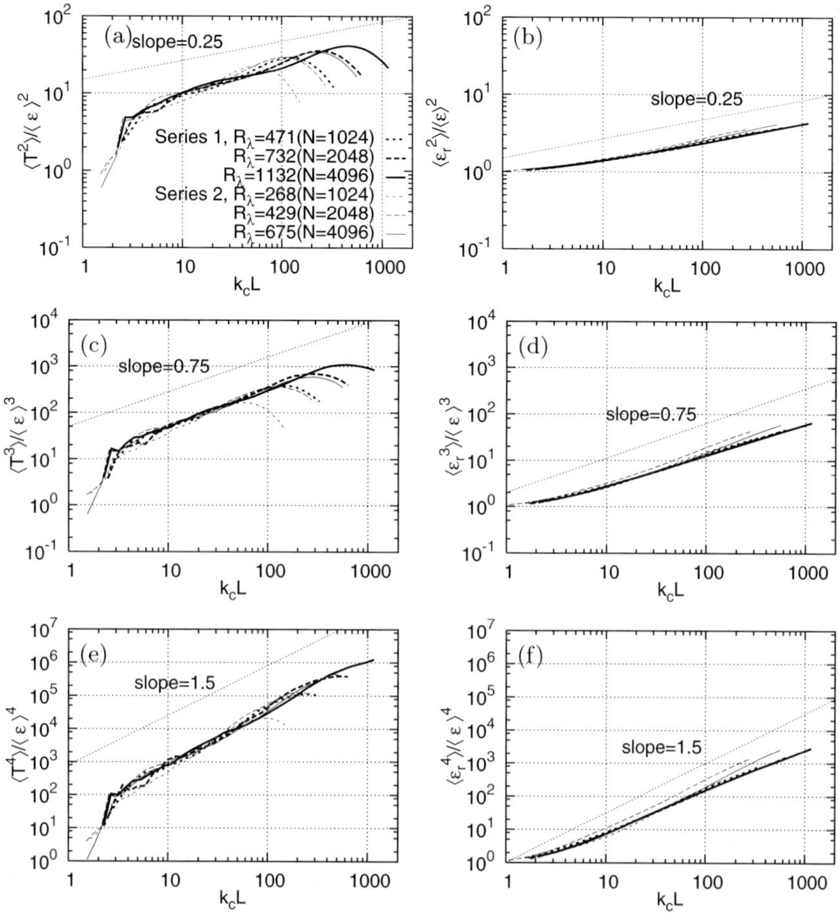

Fig. 2. Second, third and fourth order moments of T ((a), (c), (e)) and ϵ_r ((b), (d), (f)) for Series 1 and Series 2 with N =1024, 2048, and 4096. The straigh lines in (b), (d) and (f) correspond to the slope $< (\epsilon_r)^n > \propto (r/L)^{\mu n(n-1)/2}$ by the Log-normal theory

and turbulence suggests that there may be two kinds of relations in turbulence as well: one includes those characterizing the equilibrium state itself, and the other includes those characterizing the response to disturbance added to the equilibrium state.

Ishihara et al. [10] applied this idea to turbulent shear flow, and derived a simple expression for the velocity correlation spectrum $Q_{ij}(\mathbf{k})$ such that

$$\Delta Q_{ij}(\mathbf{k}) = C_{ijkm}(\mathbf{k}) S_{km},$$

where S_{ij} is the local rate of strain tensor of the mean flow, $\Delta Q_{ij}(\mathbf{k}) = Q_{ij}(\mathbf{k}) - Q^0_{ij}(\mathbf{k})$, $Q^0_{ij}(\mathbf{k})$ the isotropic spectrum in the absence of the mean flow, and $C_{ijkm}(\mathbf{k})$ a fourth-order isotropic tensor depending on $\mathbf{k}$ and determined by the equilibrium state. They showed that high-resolution DNSs of turbulence under a mean flow of a simple shearing motion are in good agreement with the theoretical conjecture. At sufficiently small scales, the turbulence is shown to be dominated by the inherent Navier-Stokes dynamics without mean shear, and the effect of mean shear may be regarded as a small perturbation added to the equilibrium homogeneous and isotropic state.

Kaneda and Yoshida[11] showed that the similar is also the case in strongly stably stratified incompressible turbulence obeying the Boussinesq equation, and derived simple expressions for the velocity correlation and density correlation spectra. They confirmed the expressions are in good agreement with DNS.

Recently the idea was extended also to magneto-hydrodynamic turbulence under strong uniform magnetic field obeying the so-called adiabatic equations. (Ishida and Kaneda, in this volume). The detail of the theory and comparisons with DNS will be presented elsewhere.

The simulations were performed on the Earth Simulator, and on the Fujitsu HPC2500 at the Information Technology Center of Nagoya University. The work was partially supported by Grant-in-Aids for Scientific Research (B)17340117 and (C)17560051 from the Japan Society for the Promotion of Science, and also by a Grant-in-Aid for the 21st Century COE "Frontiers of Computational Science".

References

1. Yokokawa M., Itakura K., Uno A., Ishihara T., & Kaneda Y. (2002) *Proc. IEEE/ACM SC2002 Conf., Baltimore, 2002*; http://www.sc-2002.org/paperpdfs/pap.pap273.pdf.
2. Kolmogorov A. N. (1941) Dokl. Akad. Nauk SSSR **30**, 301–305.
3. Kaneda Y. & Ishihara T. (2005) Journal of Turbulence, to appear.
4. Kaneda Y., Ishihara T., Yokokawa M., Itakura K., & Uno A. (2003) Phys. Fluids **15**, L21–24.
5. Ishihara T. & Kaneda Y. (2002) In Kaneda Y. & Gotoh T. (eds), Statistical Theories and Computational Approaches to Turbulence; Modern Perspectives and Applications to Global-Scale Flows, Springer Verlag, Tokyo, 177–188.
6. Ishihara T., Kaneda Y., Yokokawa M., Itakura K., & Uno A. (2005) J. Phys. Soc. Jpn., **74**, 1467–1471.
7. Aoyama T., Ishihara T., Kaneda Y., Yokokawa M., Itakura K., & Uno A. (2005) J. Phys. Soc. Jpn. **74**, to appear.
8. Kolmogorov A. N. (1962) J. Fluid Mech. **13**, 82–85.
9. Sreenivasan K. R. & Antonia R. A. (1997) Annu. Rev. Fluid Mech. **29**, 435–472.
10. Ishihara T., Yoshida K., & Kaneda Y. (2002) Physical Review Letters, **88**, No. 15, Art. No. 154501-1–4.
11. Kaneda Y. & Yoshida K. (2004) New Journal of Physics, **6**, Art. No. 34.

Numerical Investigation of Turbulent Jet Under Random Waves

Y. P. Chen[1] and C. W. Li[2]

[1] Department of Civil and Structural Engineering, The Hong Kong Polytechnic University, 02902193r@polyu.edu.hk
[2] Department of Civil and Structural Engineering, The Hong Kong Polytechnic University, cecwli@polyu.edu.hk

1 Introduction

When wastewater is discharged into a large water body, e.g. lake, estuary or ocean, it will form a turbulent jet which is often under the influence of tides and waves. Although the jet motion in stagnant ambient as well as in steady or quasi-steady flows has been extensively studied in previous works, the knowledge of jet motion in un-steady flow such as waves is rather limited. In real situation, waves are random and have various frequencies. In order to obtain a more accurate assessment of the environmental impact due to wastewater discharge, it is essential to investigate the effect of random waves on jet behaviours.

2 Numerical Schemes

In this study, a 3D vertical jet is numerically investigated by using the large eddy simulation (LES) method. Turbulence is simulated by a dynamic sub-grid scale model. The time-averaged flow characteristics and the turbulent intensity of jet motion under wavy environment are obtained by placing the jet in a fully nonlinear numerical wave tank (NWT, Fig.1).

The NWT is developed based on the solution of the σ-transformed and spatially filtered Navier-Stokes equations (LES formulation). The numerical wave is generated from the inflow boundary, where the surface elevation and/or velocity are specified by use of analytical solution or laboratory data. The Sommerfeld/Orlanski radiation condition, combined with an artificial damping zone, is applied at the outflow boundary to reduce wave reflection. A novel Lagrange-Euler Method is used to locate the free surface elevation. The NWT has been successfully verified through simulation of various kinds of waves, including random waves [1].

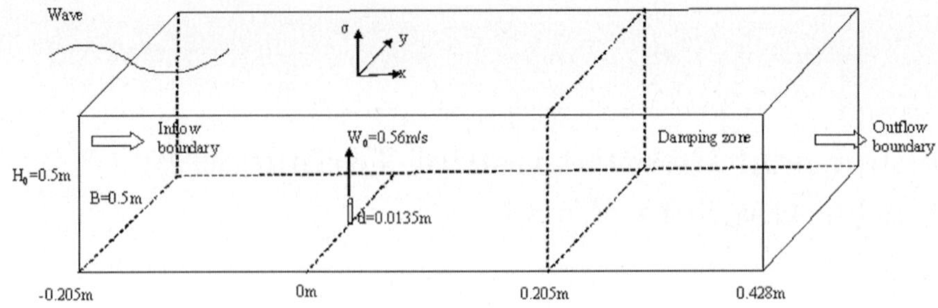

Fig. 1. Schematic diagram of numerical wave tank

3 Results & Discussions

An initial computation of a vertical jet in stagnant water is carried out in the NWT. The jet parameters used are the same as those set in the laboratory experiments (see Fig. 1). The computed flow field exhibits large vortex motions due to jet-induced shear flow instability (Fig. 2a). The time-averaged velocity and the turbulent intensity follow a self-similar relationship in the zone of established flow, and two symmetrical horizontal surface jets were formed after jet impingement on the free surface (Fig. 2b).

The random waves are also well reproduced by using the current grid system. The time history and the spectral energy density of surface elevation close to the jet exit (x=0m) agree very well with the experimental data (Figs. 3 and 4), which guarantees the wave condition can be accurately specified in the subsequent computation of jet under waves.

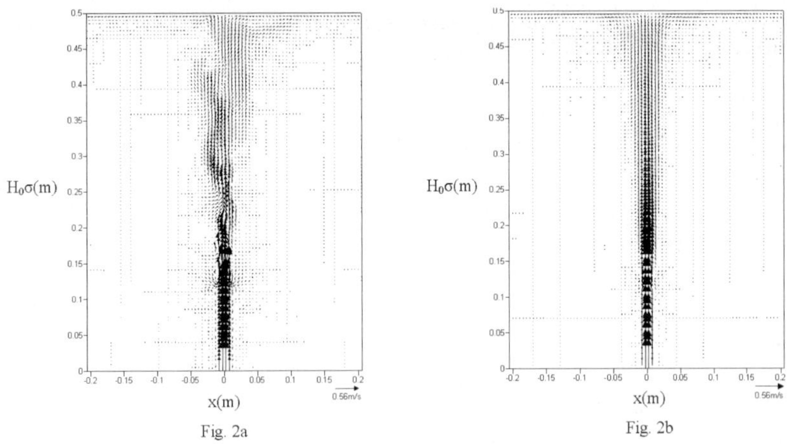

Fig. 2. Instantaneous flow field (t=10s, Fig. 2a) and mean flow field (t=5∼20s, Fig. 2b) of jet in stagnant water

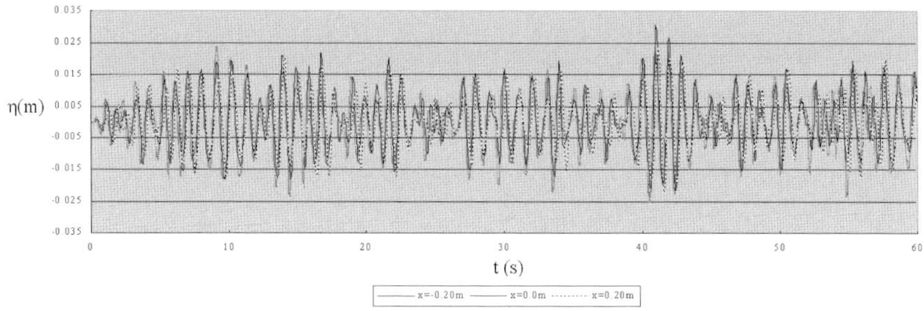

Fig. 3. Time history of random wave at x=-0.20m, 0m and 0.20m

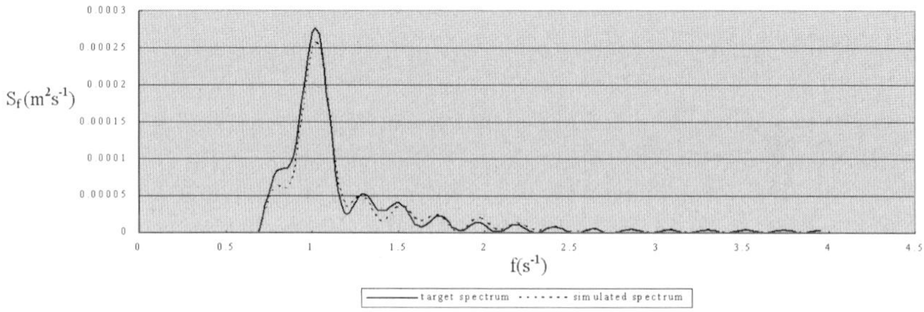

Fig. 4. Comparison of target spectrum and simulated spectrum at x=0m

Under the random wave environment, the entrainment and mixing structure of the jet are quite different from those in stagnant ambient. The so-called 'wave tractive mechanism'[2] occurs and is reproduced in the present numerical study (Fig. 5a). Although the time-averaged flow field is still symmetrical (Fig. 5b), the jet-width is apparently enlarged by wave action (Fig. 6a). The velocity along the centerline decreases rapidly away from the jet nozzle (Fig. 6b), accompanied by a significant increase of turbulent intensity. However, the turbulent intensity decreases faster in region closer to the water surface, due to the relatively weak wave-induced turbulence near the surface (Fig. 7). All the numerical results agree well with the available experimental data, which demonstrates the engineering applicability of the present numerical model.

References

1. Chen Y P, Li C W, Zhang C K(2004), Development of a fully nonlinear numerical wave tank, China Ocean Engineering, 18(4): 501-514.
2. Chyan J M, Hwung H H(1993), On the interaction of turbulent jet with waves, Journal of Hydraulic Research, 31: 791-810.

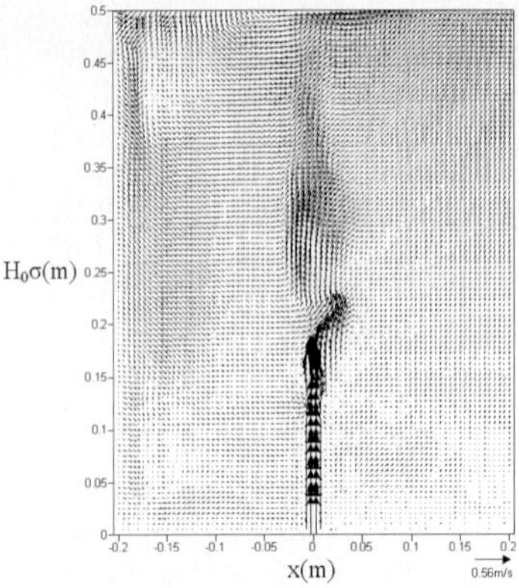

Fig. 5a

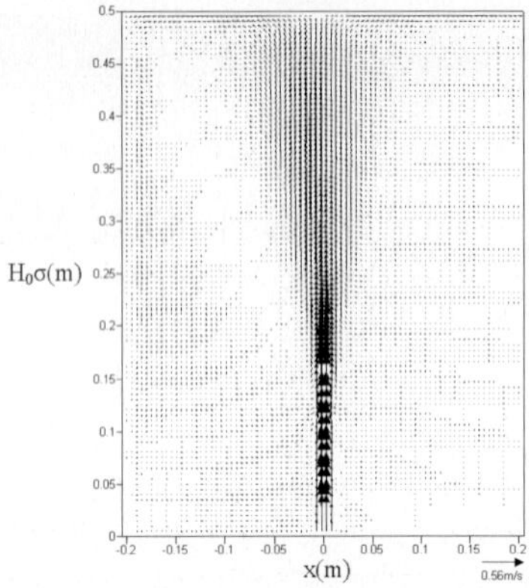

Fig. 5b

Fig. 5. Instantaneous flow field (t=41s, Fig. 5a) and mean flow field (t=10~60s, Fig. 5b) of jet under random waves

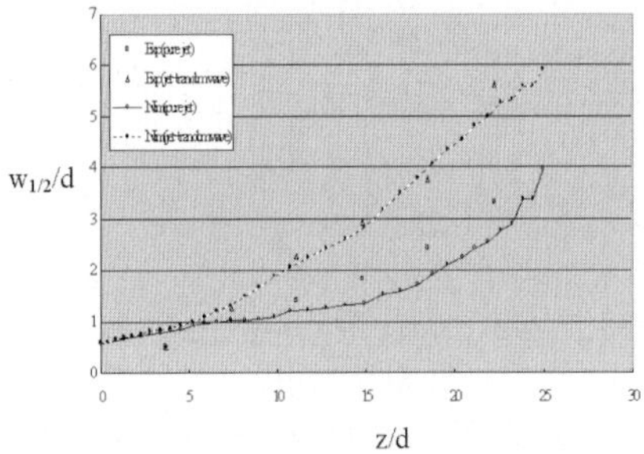

Fig. 6a

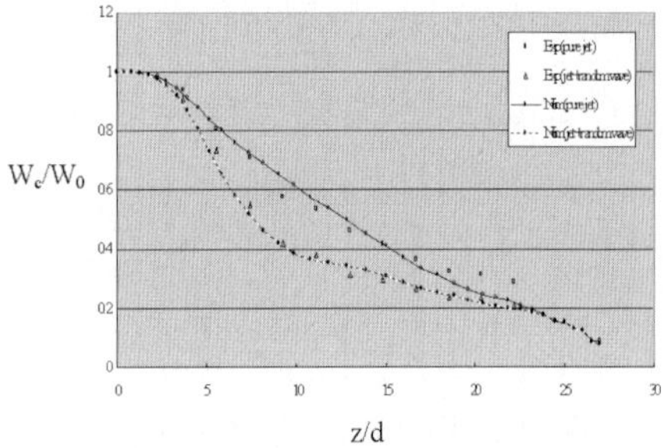

Fig. 6b

Fig. 6. Comparison of jet semi-width(Fig. 6a) and time-averaged centerline velocity(Fig. 6b) between numerical results and experimental data

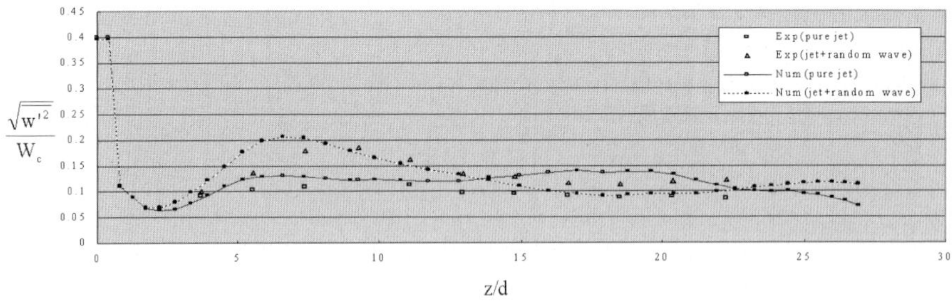

Fig. 7. Comparison of turbulent intensity between numerical results and experimental data

Scale-by-Scale Assessments of the Approach to Isotropy

Peter E. Hamlington[1], Jamison L. Szwalek[2] and Werner J.A. Dahm[1]

[1] Laboratory for Turbulence & Combustion (LTC), Department of Aerospace Engineering, The University of Michigan, Ann Arbor, MI, 48109-2140, USA
[2] Department of Naval Architecture and Marine Engineering, The University of Michigan, Ann Arbor, MI, 48109-2145, USA

The current understanding of turbulent flows is founded on the local-isotropy hypothesis of Kolmogorov's 1941 universal equilibrium theory that all turbulence attains a universal isotropic state at small scales. Substantial evidence for local isotropy has been found in both laboratory experiments and large-scale test facilities [1, 2] and it is now largely accepted that local isotropy applies in turbulent shear flows for $k \geq 10\left(S^3/\epsilon\right)^{1/2}$ when the Corrsin-Uberoi criterion is met. At the same time, other studies have reported substantial small scale departures from isotropy in the moments of various quantities [3, 4]. We thus present initial results from a study that addresses the scale-by-scale approach to isotropy in turbulent flows.

In particular, we present an anisotropy measure that reveals the effect of the mean shear rate S on the approach to isotropy in DNS of homogeneous uniformly-sheared turbulence, as well as in dual-plane stereo particle image velocimetry (DSPIV) of inhomogeneous turbulent shear flows. In recent years it has become increasingly popular to use the SO(3) decomposition to analyze isotropy, primarily in the context of structure functions. Here we do not use a structure function approach, and instead develop an extension of the classical anisotropy tensor. Using either Fourier or wavelet transforms, the field of interest, for example the velocity fluctuation field $u'_i(\mathbf{x},t)$, can be decomposed into its individual scale-components $\tilde{u}'_i(\mathbf{k})$ as indicated in Fig. 1. At every scale $k \equiv (\mathbf{k}\cdot\mathbf{k})^{1/2}$, the anisotropy can be quantified using an extension of the traditional b_{ij} anisotropy tensor as

$$b_{ij}(k) = \frac{\left\langle u'_i(k)u'_j(k)\right\rangle_N}{\left\langle u'_n(k)u'_n(k)\right\rangle_N} - \frac{1}{3}\delta_{ij} \quad . \tag{1}$$

From this scale-dependent tensor we use $\|b_{ij}(k)\| \equiv \left[b_{ij}(k)b_{ji}(k)\right]^{1/2}$ to determine the degree of anisotropy in the k-th component of the field of interest.

Note that the average in (1) is over the number N of effectively independent structures in the field at scale k. The traditional notion of isotropy,

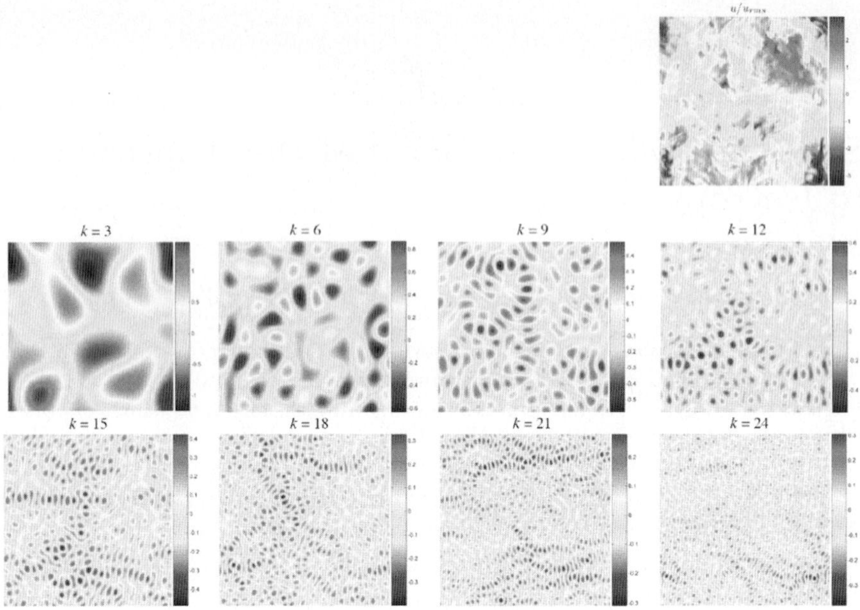

Fig. 1. Typical Fourier decomposition of an instantaneous velocity component field $u_i(x,t)$ (top right) from 512^3 DNS of homogeneous isotropic turbulence. Anisotropy at each scale k is herein analyzed via scale-by-scale anisotropy tensor $b_{ij}(k)$ in (1)

namely $b_{ij} \equiv 0$, applies only for $N \to \infty$. For finite N, tests of isotropy must thus separate the statistical convergence of $\|b_{ij}(k)\|$ from true departures from isotropy. One approach is to obtain $\|b_{ij}(k)\|$ in 'windows' sized with k so that $N(k)$ is constant. Figure 2 shows typical results for isotropic

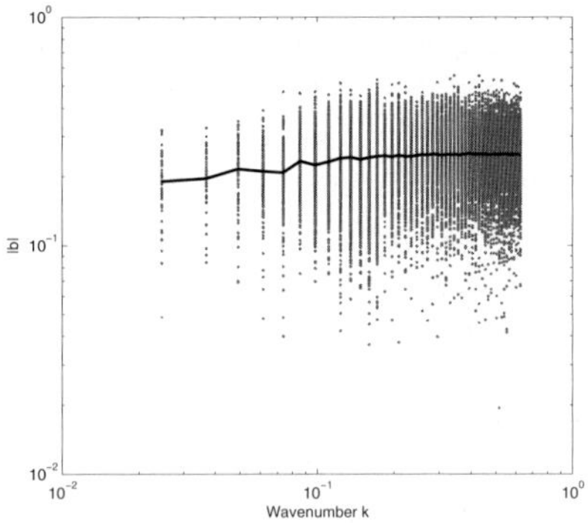

Fig. 2. Anisotropy tests for 512^2 white-noise fields, verifying average $b_{ij}(k)$ modulus *(black line)* independent of k within statistical convergence limits. Individual $b_{ij}(k)$ modulus values *(gray dots)* show results obtained from all windows at each scale k

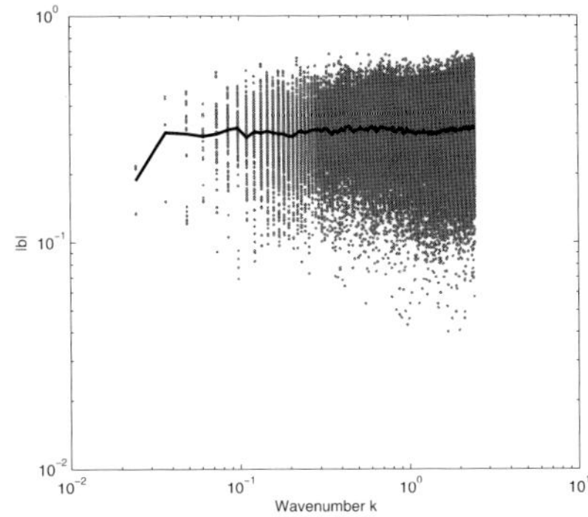

Fig. 3. Anisotropy tests for 512^2 DNS fields of homogeneous isotropic turbulence, verifying average $b_{ij}(k)$ modulus *(black line)* independent of k within statistical convergence limits. Individual $b_{ij}(k)$ modulus values *(gray dots)* are from all windows at each scale k

white noise fields, and sample results are shown in Fig. 3 from an analysis of 512^3 DNS homogeneous isotropic turbulence. The essentially k-independent $\|b_{ij}(k)\|$ reflects the largely isotropic fields that result from the periodic boundary conditions and the nominally isotropic low-k forcing.

The principal interest is in applying this approach to anisotropic sheared turbulence. We thus use $256^2 \times 129$ DNS of uniformly-sheared turbulence for mean shear rates $S = \{0, 1/\pi, 3/\pi, 6/\pi\}$ from [4]. Nominally isotropic low-k grid and shear forcing were applied throughout the domain, and periodic boundary conditions were applied on four of the six faces of the domain [4]. Slip boundary conditions were applied on the 256^2 faces to simulate inviscid top and bottom walls; this has been shown to provide improved stationarity in comparison with the more traditional approach based on fully periodic boundaries. However this introduces substantial departures from homogeneity near the slip walls and reduces the energy in the v fluctuations relative to the u and w fluctuations throughout the domain. Thus, the present analysis is based only on the u and w fluctuation fields at the domain mid-plane.

Figure 4 shows the corresponding $\|b_{ij}(k)\|$ spectra for each S case. The curves for $S = 0$, $1/\pi$ and $3/\pi$ are shown in Fig. 5 with k normalized by the Kolmogorov scale η, where it is clear that the results agree at intermediate and small scales. The same curves are shown in Fig. 6 with k normalized by the integral scale L_y, where the different levels for forcing at low-k become apparent. In particular, the increased anisotropy for the highest S case, and the decrease in anisotropy as the intermediate scales are approached, are evident.

The scale-by-scale $\|b_{ij}\|$ metric thus allows a direct investigation of the effects of S on the approach to isotropy in DNS of homogeneous uniformly-sheared turbulence, and should be applicable to inhomogeneous turbulent shear flows using data from DSPIV measurements.

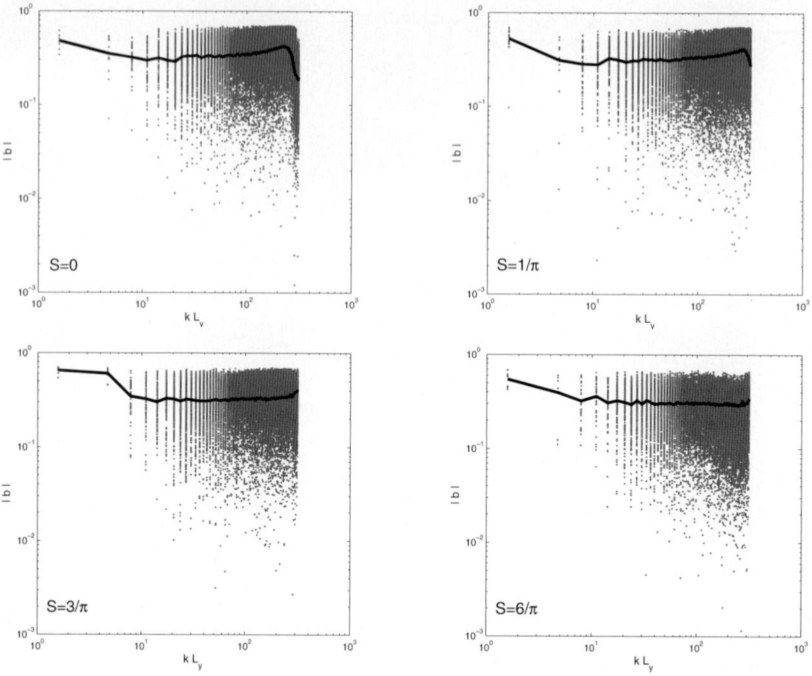

Fig. 4. Anisotropy results for DNS of homogeneous uniformly-shear turbulence for increasing shear rates S, showing average $b_{ij}(k)$ modulus *(black line)* and individual $b_{ij}(k)$ modulus values *(gray dots)* from all windows at each scale k

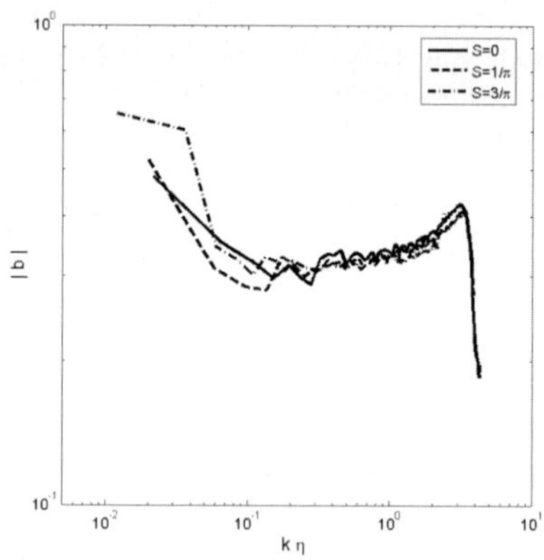

Fig. 5. Average $b_{ij}(k)$ modulus for all shear rates S from DNS of homogeneous uniformly sheared turbulence, with k normalized by η

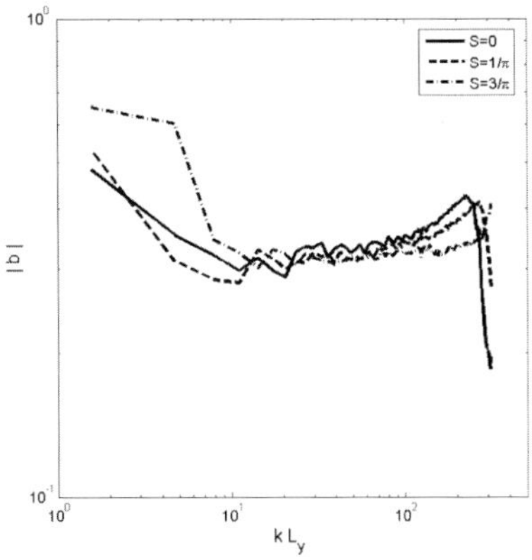

Fig. 6. Corresponding results for average $b_{ij}(k)$ modulus for homogeneous uniformly sheared turbulence, with k normalized by L_y

Acknowledgements

The DNS data were provided by Greg Burton and Jörg Schumacher; discussions with David Dowling are also acknowledged.

References

1. Saddoughi, S.G. & Veeravalli, S.V. (1994) Local isotropy in turbulent boundary layers at high Reynolds number. *J. Fluid Mech.* 268, 333-372.
2. Xu, G., Rajagopalan, S. & Antonia, R.A. (2000) Approach to isotropy in a smooth wall turbulent boundary layer. *Fluid Dyn. Res.* 26, 1-11.
3. Schumacher, J., Sreenivasan, K.R. & Yeung, P.K. (2003) Derivative moments in turbulent shear flows. *Phys. Fluids* 15, 84-90.
4. Schumacher, J. (2004) Relation between shear parameter and Reynolds number in statistically stationary turbulent shear flows. *Phys. Fluids* 16, 3094-3102.

On the Decay of Isotropic Turbulence

P.A. Davidson[1], Y. Kaneda[2] and T. Ishida[3]

[1] Dept. Engineering, University of Cambridge, Cambridge CB2 1PZ, U.K. pad3@eng.cam.ac.uk
[2] Dept. Computational Science & Engineering, Nagoya University, Nagoya 464-8603, Japan kaneda@cse.nagoya-u.ac.jp
[3] Dept. Computational Science & Engineering, Nagoya University, Nagoya 464-8603, Japan ishida@cse.nagoya-u.ac.jp

Abstract

We investigate the decay of freely-evolving, isotropic turbulence whose spectrum takes the form $E(k \to 0) \sim Ik^4$, I being Loitsyansky's integral. We report numerical simulations in a periodic domain whose dimensions, l_{box}, are much larger than the integral scale of the turbulence, l. We find that, provided $l_{box} \gg l$ and $Re \gg 1$, the turbulence evolves to a state in which Loitsyansky's integral is approximately constant and Kolmogorov's decay law, $u^2 \sim t^{-10/7}$, holds true. The approximate conservation of I in fully-developed turbulence implies that the long-range interactions between remote eddies, as measured by the triple correlations, are very weak.

1 Introduction

There are two canonical cases in freely-evolving, isotropic turbulence: $E(k \to 0) \sim Lk^2$ and $E(k \to 0) \sim Ik^4$. Here L and I are the Saffman and Loitsyansky integrals,

$$L = \int \langle \mathbf{u} \cdot \mathbf{u}' \rangle d\mathbf{r}, \qquad I = -\int \mathbf{r}^2 \langle \mathbf{u} \cdot \mathbf{u}' \rangle d\mathbf{r}, \tag{1}$$

and $\langle \mathbf{u} \cdot \mathbf{u}' \rangle = \langle \mathbf{u}(\mathbf{x}) \cdot \mathbf{u}(\mathbf{x}+\mathbf{r}) \rangle$. A key discovery by Landau (Landau & Lifshitz 1959), later pursued by Saffman (1967), is that the linear and angular impulse of a large cloud of turbulence of volume V can be related to these two integrals:

$$[\text{linear impulse}]^2 \sim V \int \langle \mathbf{u} \cdot \mathbf{u}' \rangle d\mathbf{r} + \quad (\text{correction term}, C_1)$$

$$[\text{angular impulse}]^2 \sim -V \int \mathbf{r}^2 \langle \mathbf{u} \cdot \mathbf{u}' \rangle d\mathbf{r} + \quad (\text{correction term}, C_2)$$

It turns out that the correction terms C_1 and C_2 vanish provided that $\langle \mathbf{u} \cdot \mathbf{u}' \rangle$ decays sufficiently rapidly with $|\mathbf{r}|$. In such cases the momentum conservation laws applied to a cloud of turbulence suggests that the integrals L and I should be invariants of freely-evolving turblence. (See, for example, Davidson 2000, 2004 [4], [5]) Indeed, when remote points in a turbulent flow are statistically independent, in the sense that $\langle \mathbf{u} \cdot \mathbf{u}' \rangle$ is transcendentally small for large $|\mathbf{r}|$, these suggestions may be confirmed directly using the Karman-Howarth equation. In particular, we find that the statistical independence of remote points leads to L = constant = 0 and I = constant.

Now we would expect the angular impulse (i.e. angular momentum) of a cloud of turbulence to be dominated by the large eddies and this is consistent with the integral expression for I given above. So it is common to estimate I as, $I \sim u^2 \ell^5$, where ℓ is the integral scale. Kolmogorov (1941) [2] used the alleged invariance to I, in conjunction with the energy decay law, $du^2/dt \sim -u^3/\ell$, to predict the rate of decay of energy in a turbulent cloud. The resulting expression, $u^2 \sim t^{-10/7}$, is known as Kolmogorov's 10/7 decay law.

However Batchelor & Proudman (1956) [6] have shown that, because of the transmission of information over large distances by the pressure field, $\langle \mathbf{u} \cdot \mathbf{u}\prime \rangle$ is not, in general, transcendentally small for large $|\mathbf{r}|$. This raises questions over Landau's assertion that I is proportional to the square of the angular momentum (the correction term C_2 could be of order one) as well as the formal proof (via the Karman-Howarth equation) that I is an invariant of freely evolving turbulence. Batchelor & Proudman went on to suggest that, in general, one would expect $L = 0$, as in traditional theory, but that I should be time-dependent. By implication, Kolmogorov's decay law is suspect.

Later, Saffman (1967) [3] showed that, for certain initial conditions, even more dramatic departures from the traditional theory could arise. In particular, following Birkhoff (1954) [7], he showed that it is possible to produce turbulence in which L is non-zero and constant, while I diverges. Whether one observes a Saffman, $E \sim Lk^2$, or a Batchelor, $E \sim Ik^4$, spectrum depends on the initial conditions. If the turbulence is created with sufficient linear impulse, a Saffman spectrum is obtained, otherwise we obtain a k^4 spectrum (Davidson 2004) [5]. In this paper we consider the latter case.

2 The Numerical Simulations

We have computed isotropic turbulence for a variety of domain sizes and Reynolds numbers. The direct numerical simulations employed a spectral code described in Kaneda & Ishihara (2004) [8]. The boundary conditions are periodic and the initial conditions consist of an $E \sim k^4$ spectrum with random Gaussian phases. The details of some of the simulations are tabulated below, where k_p is the wave-number at which $E(k, t = 0)$ peaks, N is the number of modes, t_{max} is normalised by the intitial turnover time, and Re is based on

N	k_p Re t_{max}
1024	80 62.5 300
1024	40 250 300
1024	40 125 300

Table 1. Details of the simulations

the integral scale $l = k_p^{-1}$. Note that, with this choice of integral scale, the initial value of l_{box}/l is simply k_p. The value of $I(t)$ was estimated by fitting the curve $E = Ik^4/24\pi^2$ to the data in the vicinity of $k = 0.08k_p$.

The results of these, and other, simulations will be reported in detail elsewhere. We merely note here that, provided $l_{box} \gg l$ and $Re \gg 1$, the turbulence evolves to a state in which I is very nearly constant (Figs. 1 and 2) and Kolmogorov's classical decay law, $u^2 \sim t^{-10/7}$, holds true (Fig. 3). The minimum value of Re needed to obtain Kolmogorov's decay law is $Re \sim 100$, while values of $l_{box}/l \sim 40$, or greater, are required to resolve the $E \sim Ik^4$ part of the spectrum. The time taken to reach the asymptotic state is around 200 turnover times.

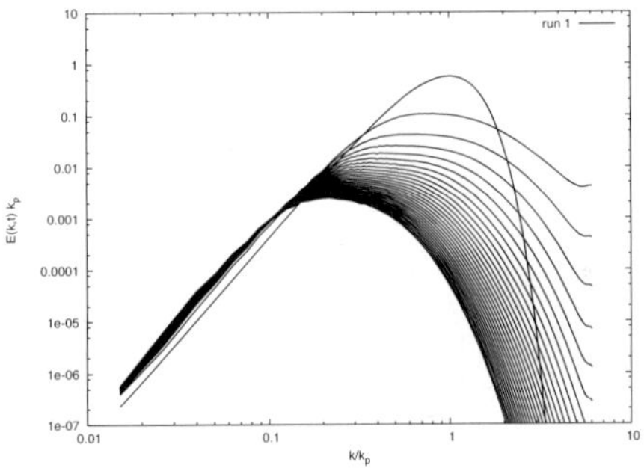

Fig. 1. $E(k,t)$ for run 1 ($k_p = 80$, $Re = 62.5$).$t = 0 - 300$ in steps of $\Delta t = 10$

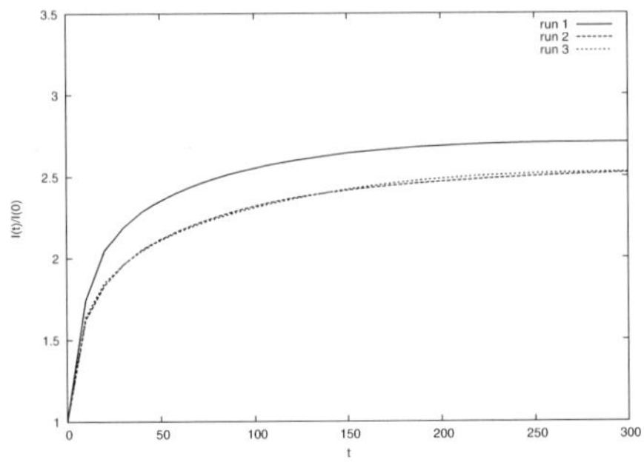

Fig. 2. $I(t)/I(0)$ versus time for run 1 ($k_p = 80, Re = 62.5$), run 2 ($k_p = 40$, $Re = 250$) and run 3 ($k_p = 40, Re = 125$)

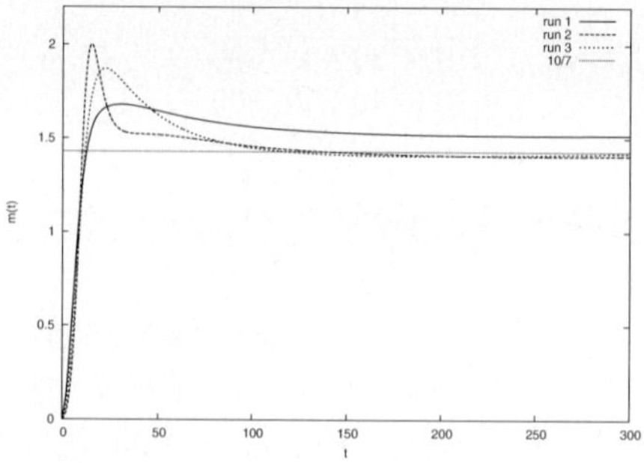

Fig. 3. $m(t)$ versus time for run 1 ($k_p = 80, Re = 62.5$), run 2 ($k_p = 40, Re = 250$) and run 3 ($k_p = 40, Re = 125$)

The approximate conservation of I in fully-developed turbulence implies that the long-range interactions between remote eddies, as measured by the triple correlations, are very weak. Interestingly, our findings are inconsistent with turbulence closure models, such as EDQNM, which predict a continual rise in I.

The authors would like to thank T. Ishihara for making available his spectral code.

References

1. Landau L D, Lifshitz E M (1959) Fluid Mechanics, 1st Edn., Pergamon
2. Kolmogorov A N (1941) Dokl. Akad. Nauk SSSR, 31(6), 538-541
3. Saffman P G (1967) J. Fluid Mech., 27, 581-593.
4. Davidson P A (2000) Jn. of Turbulence, 1.
5. Davidson P A (2004) Turbulence: An Introduction for Scientists and Engineers, Oxford University Press.
6. Batchelor G K, Proudman I (1956) Phil. Trans. Roy. Soc. A, 248, 369-405.
7. Birkhoff G (1954) Commun. Pure & Applied Math., 7, 19-44.
8. Kaneda Y, Ishihara T (2004). In: Reynolds number scaling in turbulent flows. Edited by A J Smits, Kluwer Acad. Publishers, 155-162.

Oscillatory Relaxation Towards Turbulent States

Bruno Eckhardt, Andreas Dietrich, Jörg Schumacher and Tobias M. Schneider

Fachbereich Physik, Philipps-Universität Marburg, 35032 Marburg, Germany
bruno.eckhardt at physik.uni-marburg.de

Turbulent flows are characterized by a balance, on average, between the energy supplied by external stirring and its dissipation by internal friction. In the absence of forcing the flow slows down until it comes to rest. In cases with a clear separation between the scales on which energy is fed into the fluid and those on which it is taken out, a scale resolved energy balance shows a net flux of energy from the larger to the smaller scales [1, 2]. This, however, does not imply that the energy flow on large scales is unidirectional [3] nor that the approach to a turbulent state is monotonic. Rather, various numerical studies [4, 5] and models [6, 7, 8] show an oscillatory approach to the turbulent state. Before discussing a fairly straightforward model for this phenomenon, we summarize some numerical findings.

As a first example, consider plane Couette flow in the transition region near a Reynolds number of about 400 (with the Reynolds number based on half the gap width and half the velocity difference between the plates), [4, 9]. The numerical protocol starts with a perturbation just barely strong enough to induce turbulence. The energy content of this perturbation is weaker than the one for the turbulent state, so that the energy has to increase towards the turbulent state. The turbulent state is not approached uniformly but rather as a damped oscillation. The studies over the small range of Reynolds numbers accessible then show that the relaxation rate varies weakly with the Reynolds number whereas the escape rate, the inverse of the lifetime of the turbulent state, changes dramatically.

Another example is relaxation in pipe flow. We take five different perturbations at a Reynolds number of 2500 with amplitudes close to the threshold for triggering the transition, and follow the time evolution of the energy content. All of them swing up to the turbulent state with an overshoot.

The final example is a volume driven shear flow. Varying the Reynolds number by a factor of three shows little variation in the frequencies, though perhaps a slight increase in damping. Volume driven flows have certain tech-

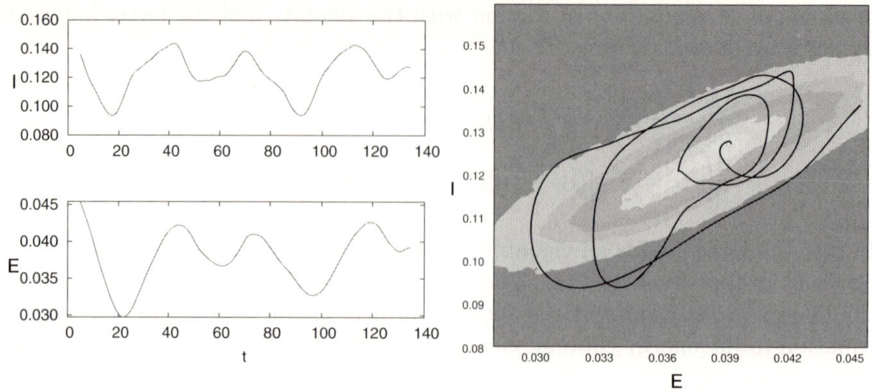

Fig. 3. Comparison of the energy model with direct numerical simulations of a shear flow. The left panels show time traces of the energy uptake I (top) and energy content E (bottom). The panel on the right shows a typical trajectory in the model relaxing to the equilibrium point and a trajectory from the numerical simulations in the projected onto the $I - E$-phase space

model and DNS and investigations of the periodically driven system will be presented elsewhere.

We thank C. R. Doering for stimulating discussions. Most of the simulations were done on the Cray T90 and Cray SV1ex supercomputers at the John von Neumann-Institut für Computing at the Forschungszentrum Jülich and we are grateful for their steady support. The work was also supported by the Deutsche Forschungsgemeinschaft.

References

1. S. B. Pope, Turbulent flow, Cambridge University Press, Cambridge 2000.
2. U. Frisch, Turbulence, Cambridge University Press, Cambridge 1995.
3. J. Schumacher and B. Eckhardt , Physica D **187**, 370–376 (2004).
4. A. Schmiegel and B. Eckhardt, Europhys. Lett. **51**, 395 (2000).
5. J. Schumacher and B. Eckhardt, Phys. Rev. E **63**, 046307 (2001).
6. V. Yakhot, Phys. Fluids **15**, L17 (2003).
7. B. Eckhardt, A. Dietrich, and J. Schumacher, in: H. I. Andersson and P.-A. Krogstad (eds.): Advances in Turbulence X, pp. 49–52. CIMNE, Barcelona, 2004.
8. R. Rubinstein and T.T. Clark, Deterministic large-eddy simulation of isotropic turbulence, Fourth International Symposium on Turbulence and Shear Flow Phenomena, JAC Humphrey, TB Gatski, JK Eaton, R Friedrich, N Kasagi and MA Leschziner (eds), Williamsburg 2005, 1067–1072.
9. B. Eckhardt, H. Faisst, A. Schmiegel and T.M. Schneider, submitted to Phil. Trans. R Soc. (London), (2005)

10. J. Schumacher and B. Eckhardt, Europhys. Lett. **52**, 627 (2000).
11. J. Schumacher, J. Fluid Mech. **441**, 109 (2001).
12. C. R. Doering, B. Eckhardt, and J. Schumacher, J. Fluid Mech. **494**, 275 (2003).
13. A. von der Heydt, S. Grossmann and D. Lohse, Phys. Rev. E **67**, 046308 (2003).
14. A. von der Heydt, S. Grossmann and D. Lohse, Phys. Rev. E **68**, 066302 (2003).

An Exact Solution for the Forced Burgers Equation

Stephan Eule[1] and Rudolf Friedrich[2]

[1] Institute of Theoretical Physics Westfälische Wilhelms Universität 48149 Münster yeahhart@uni-muenster.de
[2] Institute of Theoretical Physics Westfälische Wilhelms Universität 48149 Münster fiddir@uni-muenster.de

Abstract

We derive the exact solution for the Burgers equation with a time dependent forcing, which depends linearly on the spatial coordinate. For the case of a stochastic time dependence an exact expression for the joint probability distribution for the velocity fields at multiple spatial points is obtained. We present numerical results for fixed boundary conditions, and analyze the formation of shocks.

1 Introduction

The Burgers [1, 2, 3] has been intensively investigated in the context of turbulence. The homogeneous equation is exactly solvable by a Hopf-Cole transformation [4, 5] such that many properties of the spatio-temporal behaviour can be obtained analytically. In recent years, the stochastically forced Burgers equation has found much attention due to the work of Polyakov [6]. He considered a white noise forcing $f(x,t)$ with analytical spatial correlations $\langle f(x,t)f(x',t)\rangle = \kappa(x-x')\delta(t-t')$. This equation has been found to show interesting statistical behaviour.

In this contribution we want to present the general solution of the forced Burgers equation

$$[\frac{\partial}{\partial t} + v(r,t)\frac{\partial}{\partial r}]v(r,t) = \nu\frac{\partial^2}{\partial r^2}v(r,t) + G(t)r\,, \tag{1}$$

with the initial condition $v(r,0)$ for $-\infty < r < \infty$, where $G(t)$ is an arbitrary function of time t. As we shall discuss below, this problem is related to the problem considered by Polyakov. Investigating the special forcing, $f(r,t) = rG(t)$, has the advantage that one can find an analytic solution for an arbitrary random function $G(t)$. For the case of a white noise force $G(t)$ we are able to find a solution of the hierarchy of evolution equations for the joint probability distributions $f(v_i, r_i, t)$ for the velocities v_i at spatial points r_i. As we will show, a finite time singularity appears, which, for bounded forcing, can be interpreted as a formation of a shock. Finally, we present the negative tail of the velocity-increment probability distribution function (pdf), obtained by numerical simulations with fixed boundary conditions.

2 Exact Solution for Linear Stochastic Forcing

The general solution of (1) can be obtained by direct calculation using the ansatz

$$v(r,t) = a(t)r + \Lambda(t)w(\Lambda(t)r, \tau(t)) \ . \tag{2}$$

Thereby, the functions $a(t)$, $\Lambda(t)$, $\tau(t)$ have to fulfill the following solvability conditions:

$$\dot{a}(t) + a(t)^2 = G(t), \qquad \dot{\Lambda} = -a(t)\Lambda, \qquad \dot{\tau} = \Lambda^2 \ . \tag{3}$$

In order that the ansatz (2) obeys the forced Burgers equation, the function $w(\xi, \tau)$ has to obey the homogeneous Burgers equation,

$$[\frac{\partial}{\partial \tau} + w(\xi,\tau)\frac{\partial}{\partial \xi}]w(\xi,\tau) = \nu \frac{\partial^2}{\partial \xi^2} w(\xi,\tau) \ . \tag{4}$$

We note that the structure of the solution (2), is similar to the solution describing stretched vortices in the Navier-Stokes equation (see e.g. [7]).

From the solvability conditions it is clear that the crucial point is to obtain a solution of the nonlinear differential equation for $a(t)$. The behaviour of $a(t)$ may be visualized by introducing the potential $V(a) = \frac{1}{3}a^3$: The variable $a(t)$ then behaves like an overdamped particle moving in this potential under the influence of the time dependent force $G(t)$.

3 The Probability Distribution of the Velocity Increments

Let $G(t)$ now be a stochastic white noise force. The Fokker-Planck equation for the probability distribution $f(a,t) = \langle \delta(a - a(t)) \rangle$ then reads

$$\frac{\partial}{\partial t} f(a,t) = [\frac{\partial}{\partial a}\frac{\partial V(a)}{\partial a} + Q\frac{\partial^2}{\partial a^2}] f(a,t) \ . \tag{5}$$

Since the potential $V(a)$ is not bounded from below, there does not exist a stationary state. From (3) is evident that a kind of finite time singularity exists, where $a(t)$ tends to $a(t) \to -\infty$. As a consequence of (3) the variable $\Lambda(t)$ increases, too.

Despite the existence of a finite time singularity, it is nevertheless interesting to investigate the structure of the probability distribution of the velocity field $v(r,t)$. To this end let us introduce the following distribution $P(a, \Lambda, \tau; t)$, which obeys the Fokker-Planck equation

$$\frac{\partial}{\partial t} P = \{\frac{\partial}{\partial \Lambda} a\Lambda - \frac{\partial}{\partial \tau}\Lambda^2 + \frac{\partial}{\partial a} a^2 + \frac{\partial^2}{\partial a^2}\} P . \tag{6}$$

Using $P(a, \Lambda, \tau; t)$ the probability distribution of the velocity field $v(r,t)$, $f(v,r,t) = \langle \delta(v - v(r,t)) \rangle$ reads

$$\begin{aligned} f(v,r,t) &= \int d\Lambda d\tau da P(a, \Lambda, \tau, t)\delta(v - ar - \Lambda w(\Lambda r, \tau)) \\ &= \frac{1}{r}\int d\tau d\Lambda P\left(\frac{v - \Lambda w(\Lambda r, \tau)}{r}, \Lambda, \tau, t\right) \end{aligned} \tag{7}$$

For the joint pdf of the velocity field at multiple scales r_i one obtains

$$f(\{v_i, r_i\}, t) = \int d\Lambda d\tau da P(a, \Lambda, \tau, t) \Pi_{i=1}^N \delta(v_i - ar_i - \Lambda w(\Lambda r_i, \tau)) . \tag{8}$$

This formula demonstrates how the spatial coherence of the stochastic force governs the statistics of the multiple scales pdf of the velocity field.

Using standard methods it is straightforward to derive a hierarchy of evolution equations for the pdf of the velocity field at multiple scales $f(v_i, r_i, t)$:

$$\begin{aligned} &\frac{\partial}{\partial t} f(v_i, r_i, t) + \sum_i v_i \frac{\partial}{\partial r_i} f(v_i, r_i, t) + \sum_i \frac{\partial}{\partial r}\int_{-\infty}^{v_i} dv_i' f(v_i', r_i, t) \\ &\quad = -\sum_i \nu \frac{\partial}{\partial v_i}\int dv' \int dr' \delta(r - r') v' \frac{\partial^2}{\partial r'^2} f(v', r'; v_i, r_i; t) \\ &\quad + \sum_{i,j} Q(r_i, r_j) \frac{\partial^2}{\partial v_i \partial v_j} f(v_i, r_i, t) \end{aligned} \tag{9}$$

Here, the matrix $Q(r_i, r_j)$ is given by $Q(r_i, r_j) = Qr_i r_j$.

We present this hierarchy, since Polyakov [6] considered the same hierarchy for the case of a translational invariant white noise forcing. Here, we could obtain a solution to this chain of equations by a direct calculation in terms of (8).

We now consider the case, where the velocity-increment is held fixed at the boundary, e.g. $v(L,t) = 0$. This correspondends to a bounded forcing and allows to perform numerical calculations for the pdf of the increments. The emergence of the singularity in the case of unbounded forcing is in fact the formation of a new shock when the value on the boundary is fixed. Obviously the increment-profile is dominated by the shocks so that after the building of

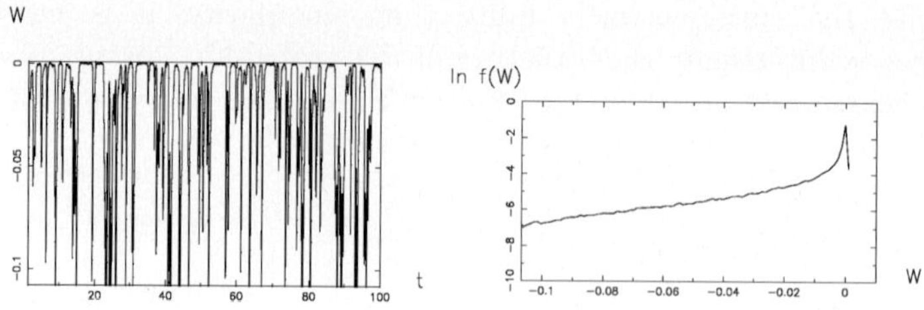

Fig. 1. Time Series and negative tail of the pdf of $W(t)$

a shock the random force is almost negligible for the evolution of $v(r,t)$. The graphs show the behavior of the velocity-gradient at the origin $W = \frac{dv}{dr}|_{r=0}$.

It seems that this behavior can be considered as a kind of excitability. The random force generates a velocity profile with negative slope and sets the initial condition for the formation of a shock. This shock eventually approaches the state $v = 0$ due to the influence of the boundary conditions. Now the linear forcing becomes important again leading to the continous generation of shocks at the origin $r = 0$.

4 Conclusions

Summarizing we have considered a forced Burgers equation with a forcing linear in the spatial coordinate. We have shown that in the one dimensional case this problem can be solved completely. We have discussed the relationship to the velocity increment statistics of the Burgers equation with translationally invariant forcing. Numerical results for the case of fixed boundary conditions have beeen presented and we proposed to consider shock formation as a kind of excitability.

References

1. J.M. Burgers, *The Nonlinear Diffusion Equation*, D. Reidel, Dordrecht 1974.
2. U. Frisch and J. Bec, Les Houches 2000: *New Trends in Turbulence*, M Lesieur, ed., Springer, Berlin (2001)
3. W.A. Woyczyński, *Burgers-KPZ Turbulence*, Springer, Berlin (1998)
4. E. Hopf, *The partial differential equation* $u_t + uu_x = u_{xx}$, Comm. Pure Appl. Math. **3** (1950), 201–230.
5. J.D. Cole, *On a quasi-linear parabolic equation occurring in aerodynamics*, Quart. Appl. Math. **9** (1951), 225–236.
6. A.M. Polyakov, *Turbulence without pressure*, Phys. Rev. E **52** (1995), 6183–6188.
7. D.I. Pullin, P.G. Saffman, *Vortex dynamics in turbulence*, Anu. Rev. Fluid Mech. 30, 31 (1998)

Consistent Invariant Modelling of Axially Rotating Pipe Flow

S. Guenther and M. Oberlack

Hydromechanics and Hydraulics Group, Technische Universität Darmstadt, Petersenstraße 13, 64287 Darmstadt, Germany,
guenther@hyhy.tu-darmstadt.de oberlack@hyhy.tu-darmstadt.de

1 Introduction

In RANS modelling it is common practice to use classical canonical flow cases such as the isotropic decay, the logarithmic law of the wall or homogeneous shear flows for calibrating the model constants. With the help of Lie group analysis a broad variety of invariant solutions (scaling laws) can be derived comprising the latter classical solutions as well as a broad variety of new solutions which have so far not been used for model calibration or development. The symmetry methods provide therefore a very useful tool for the improvement of existing turbulence models or may be a guideline for the development of new models. In the following we will exemplary investigate linear and nonlinear eddy viscosity models as well as Reynolds stress transport models for their capability to reproduce the scaling laws derived from symmetry methods for the fully developed turbulent rotating pipe flow.

2 Symmetry Analysis

In Oberlack (1999) new scaling laws for high-Reynolds-number turbulent pipe flow are derived using symmetry methods and validated using experimental and numerical data. For the analysis an infinite Reynolds-number was assumed and hence viscosity has been neglected. Thus only large-scale quantities such as the mean velocities are determined. Thereby two cases have been distinguished.

The first case is the most general case since no symmetry breaking is imposed on the flow, giving an algebraic scaling law for the axial and azimuthal mean velocity profile. Thus Oberlack (1999) received for the axial mean velocity the velocity defect law

$$\frac{\bar{u}_c - \bar{u}_z}{u_\tau} = \chi\left(\frac{u_w}{u_\tau}\right)\left(\frac{r}{R}\right)^{\psi} , \tag{1}$$

with χ being a function of the velocity ratio u_w/u_τ and since $u_w = R\Omega$ therewith dependends on the rotation rate Ω. This dependence has also been derived in Guenther and Oberlack (2005) analysing second moment closure models.
In Oberlack (1999) it was also found that the algebraic scaling law for the azimuthal velocity component can be rewritten as

$$\frac{\bar{u}_\phi}{u_w} = \zeta \left(\frac{r}{R}\right)^\psi . \tag{2}$$

The second test case is derived for a special combination of group parameters which applies if an external velocity scale acts on the flow. For this case a logarithmic mean velocity profile for the axial velocity is obtained in Oberlack (1999) with the singularity appearing on the pipe axis, not at the wall like in the classical law of the wall:

$$\frac{\bar{u}_z}{\bar{u}_w} = \lambda \ln\left(\frac{r}{R}\right) + \omega . \tag{3}$$

Oberlack found that (3) applies in some section of the radius for rapidly rotating pipes in which the wall velocity dominates the friction velocity u_τ and is therefore the symmetry-breaking velocity scale. The corresponding azimuthal velocity is given by

$$\bar{u}_\phi = \gamma , \tag{4}$$

with γ being a constant.

3 Model Implications

3.1 Linear Eddy Viscosity Models

Investigating plane and axisymmetric parallel shear flows with rotation Oberlack (2000) found that the standard $K-\epsilon$ model (Hanjalic and Launder, 1976) as well as other two-equation models have too many symmetries, leading to non-physical behavior under certain flow conditions such as rotation or stream-line curvature. This is due to the fact that the $K-\epsilon$ model equations do not contain Coriolis terms for any type of flow so that no symmetry breaking of scaling of time is possible. A complete group analysis of the $K-\epsilon$ equations in cylinder coordinates discloses an additional symmetry of the form:

$$r^* = r , \quad \bar{u}_z^* = \bar{u}_z , \quad \bar{u}_\phi^* = \bar{u}_\phi + br , \quad K^* = K , \quad \epsilon^* = \epsilon , \tag{5}$$

where b represents the group parameter. This additional symmetry allows to add a solid body rotation to the azimuthal velocity without any change to the remaining flow quantities. Obviously this is unphysical since turbulence is highly sensitive to rotation.

3.2 Non-linear Eddy Viscosity Models and Reynolds Stress Models

In recent years many non-linear stress-strain relations have been proposed to extend the applicability of linear eddy viscosity models to streamline curvature and rotation at modest computational costs. Thus it seems to be promising to apply a non-linear eddy viscosity model to the fully developed rotating pipe flow. The basic assumption behind a non-linear eddy viscosity model is that the Reynolds stresses are uniquely related to the rates of strain, the rate of rotation and scalar quantities. The two parameters K and ϵ are usually used to normalise the latter quantities as follows:

$$b_{ij} = \frac{\overline{u_i' u_j'}}{2K} - \frac{1}{3}\delta_{ij}, \tag{6}$$

$$S_{ij}^* = \frac{K}{\epsilon} S_{ij} = \frac{1}{2}\frac{K}{\epsilon}\left(\frac{\partial \bar{u}_i}{\partial x_j} + \frac{\partial \bar{u}_j}{\partial x_i}\right), W_{ij}^* = \frac{K}{\epsilon} W_{ij} = \frac{1}{2}\frac{K}{\epsilon}\left(\frac{\partial \bar{u}_i}{\partial x_j} - \frac{\partial \bar{u}_j}{\partial x_i}\right).$$

Owing to the Cayley-Hamilton theorem, the number of independent invariants and linearly dependent second order tensor forms built from the strain and rotation tensor to model the Reynolds stresses is finite. In general form the stress-strain relationship might be written in the closed tensor polynomial:

$$\boldsymbol{b} = \sum_{\lambda=1}^{10} G_\lambda \boldsymbol{T}^{*\lambda}. \tag{7}$$

Thereby the coefficients G_λ are functions of a finite number of scalar invariants. In the general, three-dimensional case there are ten tensors and five scalar invariants:

$$\begin{aligned}
\boldsymbol{T}^{*1} &= \boldsymbol{S}^*, & \boldsymbol{T}^{*6} &= \boldsymbol{W}^{*2}\boldsymbol{S}^* + \boldsymbol{S}^*\boldsymbol{W}^{*2} - \tfrac{2}{3}\delta\{\boldsymbol{S}^*\boldsymbol{W}^{*2}\},\\
\boldsymbol{T}^{*2} &= \boldsymbol{S}^*\boldsymbol{W}^* - \boldsymbol{W}^*\boldsymbol{S}^*, & \boldsymbol{T}^{*7} &= \boldsymbol{W}^*\boldsymbol{S}^*\boldsymbol{W}^{*2} - \boldsymbol{W}^{*2}\boldsymbol{S}^*\boldsymbol{W}^*,\\
\boldsymbol{T}^{*3} &= \boldsymbol{S}^{*2} - \tfrac{1}{3}\delta\{\boldsymbol{S}^{*2}\}, & \boldsymbol{T}^{*8} &= \boldsymbol{S}^*\boldsymbol{W}^*\boldsymbol{S}^{*2} - \boldsymbol{S}^{*2}\boldsymbol{W}^*\boldsymbol{S}^*,\\
\boldsymbol{T}^{*4} &= \boldsymbol{W}^{*2} - \tfrac{1}{3}\delta\{\boldsymbol{W}^{*2}\}, & \boldsymbol{T}^{*9} &= \boldsymbol{W}^{*2}\boldsymbol{S}^{*2} + \boldsymbol{S}^{*2}\boldsymbol{W}^{*2} - \tfrac{2}{3}\delta\{\boldsymbol{S}^{*2}\boldsymbol{W}^{*2}\},\\
\boldsymbol{T}^{*5} &= \boldsymbol{W}^*\boldsymbol{S}^{*2} - \boldsymbol{S}^{*2}\boldsymbol{W}^*, & \boldsymbol{T}^{*10} &= \boldsymbol{W}^*\boldsymbol{S}^{*2}\boldsymbol{W}^{*2} - \boldsymbol{W}^{*2}\boldsymbol{S}^{*2}\boldsymbol{W}^*
\end{aligned} \tag{8}$$

and

$$I_1^* = \{\boldsymbol{S}^{*2}\},\ I_2^* = \{\boldsymbol{W}^{*2}\},\ I_3^* = \{\boldsymbol{S}^{*3}\},\ I_4^* = \{\boldsymbol{W}^{*2}\boldsymbol{S}^*\},\ I_5^* = \{\boldsymbol{W}^{*2}\boldsymbol{S}^{*2}\}. \tag{9}$$

Hereby $\boldsymbol{S}^{*2} = S_{ik}^* S_{kj}^*$, $\boldsymbol{W}^*\boldsymbol{S}^{*2} = W_{ik}^* S_{kl}^* S_{lj}^*$ etc. and $\{\cdot\}$ denotes the trace.

Special attention is turned in the following to constraints which break the unphysical symmetry (5) which is admitted by linear two-equation models in order to be sensitive to rotation in a physically correct manner.

For a fully developed, rotating, turbulent pipe flow the production term reduces to

$$P = -\overline{u'_r u'_z}\frac{\partial \bar{u}_z}{\partial r} = b_{rz}2K(S_{rz} - W_{rz}). \qquad (10)$$

Thus only the $[r, z]$ component of the tensors $\boldsymbol{T}^1 - \boldsymbol{T}^{10}$ gives a contribution. For the given flow case only the tensors $\boldsymbol{T}^1, \boldsymbol{T}^5, \boldsymbol{T}^6$ and $\boldsymbol{T}^{10}$ have a $[r, z]$ component which is unequal to zero. Considering these four tensors it has been found that the linear solution for the azimuthal velocity is not admitted by $\boldsymbol{T}^5, \boldsymbol{T}^6$ and $\boldsymbol{T}^{10}$. Only these tensors contain a term of the form $r\frac{d\bar{u}_\phi}{dr} + \bar{u}_\phi$ which is not invariant under (5) and hence make them sensitive to rotation. Interesting enough models containing $\boldsymbol{T}^5$, $\boldsymbol{T}^6$ and $\boldsymbol{T}^{10}$ may be used to derive the unknown rotation rate dependence of (1) due to $\chi(u_w/u_\tau)$. Using $\overline{u'_r u'_\phi} = 0$, what comes due to the simplification of the Reynolds stress equations for the given flow case, we solely consider the $[r, \phi]$ component of (7). Introducing into this equation the invariant solutions it reduces to an algebraic equation, which can be solved easily for the coefficient of the axial velocity component $\chi(u_w/u_\tau)$. If we now introduce this solution into the expression for the axial velocity component (1), we receive the dependence of the axial velocity component on the rotation rate:

$$\frac{u_c - \bar{u}_z}{u_\tau} = \frac{\sqrt{C_1 R^2 \Omega^2 + C_2 u_b^2}}{u_\tau}\left(\frac{r}{R}\right)^\psi, \qquad (11)$$

with C_1 and C_2 being constants. At the derivation of (11) $u_w = R\Omega$ has been used. The same dependence of the axial velocity component on the rotation rate (11) can be derived for Reynolds stress models using the EARSM procedure. This finding is also confirmed by experiments since the profile for $\bar{u}_z$ becomes steeper or in other words $\chi(u_w/u_\tau)$ increases for increasing rotation rates. Using the experiments from Kikuyama et al. (1983) the coefficients may be fitted to $C_1 = -0.59$, $C_2 = 0.43$ and $\psi = 2$.

References

Guenther, S., Oberlack, M.: Symmetry methods in modelling rotating, turbulent pipe flow, *submitted to J. Fluid Mech.*,(2005b).

Hanjalić, K., Launder, B.E.: Contribution towards a Reynolds stress closure for low Reynolds number turbulence, *J. Fluid Mech.* **74**, (1976), 593-610.

Kikuyama, K., Murakami, M., Nishibori, K., Maede, K.: Flow in axially rotating pipe, *Bull. JSME*, **26**, (1983), 506-513.

Oberlack M.: Similarity in non-rotating and rotating turbulent pipe flow, *J. Fluid Mech.*, **379**, (1999), 1-22.

Oberlack M.: Symmetrie, Invarianz und Selbstähnlichkeit in der Turbulenz, *Habilitationsschrift, Shaker Verlag*, (2000).

Small-Scale Anisotropy in MHD Turbulence Under Strong Uniform Magnetic Field

Takaki Ishida[1] and Yukio Kaneda[2]

Department of Computational Science and Engineering, Nagoya University, Nagoya, 464-8603, Japan
[1]ishida@fluid.cse.nagoya-u.ac.jp
[2]kaneda@cse.nagoya-u.ac.jp

The anisotropy at small scales of incompressible magneto-hydrodynamic (MHD) turbulence under a strong uniform magnetic field at low magnetic Reynolds number is studied theoretically and numerically on the basis of the quasi-static approximation. A simple expression is derived for the wavenumber dependence of the anisotropy of the velocity correlation spectrum. It is in good agreement with direct numerical simulation (DNS). The relaxation of the velocity spectrum to isotropy after the switching off the uniform magnetic field is also studied by the DNS.

1 Introduction

It is known in the statistical mechanics of thermal equilibrium system that there are not only universal relations characterizing the equilibrium state itself, but also those characterizing the response of the system to disturbances added to the equilibrium state, see for example [1]. The similar idea has been applied to the study of anisotropy at small scales of turbulent shear flow by Ishihara et al. [2], and to that of stably stratified turbulence by Kaneda & Yoshida [3].

In this paper we consider incompressible magneto-hydrodynamic(MHD) turbulence that is known to be important in industrial applications, such as electromagnetic processing of materials(EPM) in metallurgical industry. We confine ourselves here to MHD turbulence under a strong uniform magnetic field at low magnetic Reynolds number($Re_{\mathrm{m}} \ll 1$), for which one may apply the so-called quasi-static approximation [4, 5]. Such MHD turbulence at low Re_{m} occurs, for example, in electrically conducting liquid under a strong uniform magnetic field, and has been the subject of studies by direct numerical simulation (DNS) [6, 7, 8].

Under the presence of the uniform magnetic field, the turbulence cannot be isotropic. By applying the idea developed by Ishihara et al.[2] and Kaneda

& Yoshida [3], we derive a simple expression for the anisotropy of velocity correlation spectrum at small scales. The theoretical conjecture has been examined by comparison with DNS with 512^3 grid points, and shown to be in good agreement with the DNS.

The analogy of turbulence with thermal equilibrium system suggests also that if the external uniform magnetic field is switched off, then the turbulence would relax to an isotropic equilibrium state. The relaxation process is also studied by the DNS. It suggests that the relaxation to isotropy is faster at smaller scales, as would be expected.

2 Basic Equations and Theoretical Conjecture

In the limit of low Re_{m}, the motion of MHD turbulence of incompressible fluid can be approximated by the so-called quasi-static approximation [4, 5],

$$\frac{\partial \boldsymbol{u}}{\partial t} + (\boldsymbol{u} \cdot \mathrm{grad})\, \boldsymbol{u} = -\mathrm{grad} p' + \nu \triangle \boldsymbol{u} + \boldsymbol{F}, \tag{1}$$

$$\mathrm{div} \boldsymbol{u} = 0, \qquad \boldsymbol{F} = -\frac{\sigma_{\mathrm{e}}}{\rho} \triangle^{-1} (\boldsymbol{B}_0 \cdot \mathrm{grad})^2\, \boldsymbol{u}, \tag{2}$$

where $\boldsymbol{u}$, $\boldsymbol{B}_0$, ν, ρ and σ_{e} are the velocity, applied magnetic field, kinematic viscosity, constant fluid density and electric conductivity, respectively. The external force $\boldsymbol{F}$ represents the rotational part of the Lorentz force due to the magnetic field, and the modified pressure p' includes the fluid pressure and magnetic pressure. The symbol $\triangle^{-1}$ denotes the inverse Laplace operator.

At small scales, the force $\boldsymbol{F}$ may be regarded as a disturbance added to the equilibrium turbulence state governed by the inherent Navier-Strokes dynamics in the absence of the magnetic field. With this in mind, let us write the velocity correlation spectrum Q defined by

$$Q_{ij}(\boldsymbol{k}, t) \equiv (2\pi)^{-3} \int \mathrm{d}^3 \boldsymbol{r}\, \langle u_i(\boldsymbol{x} + \boldsymbol{r}, t) u_j(\boldsymbol{x}, t) \rangle\, e^{-i\boldsymbol{k}\cdot\boldsymbol{r}},$$

as

$$Q_{ij}(\boldsymbol{k}, t) = Q_{ij}^{(0)}(\boldsymbol{k}, t) + \Delta Q_{ij}(\boldsymbol{k}, t), \tag{3}$$

where $Q_{ij}^{(0)}(\boldsymbol{k})$ is the equilibrium isotropic spectrum in the absence of the magnetic field, and $\Delta Q_{ij}(\boldsymbol{k})$ the response of the equilibrium state to the magnetic field $\boldsymbol{F}$. A perturbation expansion method based on the smallness of effect of $\boldsymbol{F}$ yields

$$\Delta Q_{ij}(\boldsymbol{k}, t) \sim \sum_{m,n} C_{ijmn}(\boldsymbol{k}, t) \alpha_{mn}, \tag{4}$$

for high wavenumbers. Here $\alpha_{mn} = (B_0)_m (B_0)_n \sigma_{\mathrm{e}}/\rho$, and C_{ijmn} is given by

$$C_{ijmn}(\boldsymbol{k}, t) = \frac{1}{2} q^{(1)}(k, t) \left[P_{im}(\boldsymbol{k}) P_{jn}(\boldsymbol{k}) + P_{jm}(\boldsymbol{k}) P_{in}(\boldsymbol{k}) \right]$$
$$+ q^{(2)}(k, t) P_{ij}(\boldsymbol{k}) \frac{k_m k_n}{k^2} + q^{(3)}(k, t) P_{ij}(\boldsymbol{k}) \delta_{mn}, \tag{5}$$

where $P_{ij}(\boldsymbol{k}) = \delta_{ij} - (k_i k_j)/k^2$, and $q^{(i)}(k,t)$ depends on $\boldsymbol{k}$ only through $k = |\boldsymbol{k}|$.

3 DNS Method and Results

To examine the theoretical predictions, we performed DNS of incompressible forced MHD turbulence obeying 1 and 2 in a periodic box. The DNS is based on a spectral method free from alias error. The number of grid points is 512^3, and the maximum wavenumber $k_{\max}$ of the retained Fourier modes is approximately 241. The time marching was accomplished by a fourth order Runge-Kutta method. The total kinetic energy E is kept almost constant (≈ 0.5), by introducing negative viscosity [9]at the low wavenumber range $k < 2.5$. The initial incompressible velocity field is generated by random numbers with given energy spectrum $E(k)$ with the peak wavenumber at $k = 2$ and a power law dependence $E(k) \propto k^4$ at low wavenumber. [10].

Prior to the application of the magnetic field, we simulated turbulence field without the magnetic field until a certain time, say $t = t_I$, by which the turbulence field attains at a statistically quasi-stationary state. Then we switched on a uniform magnetic field $\boldsymbol{B}_0$ in (2) at $t = t_I$, and continued further the simulation until $t = t_F$. During the period between $t = t_I$ and $t = t_F$, $\boldsymbol{B}_0$ is kept constant. In the run reported below, we put $B_0 = |\boldsymbol{B}_0|$ as $\sigma_{\mathrm{e}}(B_0)^2/\rho = u_0/L$ at $t = t_I$, $t_I/T = 16$ and $t_F/T = 32$, and kinematic viscosity $\nu = 2.8 \times 10^{-4}$, where T is the eddy turn over time defined by $T \equiv L/u_0$, L the integral scale and $3u_0^2/2 = E$. The Reynolds number $Re_\lambda (\equiv u_0\lambda/\nu) = 525$ at $t = t_F$, where λ is the Taylor microscale.

The theoretical conjecture (3) with (4) was examined by comparison with the DNS. The comparison shows that angular dependence as well as the wavenumber dependence predicted by (3) and (4) are in good agreement with the DNS at high wavenumbers. The details of the comparison will be presented elsewhere.

The analogy with the linear response theory in statistical mechanics of thermal equilibrium system [1] suggests that the turbulence would relax to an isotropic equilibrium system if the external magnetic field is switched off, and this relaxation process may also reflect a certain kind of universality characterizing the equilibrium state. As a first step toward the understanding of such a relaxation process, we continued further our DNS after the switching off the magnetic filed $\boldsymbol{B}_0$ at time $t = t_F$, and observed the time dependence of anisotropy at small scales.

Let $\boldsymbol{B}_0$ be parallel to the x_3 axis. Then (3) and (4) give

$$p(\boldsymbol{k},t) \equiv \frac{k^2 + k_3^2}{k^2 - k_3^2} Q_{33}(\boldsymbol{k},t) - Q_{11}(\boldsymbol{k},t) - Q_{22}(\boldsymbol{k},t) = q^{(1)}(k,t)\alpha_{33}\sin^2\theta,$$

and $p(\boldsymbol{k},t) = 0$ if the field is isotropic, so that $p(k,t) \equiv [p(\boldsymbol{k},t)]_k$ gives a measure of anisotropy due to the disturbance $\boldsymbol{B}_0$ added to the equilibrium

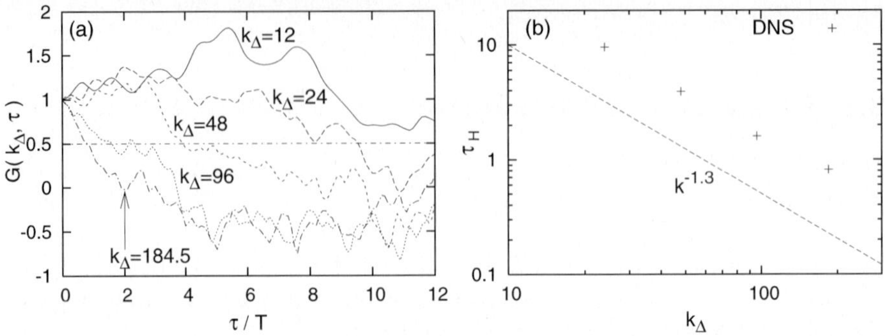

Fig. 1. (a) $G(k_\Delta, \tau)$ vs. $\tau/T \equiv (t - t_F)/T$ for $k_\Delta = 12, 24, 48, 96$ and 184.5, (b) τ_H vs. wavenumber k_Δ. The straight shows the slope $k_\Delta^{-1.3}$

state. Here θ is the angle between $\boldsymbol{k}$ and the third axis, and $[f(\boldsymbol{k})]_k$ is the average of $f(\boldsymbol{q})$ over the spherical surface given $|\boldsymbol{q}| = k$.

Figure 1(a) shows the time dependence of the DNS value of $[p(k_\Delta, t)]'$ normalized by $[p(k_\Delta, t_F)]'$, i.e.

$$G(k_\Delta, \tau) \equiv [p(k_\Delta, \tau + t_F)]'/[p(k_\Delta, t_F)]', \quad \text{for } \tau \geq 0,$$

where $[f(k_\Delta)]'$ is the average of $f(k)$ over $\Delta \equiv [k_0, k_1)$ with $k_\Delta = (k_0 + k_1)/2$. In Fig. 1 (a), we put $(k_0, k_1) = [8, 16)$,$[16, 32)$,$[32, 64)$,$[64, 128)$ and $[128, 241)$. The decay rate may be characterized by the time τ_H such that $G(k_\Delta, \tau_H) = 1/2$. Figure 1 (b) shows τ_H vs. k_Δ. These figures suggest that the decay of $G(k_\Delta, \tau)$ with respect to τ is faster at higher wavenumbers, i.e. at small scales.

The authors would like to thank T. Ishihara for making available his pseudo-spectral code. The simulations were performed on the Fujitsu HPC2500 at the Information Technology Center of Nagoya University. The work was partially supported by Grant-in-Aids for Scientific Research (B)17340117 from the Japan Society for the Promotion of Science, and also by a Grant-in-Aid for the 21st Century COE "Frontiers of Computational Science".

References

1. Kubo R (1966) Rep Prog Phys 29:255–284
2. Ishihara T, Yoshida K, Kaneda Y (2002) Phys Rev Lett 88:154501-1–4
3. Kaneda Y, Yoshida K (2004) New J Phys 6:Art. No. 34
4. Roverts P (1967) An Introduction to Magnetohydrodynamics. American Elsevier
5. Moreau R (1990) Magnetohydrodynamics. Kluwer
6. Schumann U (1976) J Fluid Mech 74:31–58
7. Zikanov O, Thess A (1998) J Fluid Mech 358:299–233
8. Knaepen B, Kassinos S, Carati D (2004) J Fluid Mech 513:199–220
9. Jimenez J, Wray A A, Saffman P G, Rogallo R S (1993) J Fluid Mech 255:65–90
10. Chasnov J R (1995) Phys Fluids 7:600–605

Scaling in an Ensemble of Stochastic Forced Point Vortices

Oliver Kamps[1] and Rudolf Friedrich[1]

Institute of Theoretical Physics Westfälische Wilhelms Universität 48149 Münster
okamp@uni-muenster.de
fiddir@uni-muenster.de

Abstract

Onsager's point vortex model of two dimensional turbulence is extended by the inclusion of time dependent vortex circulations. If the time dependence of the circulations is governed by statistically independent Onstein-Uhlenbeck processes we observe the emergence of scaling regimes for the structure functions of the Eulerian and the Lagrangian velocity increments.

Fully developed turbulent flows are flux equilibrium systems leading to selfsimilarity and scaling behaviour of correlation functions. The probability distributions corresponding to the simplest case of idealized, i.e. homogeneous, isotropic, and stationary fully developed flows are unknown although the underlying fluid dynamical equations and its statistical counterparts are well-established [1]. Fluid motions can be treated either from an Eulerian or a Lagrangian point of view. Most analytical theories have been formulated in the Eulerian framework. Point vortex models (see e.g. [2]), which have been extensively investigated especially for the case of two dimensional flows, essentially make use of a Lagrangian description.

Since the point vortex equations of an ideal two dimensional fluid, which is not stirred, are of Hamiltonian nature, a statistical treatment based on the microcanonical ensemble can be established. This has been discussed for the first time by Onsager [3]. The purpose of the present Letter is to show that an extension of Onsager's point vortex model, which allows for fluctuating circulations of the point vortices, leads to a state which shows scaling behaviour of the structure functions.

1 The Point Vortex Model

The behaviour of two dimensional fluid motion is governed by the vorticity equation:

$$\frac{\partial}{\partial t}\omega(\mathbf{x},t) + \mathbf{u}(\mathbf{x},t)\cdot\nabla\omega(\mathbf{x},t) = \nu\Delta\omega(\mathbf{x},t) - \gamma\omega(\mathbf{x},t) + F(\mathbf{x},t) \ . \tag{1}$$

The fluctuating force $F(\mathbf{x},t)$ is assumed to be Gaussian and δ-correlated in time. Furthermore, the spatial correlations should decay rapidly in space such that the energy injection appears on small scales. The damping term $-\gamma\omega$ extracts energy at the largest scales and avoids the generation of large scale flows. From both, experiment [4] and numerical calculations [5] the model is known to exhibit scaling.

The vorticity equation serves as a motivation for the formulation of the following point vortex model. The vorticity $\omega(\mathbf{x},t)$ is represented as a superposition of point vortices

$$\omega(\mathbf{x},t) = \sum_i \Gamma_i\delta(\mathbf{x}-\mathbf{x}_i(t)) \ . \tag{2}$$

Neglecting viscosity and assuming the driving forces to be located exactly at the point vortices,

$$F(\mathbf{x},t) = \sum_i^N \sqrt{Q}F_i(t)\delta(\mathbf{x}-\mathbf{x}_i(t)) \ , \tag{3}$$

the following equations correspond to a weak solution of the vorticity equation:

$$\begin{aligned}\frac{d}{dt}\mathbf{x}_i(t) &= \sum_{i\neq j}^N \omega_i(t)\mathbf{u}(\mathbf{x}_i(t)-\mathbf{x}_j(t)) \ , \\ \frac{d}{dt}\Gamma_i(t) &= -\gamma\Gamma_i(t) + \sqrt{Q}F_i(t) \ .\end{aligned} \tag{4}$$

The fluctuating forces are assumed to possess the usual white noise characteristics:

$$< F_i(t) >= 0 \quad , \quad < F_i(t)F_j(t') >= \delta(t-t')\delta_{ij} \ . \tag{5}$$

In the next step the velocity field of a point vortex $\mathbf{u}(\mathbf{x},t) = \frac{1}{2\pi}\mathbf{e}_z \times \frac{\mathbf{x}}{\mathbf{x}^2}$ is replaced by the velocity field of a vortex blob

$$\mathbf{u}(\mathbf{x}) \sim a^2\mathbf{e}_z \times \frac{\mathbf{x}}{\mathbf{x}^2+a^2} \ . \tag{6}$$

The vortex model is investigated in a periodic box of lenght L. The radius a is taken to be $a = L/\sqrt{N}$. After nondimensionalizing one can see that system (4) depend only on the parameters N and $\epsilon := \frac{Q}{\gamma^3}$. For the following calculations we choose $N = 10000$ and $\epsilon = 400$.

2 Eulerian Statistics

We have performed numerical studies of the stochastic point vortex model (4). In order to make contact with the Eulerian treatment of turbulent flows the Eulerian longitudinal structure functions of order n

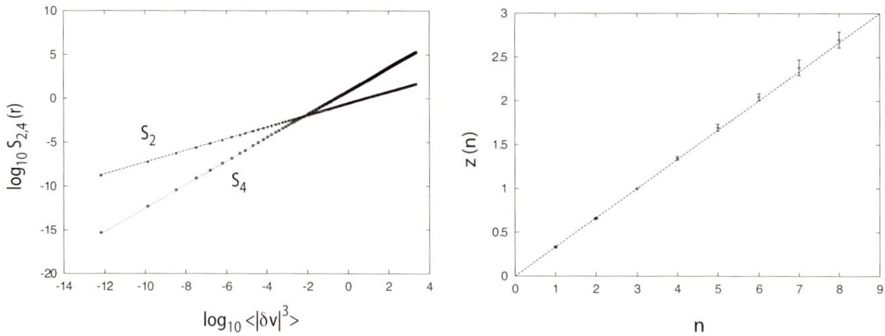

Fig. 1. ESS-plot of S_2 and S_4 (left) and ESS-scaling exponents (right)

$$S_n(r) = \langle [\delta v_r]^n \rangle = \langle [\delta \mathbf{v}_\mathbf{x} \cdot \mathbf{e}_r]^n \rangle. \tag{7}$$

Using the method of extended selfsimilarity (ESS) [6] the scaling behaviour can be extracted most efficiently. Due to ESS the modified structure functions $S_n^* = \langle |\delta v_r|^n \rangle$ are related to the modified third order structure functions $S_3^* = \langle |\delta v_r|^3 \rangle$ according to

$$S_n^* = (S_3^*)^{\zeta(n)}. \tag{8}$$

As it turns out structure functions of the order n scale like $r^{n/3}$ in the inertial range. In left part of the figure on this page the logarithms of S_2 and S_4 from are plotted as a function of the logarithm of S_3^*. One clearly sees scaling behaviour and the indices $\zeta(n) = \frac{n}{3}$ can be extracted with good accuracy (see Fig. 1). This shows that the ESS - exponents of the generalized Onsager models are in agreement with Kolmogorov's similarity theory. Intermittency is not observed in the point vortex model.

3 Lagrangian Statistics

We have also investigated the velocity increment statistics $\delta v_\tau = \dot{x}_1(t+\tau) - \dot{x}_1(t)$ of a single Lagrangian particle along the same lines as described by Mordant et al. [13] for three dimensional turbulence. ($x_1(t)$ denotes the x-coordinate of a point vortex). Using ESS we have considered scaling behaviour of the structure functions $< |\delta v_\tau|^n > \approx < |\delta v_\tau|^2 >^{\eta(n)}$. In contrast to the Eulerian case these functions exhibit multifractal scaling (see Fig. 2). The exponents $\eta(n)$ for small values of n can be well approximated by the formula $\eta(n) = n/2 - \mu n(n-2)$ with $\mu = .015$

4 Conclusion

Summarizing, we have indicated that by introducing fluctuating circulations into a point vortex model of two dimensional turbulence a flux equilibrium condition can be established. By the method of ESS we have investigated

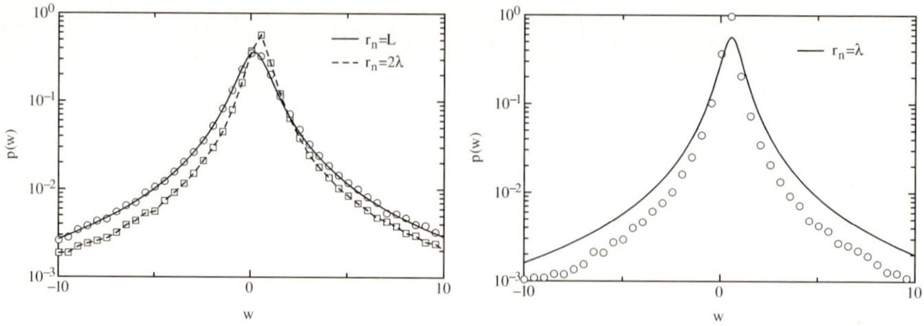

Fig. 1. Both figures show the Cauchy distribution calculated according to (12) (lines) in comparision to the data (symbols). Left: $r_1 = L$, $r_2 = L/2$ (straight line), $r_1 = 2\lambda$, $r_2 = \lambda$ (dotted line). Right: $r_1 = \lambda$, $r_2 = \lambda/2$

3 Conclusion

To summarize, we have derived the Cauchy distribution from a phenomenological Fokker-Planck equation under the assumption that the skewness and the intermittency can be neglected. Taking the estimated drift and diffusions coefficients from the Fokker-Planck equation, the Cauchy distribution (12) coincides with experimental data, see Fig. 1. Thus we have a closed expression for the multiplier's distribution which is valid from the integral length scale down to the Taylor length scale. An interesting point is that in contrast to other methods, like also the proper Fokker-Planck analysis [3], the multiplier's distribution does not grasp intermittency effects or are not sensitive to it, respectively. This is also in accordance with the findings of [1] that the multiplier statistics can be generated by a fractional brownian motion.

For helpful discussion we want to thank B. Castaing.

References

1. Q. Chen, S. Chen, G. L. Eyink, and K. R. Sreenivasan. Kolmogorov's third hypothesis and turbulent sign statistics. *Phys. Rev. Lett.*, 90:art. no. 254501, 2003.
2. A.N. Kolmogorov. A refinement of previous hypotheses concerning the local structure of turbulence in a viscous incompressible fluid at high reynolds number. *J. Fluid Mech.*, 13:82–85, 1962.
3. C. Renner, J. Peinke, and R. Friedrich. Experimental indications for Markov properties of small-scale turbulence. *J. Fluid Mech.*, 433:383–409, 2001.

DNS and Scaling Laws of a Turbulent Channel Flow with Streamwise Rotation

Tanja Weller and Martin Oberlack

Fluid- and Hydromechanics Group, Technische Universität Darmstadt, Petersenstraße 13, 64287 Darmstadt, Germany, weller@hyhy.tu-darmstadt.de, oberlack@hyhy.tu-darmstadt.de

Summary. In this work a turbulent channel flow rotating about the streamwise direction is presented. The theory is based on the investigations of [4] employing the symmetry group theory. It was found that a cross flow in the spanwise direction is induced. Statistical evaluations have shown that all six components of the Reynolds stress tensor are non zero. In this paper the results of a Lie group analysis and a corresponding direct numerical simulations (DNS) are presented.

1 Introduction

Rotating turbulent flows play more a major role in engineering applications such as in gas turbine blade passages, pumps and rotating heat exchangers to name only a few. In these cases secondary flows are induced caused by centrifugal or Coriolis forces.

Investigations of [4] using symmetry theory showed that there is a new turbulent scaling law related to the turbulent channel flow rotating about the mean flow direction. Figure 1 depicts the flow geometry.

The flow has several common features with the classical rotating channel flow rotating about the spanwise direction [2] but also has some different characteristics. The induction of a mean velocity in x_3-direction [5] is the most

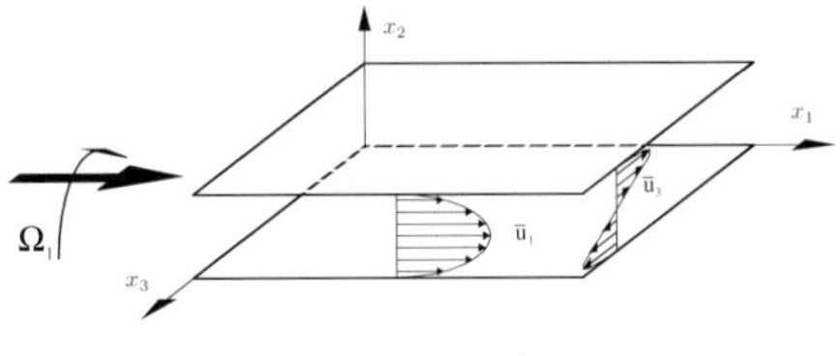

Fig. 1. Sketch of the flow geometry of a turbulent channel flow rotating about the mean flow direction

obvious difference compared to the classical case. In this paper the results of a Lie group analysis and a corresponding DNS at Reynolds number Re=180 for different rotation rates are presented and discussed. The main objective of the present paper is to validate the results from the Lie group analysis.

2 Lie Group Analysis

Lie group analysis is used to find symmetries and invariant solutions of the two-point correlation (TPC) equations. The procedure is divided into three parts. The first step is the computation of the determining system of the TPC equations. An overdetermined system with more than 500 linear partial equations is derived. Its solutions are determined using the software package [1]. As a result eight symmetry groups were obtained. The second step is to find the global transformations. They (1) can be derived employing Lie's first theorem.

$$\begin{aligned}
G_{1-3}: &\quad x_2^*{=}x_2,\, r_i^*{=}r_i{+}a_i,\, \bar{u}_1^*{=}\bar{u}_1,\, \bar{u}_3^*{=}\bar{u}_3,\, R_{ij}^*{=}R_{ij},\, P_i^*{=}P_i \\
G_4: &\quad x_2^*{=}x_2{+}a_4,\, r_i^*{=}r_i,\, \bar{u}_1^*{=}\bar{u}_1,\, \bar{u}_3^*{=}\bar{u}_3,\, R_{ij}^*{=}R_{ij},\, P_i^*{=}P_i \\
G_5: &\quad x_2^*{=}x_2,\, r_i^*{=}r_i,\, \bar{u}_1^*{=}\bar{u}_1{+}a_5,\, \bar{u}_3^*{=}\bar{u}_3,\, R_{ij}^*{=}R_{ij},\, P_i^*{=}P_i \\
G_6: &\quad x_2^*{=}x_2,\, r_i^*{=}r_i,\, \bar{u}_1^*{=}\bar{u}_1,\, \bar{u}_3^*{=}\bar{u}_3{+}a_6,\, R_{ij}^*{=}R_{ij},\, P_i^*{=}P_i \\
G_7: &\quad x_2^*{=}x_2,\, r_i^*{=}r_i,\, \bar{u}_1^*{=}\bar{u}_1,\, \bar{u}_3^*{=}\bar{u}_3,\, R_{ij}^*{=}R_{ij}{\cdot}e^{a_7},\, P_i^*{=}P_i{\cdot}e^{a_7} \\
G_8: &\quad x_2^*{=}x_2{\cdot}e^{a_8},\, r_i^*{=}r_i{\cdot}e^{a_8},\, \bar{u}_1^*{=}\bar{u}_1{\cdot}e^{a_8},\, \bar{u}_3^*{=}\bar{u}_3{\cdot}e^{a_8}, \\
&\quad R_{ij}^*{=}R_{ij}{\cdot}e^{a_8},\, P_i^*{=}P_i{\cdot}e^{a_8}
\end{aligned} \tag{1}$$

The variables r_i, x_2, $\bar{u}_1$, $\bar{u}_3$, R_{ij} and P_i are respectively the correlation radius, wall-normal coordinate, mean velocities in streamwise and spanwise directions, velocity-pressure correlations and TPC functions. $a_1 - a_8$ are the group parameters of the transformations. The first four groups are translation groups where $G_1 - G_3$ represent the translation in correlation space, they have no physical meaning and G_4 the translation in wall-normal direction. G_5 and G_6 correspond to the Galilean invariances of the velocity in streamwise and spanwise directions. G_7 and G_8 are the scaling groups where G_7 means scaling for TPC functions and pressure. G_8 means the scaling for all variables. In the last step the invariant solutions (3) can be obtained from the characteristics equation which is given by

$$\frac{dx_2}{a_8 \cdot x_2{+}a_4} = \frac{d\bar{u}_1}{a_8 \cdot \bar{u}_1{+}a_5} = \frac{d\bar{u}_3}{a_8 \cdot \bar{u}_3{+}a_6} = \frac{dR_{ij}}{a_7 \cdot R_{ij}} = \cdots \tag{2}$$

$$\bar{u}_1 = c_1 \cdot \Omega_1 \cdot x_2{+}c_2, \quad \bar{u}_3 = c_3 \cdot \Omega_1 \cdot x_2{+}c_4 \tag{3}$$

The invariant solutions for $\bar{u}_1$ and $\bar{u}_3$ are linear functions of the wall-normal coordinate x_2. To verify the invariant solutions (scaling laws) a DNS was performed.

3 Direct Numerical Simulation

The numerical technique which was chosen is a standard spectral method with Fourier decomposition in streamwise and spanwise direction as well as Chebyshev decomposition in wall-normal direction. The numerical code for channel flow was developed at KTH/Stockholm [3]. Additional features such as the streamwise rotation and statistics were added during the project. After the simulation were finalized all flow quantities were normalized on the friction velocity.

Four computations at at rotation numbers Ro=2, Ro=5, Ro=7 and Ro=10 and Reynolds number $Re = 180$ have been conducted. The size of the domain used in the x_1, x_2, and x_3 directions are respectively 4π, 2, and 2π on a $128 \times 129 \times 128$ grid at rotating numbers Ro=2 and Ro=5. At rotating number Ro=7 and Ro=10 the length of the domain was doubled in x_1 and x_3 direction and the number of grid points were increased up to 256 only in x_1-direction. All computations were run for $10000 \frac{h/2}{u_c l}$ time units and the statistics accumulation was performed for the last 5000 time units.

In Figs. 2 and 3 the mean velocity profiles are presented. The velocity profile in streamwise direction decreases invariably with increasing rotation numbers. For smaller rotation numbers faster as for the higher rotation numbers. Each mean velocity profile (for example see Fig. 4) has a linear region on each side of the centerline in the core of the flow. The near-wall regions up to $x_2 = \pm 0.9$ are only marginally perturbed. In the DNS the theoretically expected cross flow could be verified. In general the spanwise velocity

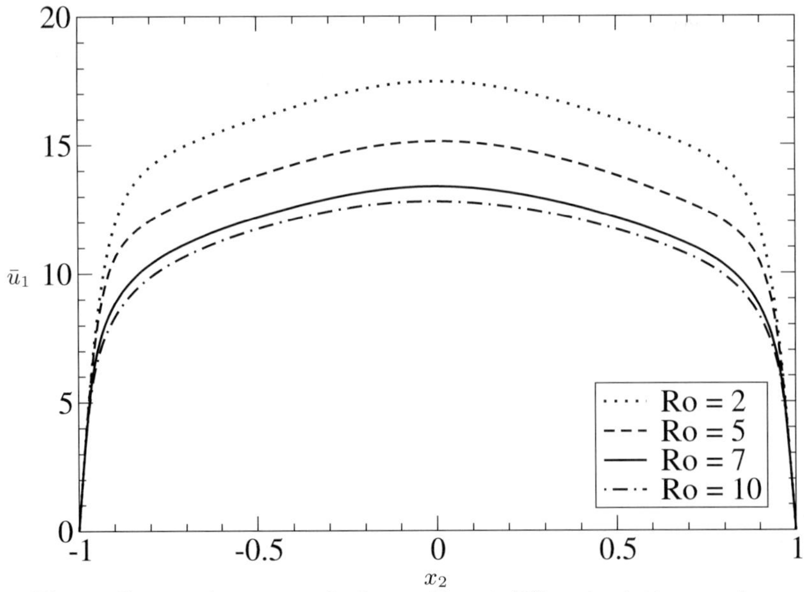

Fig. 2. Streamwise mean velocity profiles at different rotation numbers

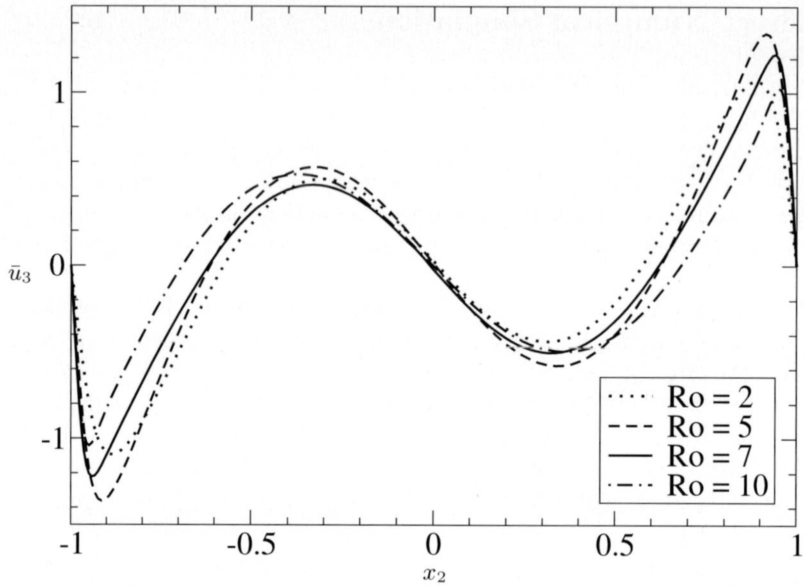

Fig. 3. Spanwise mean velocity profiles at different rotation numbers

profiles are skew-symmetric about the centerline and the predicted linear profiles are marginally visible (see Fig. 5). For small rotation rates up to Ro=5 the spanwise mean velocity profiles (Fig. 3) increase. At rotation number Ro=7 this effect appears to reverse. As predicted from the lie group analysis the DNS shows also that all six components of the Reynolds stress tensor

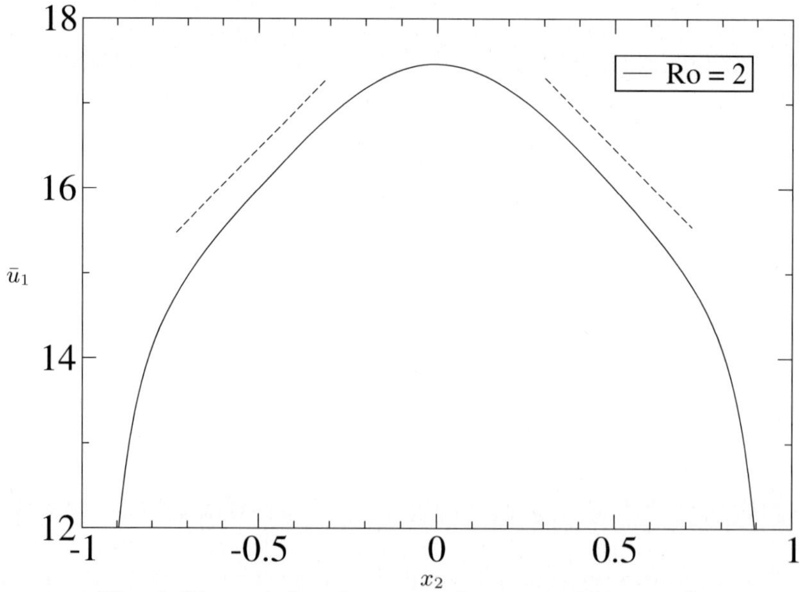

Fig. 4. Linear regions in streamwise mean velocity profile

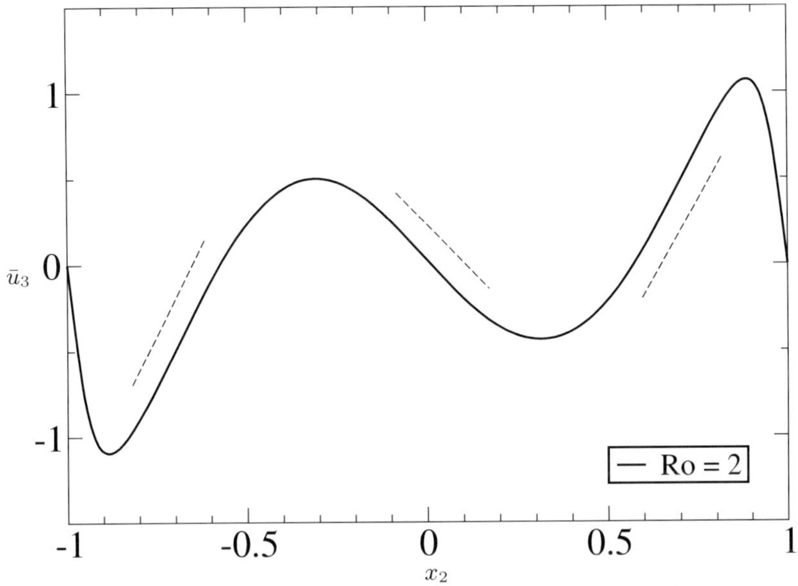

Fig. 5. Linear regions in spanwise mean velocity profile

(not shown here, see [6]) are non zero and all statistical curves are symmetric or skew-symmetric about the centerline.

4 Conclusions and Summary

The results of the DNS agree with the results of the symmetry theory. Symmetries and scaling laws provide the theoretical results for turbulence modeling which can quantitatively be adjusted employing the DNS data. With the DNS the induced phenomena of a cross flow in spanwise direction has been computed for different rotation rates. It has been confirmed by DNS that there are linear regions in both the streamwise and spanwise mean velocity. Furthermore, it could be verified that all components of the Reynolds stress tensor are non zero.

References

1. Carminati, J. , Vu, K. (2000): Symbolic computation and differential equations: Lie symmetries. J. Symbolic Comp.,29:95-116.
2. Johnston, J. P. and Halleen, R. M. and Lazius, D. K. (1972): Effects of spanwise rotation on the structure of two-dimensional fully developed turbulent channel flow. J. Fluid Mech. 56:533–557.
3. Lundbladh, A., Henningson, D., Johanson, A. (1992): An efficient spectral integration method for the solution of the Navier-Stokes eqautions. FFA-TN 1992-28, Aeronautical Research Institute of Sweden, Bromma.

4. Oberlack, M., Cabot, W., Rogers, M. M. (1998): Group analysis, DNS and modeling of a turbulent channel flow with streamwise rotation. Studying Turbulence Using Numerical Datebase - VII, Center for Turbulence Research, Stanford University/NASA Ames, 221–242.
5. Oberlack, M. (2001): A unified approach for symmetries in plane parallel turbulent shear flows. J. Fluid Mech., 427:299–328.
6. Weller, T., Oberlack, M. (2005): DNS of a Turbulent Channel Flow with Streamwise Rotation - Investigation on the Cross Flow Phenomena, accepted for publication in Proc. of DLES6.

Part II

Experiments

experiment has been obtained for the inertial range and for the evolution time of the structure T_v.

In this paper, we present new results on this experiment using PIV measurements, that give direct access to the spatial velocity fields. The procedure used allows phase averaging of the quantities extracted from the PIV velocity fields, so that real ensemble average for different times of the flow cycle can be performed.

1.1 Experimental Set Up

A stretched vortex is generated in a low velocity hydrodynamic channel. A small step added to a laminar boundary layer profile in the bottom wall produces the initial vorticity that is strongly enhanced by the stretching produced by sucking the flow through slots on each lateral wall (Fig. 1). Varying the experimental parameters, i.e. the suction flow-rate Q_2, and the downstream flow-rate Q_1, makes two regimes occur: in the first one the vortex is stable, in the second one the vortex is experiencing periodical bursts (Fig. 2). The flow is characterized by the following experimental values: $R = 3$ cm (lateral extension of the burst), $Re = [4000 - 6000]$, with $Re = \frac{\Gamma}{\nu}$ and Γ the vortex circulation for $r = R$), a coarse estimation of the stretching $a = \frac{\partial U_z}{\partial z}$, where z is the axis of stretching as shown in Fig. 1, is given by $a = 1 - 10$ s^{-1}.

We focus on the second regime, where the velocity field of the flow is measured via synchronized hot film velocity measurements and PIV measurements. The PIV measurements are performed in a cross section of the vortex at the middle of the channel ($(z = 0)$ plane) (Fig. 3). Our PIV system is composed of a high resolution camera (1280 × 1024 pixels) capturing images

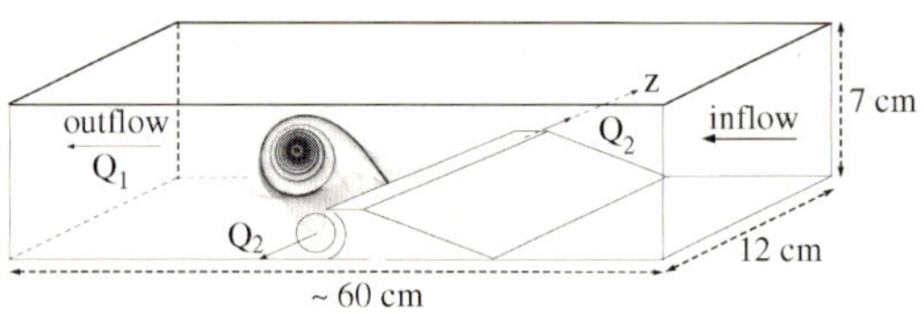

Fig. 1. Experimental setup

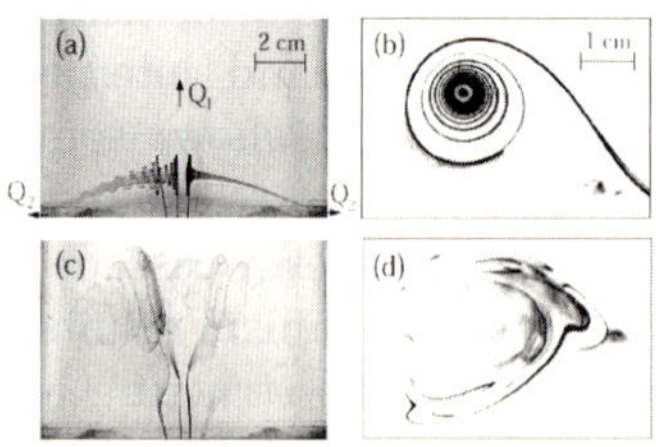

Fig. 2. Flow visualizations (a-b), laminar state of the flow, (a) top view (b) cross section, (c-d) turbulent state of the flow , (c) top view, (d) cross section

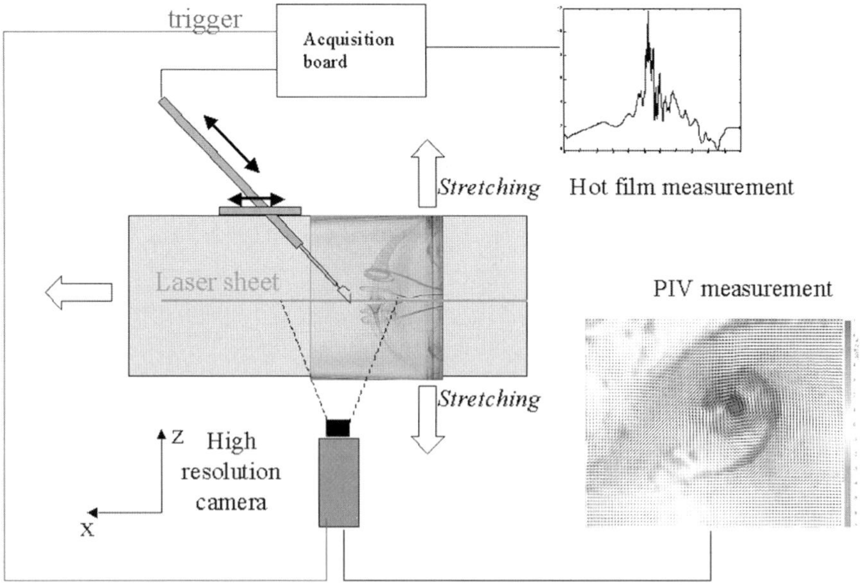

Fig. 3. Experimental setup for synchronized PIV and hot film acquisition

at a frequency of 4 Hz and a double pulsed Nd:Yag laser delivering 12 mJ at each pulse. We use a measurement areas of 5 cm × 4 cm. The interrogation window W chosen is 32 × 32 pixels [2]. The spatial resolution obtained is 86 × 64 vectors.

The PIV measurement is synchronized with a hot film measurement. The beginning of the hot film acquisition is triggered by the first PIV measurement and is continuous afterwards. The hot film probe is located in the middle of the channel and is set parallel to the z axis so that it measures: $U = \sqrt{U_r^2 + U_\theta^2}$, where U_r is the radial velocity and U_θ is the azimuthal velocity.

2 Data Processing

A typical velocity signal from the hot film shows the periodicity of the flow (a period of duration $T \simeq 10$ s is formed of a laminar phase and a turbulent phase). Times t_n for the beginning of the nth cycle are determined via an *ad-hoc* post processing of this signal detailed in [2]. Simultaneously to the hot film acquisition, a large number of PIV velocity fields are recorded ($N \simeq 4000$ corresponding to about 200 cycles). The acquisition time t_a of each velocity field is first rescaled using $t = t_a - t_n$, with $0 \le t \le T$. Hence, all velocity fields, which initially come from different cycles, are indiscriminately time-located (with time t) in a unique (virtual) cycle. The whole cycle $[0, T]$ is divided into short sub-intervals $I_i = [t_i, t_i + \Delta t[$, with $\Delta t \ll T$ but large enough to

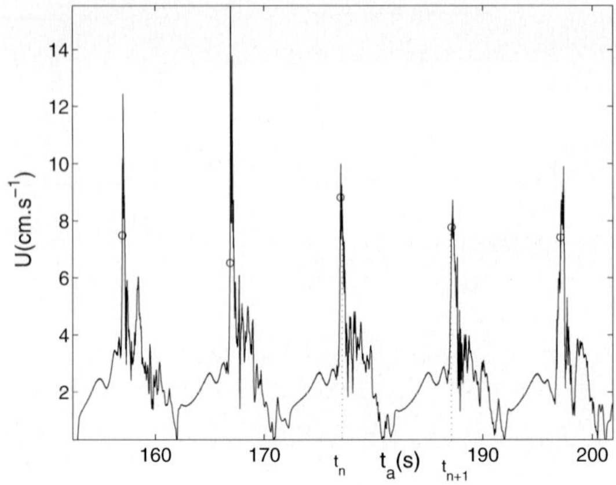

Fig. 4. Hot film signal over 5 cycles, times t_n indicate the beginning of the turbulent part for each cycle

have several velocity fields in each interval. With $\Delta t = 0.1$ s, approximately 40 velocity fields are recorded at time $t \in I_i$. These velocity fields are denoted $\mathbf{U}_p(\mathbf{r}, t)$, with $\mathbf{r}$ the vector position and and $p = 1, \ldots, N_p \sim 40$ the velocity field record index. It is now possible to define the ensemble average at time t_i

$$\langle \mathbf{U} \rangle (\mathbf{r}, t_i) \equiv \frac{1}{N_p} \sum_p \mathbf{U}_p(\mathbf{r}, t \in I_i). \tag{1}$$

and to decompose each velocity field between a fluctuating velocity and a mean velocity:

$$\mathbf{U}_p(\mathbf{r}, t_i) = \langle \mathbf{U} \rangle (\mathbf{r}, t_i) + \mathbf{u}'_p(\mathbf{r}, t_i) \tag{2}$$

From mean and fluctuating velocity fields, we can compute mean $\langle \omega \rangle (\mathbf{r}, t_i)$ and fluctuating vorticity fields $\omega'_p(\mathbf{r}, t_i)$. An example of such a decomposition between mean and fluctuating vorticity at $t_i = 0\ s$ is shown on Fig. 5. This

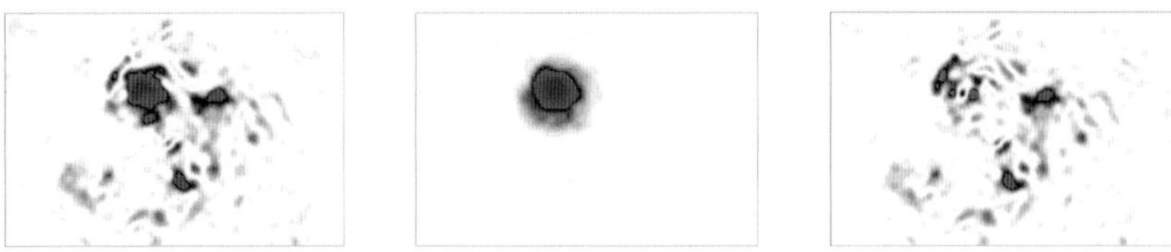

Fig. 5. Examples of decomposition between mean vorticity and fluctuating vorticity at $t_i = 0\ s$. From left to right: the whole vorticity field $\omega_p(\mathbf{r}, t)$, the mean vorticity field $\langle \omega \rangle (\mathbf{r}, t_i = 0)$ and the fluctuating vorticity field $\omega'_p(\mathbf{r}, t)$

decomposition clearly exhibits a fluctuation part composed of vorticity patches structures surrounding a nearly axisymmetric averaged vorticity field $\langle\omega\rangle$.

3 Characterization of the Vortex in the Spectral Space

An important issue of the present study is the characterization of the build-up of the turbulent energy spectrum with time. In order to characterize the temporal evolution of the burst in the spectral space, we have defined in [1, 2] quasi-instantaneous energy spectra and cumulative spectra from the hot film measurements, which provide respectively the quasi-instantaneous and time integrated spectral signature of the turbulent vortex. However two difficulties arise in the computation of these spectra from the hot film measurements. First, the correspondence between temporal velocity fluctuations and spatial velocity fluctuations is achieved using a local Taylor hypothesis in a critical condition. Second, to keep a minimal spectral resolution, spectra are computed over a minimal time interval Δt, independent of the number of recorded cycles.

The use of PIV measurements overcomes these two difficulties. We can directly write the instantaneous bidimensional velocity PSD $e(k,t)$ as:

$$e(k,t_i) \equiv \langle k \int_0^{2\pi} |\widehat{\mathbf{u}'_p}(\mathbf{k},t_i)|^2 d\theta_k \rangle \tag{3}$$

Where $\mathbf{k} \equiv (k,\theta_k)$ in polar coordinates, and $\widehat{\mathbf{u}'_p}$ is the fourier transform of $\mathbf{u}'_p$.

Then, from the instantaneous spectra we can define the cumulative spectrum $E^c(k,t_i)$ representing the time integrated spectrum of the flow as:

$$E^c(k,t_i) = \frac{1}{i}\sum_{j=1}^{i} e(k,t_j) \tag{4}$$

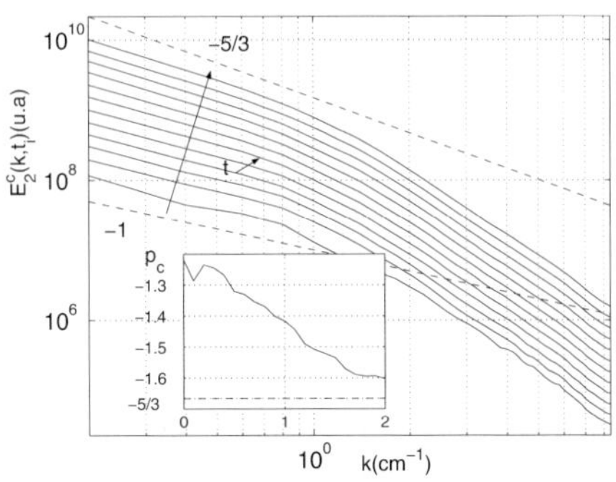

Fig. 6. Evolution of the instantaneous PIV energy spectra exponents with time

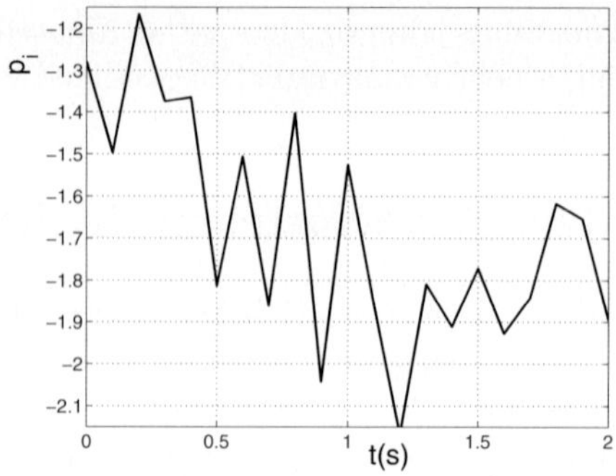

Fig. 7. PIV Energy spectra integrated over increasing times, the final spectra integrated over T_v shows a -5/3 slope. The inset shows the evolution of the corresponding spectra slopes

The evolution of the instantaneous energy spectra is characterized through the evolution of the spectral exponent p_i computed over the inertial range. A general evolution trend from -1 to -2 of the spectral slope is observed as shown in Fig. 6.

Cumulative spectra for different times t_i and corresponding evolution of the spectral slope p_i are represented on Fig. 7. The -5/3 behavior appears to result from the time integration of the instantaneous spectra over a time $T_v = 2$ s. Previous results obtained from hot film measurements in [1, 2] are confirmed without uncertainties associated with hot film measurements.

4 Conclusion

The results reported in this article show the ability of PIV phase averaged measurements to resolve temporal and spatial scales associated with the dynamic evolution of the burst.

In [1, 2], the transition between -1 and -2 was interpreted as the shearing of vortex patches into vorticity layers. A work in progress is to use the PIV data to model the experimental flow as a particular Lundgren solution in order to test this hypothesis. The decomposition of the vorticity field between a fluctuating and a mean field obtained from the PIV data is useful for this prospect.

References

1. Cuypers Y, Maurel A, Petitjeans P (2003). Vortex burst as a source of turbulence. Phys. Rev. Lett 91:19450.
2. Cuypers Y, Maurel A, Petitjeans P (2004). Comparison between an experimental turbulent vortex and the Lundgren vortex *J. Of Turb.* 5:30.
3. Lundgren T S (1982). Strained spiral vortex model for tubulent fine structures. Phys. Fluids 25:2193-2203.

Experimental Detection of the New Phenomenon of Turbulent Thermal Diffusion

A. Eidelman, T. Elperin, N. Kleeorin, A. Markovich, I. Rogachevskii

Department of Mechanical Engineering, Ben-Gurion University of the Negev, POB 653, 84105 Beer-Sheva, Israel; elperin@bgu.ac.il

The main goal of this communication is to review our recent experimental studies of a new phenomenon of turbulent thermal diffusion which was predicted theoretically in [1]. The essence of this phenomenon is the appearance of a nondiffusive mean flux of particles in the direction of the mean heat flux. This phenomenon causes formation of large-scale inhomogeneities in the spatial distribution of particles that accumulate in the regions of minimum mean temperature of the surrounding fluid.

The mechanism of the phenomenon of turbulent thermal diffusion for $\rho_p \gg \rho$ is as follows (see [1, 2]), where ρ_p is the material density of particles and ρ is the density of the fluid. The inertia causes particles inside the turbulent eddies to drift out to the boundary regions between eddies (i.e. regions with low vorticity or high strain rate and maximum of fluid pressure, [3]). Therefore particles are accumulated in regions with maximum pressure of the turbulent fluid. Similarly, there is an outflow of particles from regions with minimum pressure of fluid. In homogeneous and isotropic turbulence without large-scale external gradients of temperature a drift from regions with increased (decreased) concentration of particles by a turbulent flow of fluid is equiprobable in all directions, and pressure (temperature) of the surrounding fluid is not correlated with the turbulent velocity field. Thus, there exists only turbulent diffusion of particles.

The situation drastically changes in a turbulent fluid flow with a mean temperature gradient, whereby the turbulent heat flux $\langle \mathbf{u}'\theta' \rangle$ is not zero, i.e. fluctuations of temperature θ' and velocity $\mathbf{u}'$ of the fluid are correlated. Fluctuations of temperature cause fluctuations of pressure of the fluid, and the pressure fluctuations result in fluctuations of the number density of particles. Increase (decrease) of the pressure of the surrounding fluid is accompanied by accumulation (outflow) of the particles. Therefore, direction of the additional turbulent flux of particles coincides with that of turbulent heat flux, i.e. $\langle \mathbf{u}'n' \rangle \propto \langle \mathbf{u}'\theta' \rangle \propto -\boldsymbol{\nabla} T$, where T is the mean fluid temperature and n' are the fluctuations of particle number density. The additional turbulent flux of particles is directed towards the location with the minimum mean temperature, and the particles are accumulated in this region.

The additional effective velocity of particles caused by the phenomenon of turbulent thermal diffusion is given by $\mathbf{V}_{\rm eff} = -\tau \langle \mathbf{v}' \operatorname{div} \mathbf{v}' \rangle = -D_T(1 + \kappa(a))\boldsymbol{\nabla} T/T$ ([1,2]), where $\mathbf{v}'$ are the fluctuations of particle velocity, D_T is the turbulent diffusion coefficient, τ is the correlation time of the turbulent velocity field, the coefficient $\kappa(a)$ depends on a particle size a and the Reynolds number. The additional turbulent flux of particles, $N\,\mathbf{V}_{\rm eff}$, is caused by turbulent thermal diffusion, where N is the mean number density of particles. The velocity $\mathbf{V}_{\rm eff} = -\tau \langle \mathbf{v}' \operatorname{div} \mathbf{v}' \rangle$ was also derived by other methods in [4,5].

First experiments in oscillating grids turbulence that demonstrated the phenomenon of turbulent thermal diffusion, were discussed in [6,7]. Later, the phenomenon of turbulent thermal diffusion was observed in multi-fan turbulence generator [8]. Observation time in these devices is almost not limited due to the small mean velocity of the flow. The phenomenon of turbulent thermal diffusion in strongly inhomogeneous three-dimensional temperature fields was recently studied in a turbulence generated by oscillating grids in the air flow. In all these experiments we used Particle Image Velocimetry to determine a turbulent velocity field, an Image Processing Technique to determine the spatial distribution of particles, and a temperature probe with twelve thermocouples to measure the temperature field. It was demonstrated that even in strongly inhomogeneous temperature fields particles in non-isothermal turbulent fluid flow accumulate at the regions with minimum of mean temperature of surrounding fluids due to the phenomenon of turbulent thermal diffusion.

Let us discuss recent experiments which were carried out in a turbulence generated by oscillating grids in the air in strongly inhomogeneous three-dimensional temperature fields. The test section of the oscillating grids turbulence generator was constructed as a rectangular chamber of dimensions $29 \times 58 \times 29$ cm^3. Pairs of vertically oriented grids with bars arranged in a square array (with a mesh size 5 cm) are attached to the right and left horizontal rods. Both grids are driven independently with speed-controlled motors and are positioned at a distance of two grid meshes from the chamber walls parallel to them. This two grids system is capable of oscillating at a controllable frequency up to 20 Hz.

A mean temperature gradient in the turbulent flow was formed with two aluminium heat exchangers attached to the bottom and top walls of the chamber. We performed experiments with two directions of the mean temperature gradient, for stably stratified fluid flow (the cooled bottom and heated top walls of the chamber) and for unstably stratified fluid flow (the heated bottom and cooled top walls of the chamber). In order to improve heat transfer in the boundary layers at the walls we used a heat exchanger with rectangular fins ($3 \times 3 \times 15$ mm) which allowed us to form a mean temperature gradient up to 200 K/m for the stably stratified flow and up to 110 K/m for the unstably stratified flow at a mean temperature of about 300 K. The temperature field was measured with a temperature probe equipped with twelve thermocouples.

The turbulent velocity field was measured using a digital Particle Image Velocimetry (PIV) system with LaVision Flow Master III. A double-pulsed light sheet is provided by a Nd-YAG laser (Continuum Surelite 2×170 mJ). We employed a progressive-scan 12 Bit digital CCD camera (pixels with a size 6.7μm $\times 6.7\mu$m each) with dual frame technique for cross-correlation processing of captured images. An incense smoke with sub-micron particles as the tracer was used for PIV measurements. Smoke (with $\rho_p/\rho \sim 10^3$ and the mean particle diameter is 0.7μm) was produced by high temperature sublimation of solid incense particles.

We determined the mean and the r.m.s. velocities, two-point correlation functions and an integral scale of turbulence from the measured velocity fields. Series of 130 pairs of images acquired with a frequency of 4 Hz were stored for calculating velocity maps and for ensemble and spatial averaging of turbulence characteristics. We measured velocity for flow areas from 60×60 mm^2 up to 212×212 mm^2 with a spatial resolution of 1024×1024 pixels. The mean and r.m.s. velocities for every point of a velocity map (1024 points) were calculated by averaging over 130 independent maps, and then they were averaged over 1024 points. An integral scale l of turbulence was determined from the two-point correlation functions of the velocity field.

Particle spatial distribution was determined using digital Particle Image Velocimetry (PIV) system. In particular, the effect of Mie light scattering by particles was used to determine the particle spatial distribution in the flow. In order to characterize the spatial distribution of the mean particle number density $N \propto E^T/E$ in the non-isothermal flow, the distribution of the scattered light intensity E measured in the isothermal case was used for the normalization of the scattered light intensity E^T obtained in a non-isothermal flow under the same conditions. The scattered light intensities E^T and E in each experiment were normalized by corresponding scattered light intensities averaged over the vertical coordinate.

The turbulent flow parameters in the oscillating grids turbulence generator are as follows: the r.m.s. velocity is $\sqrt{\langle \mathbf{u}'^2 \rangle} = 3.6 - 14$ cm/s depending on the frequency of grid oscillations, the integral scale of turbulence is $l = 1.6 - 2.3$ cm, the Kolmogorov length scale is $\eta = 380 - 660\,\mu$m. The measured r.m.s. velocity was several times higher than the characteristic mean velocity in the core of the flow. The Reynolds numbers and the turbulent diffusion coefficient in the core of the flow are almost independent of the distances from the grid.

The measured mean temperature fields and the particle number density distributions for the stably and unstably stratified turbulent fluid flows generated by two grids oscillating with the different frequencies ($f = 2 - 15$ Hz) demonstrated that particles are redistributed in the non-isothermal flow. In particular, they are accumulated in regions with minimum mean temperature, and there is outflow of particles from the regions where the mean temperature is larger.

In the unstably stratified flow there are two sources of turbulence, the external mechanical forcing produced by oscillating grids and the buoyancy effect. The buoyancy effect causes a strong change in spatial distribution of the mean temperature. Nevertheless, even in strongly inhomogeneous three-dimensional mean temperature fields, the locations of regions with the minimum mean temperature are strongly correlated with the locations of the regions with the maximum particle number density due to the phenomenon of turbulent thermal diffusion.

In these experiments the contribution of the mean fluid flow to the spatial distribution of particles is negligibly small. In particular, the normalized distribution of the scattered light intensity measured in the different vertical strips in the flow in the regions where the mean flow velocity and coefficient of turbulent diffusion vary strongly are practically identical (the difference is about 1 %). The number density of particles in our experiments was of the order of 10^{10} particles per m^3, so that their collision rate is negligibly small. The effect of the gravitational settling of small particles ($0.5 - 1\,\mu$m) is negligibly small since the terminal fall velocity of these particles is less than 0.01 cm/s. The accuracy of the measurements in these experiments ($\sim 0.5\%$) is considerably higher than the magnitude of the observed effect ($\sim 10\%$). Therefore, our experiments [6–8] detected the effect of turbulent thermal diffusion. This phenomenon can cause formation of the large-scale aerosol layers in the atmospheric turbulence with temperature inversions [9].

We are indebted to H. J. S. Fernando, E. Kit, V. L'vov, A. Tsinober and V. Yakhot for illuminating discussions. We thank A. Krein and I. Sapir-Katiraie for his assistance in construction of the experimental set-up and performing the experiments. This work was partially supported by the German-Israeli Project Cooperation (DIP) administrated by the Federal Ministry for Education and Research (BMBF) and by the Israel Science Foundation governed by the Israeli Academy of Science.

References

1. Elperin T, Kleeorin N, Rogachevskii I (1996) Phys Rev Lett 76: 224-228
2. Elperin T, Kleeorin N, Rogachevskii I, Sokoloff D (2001) Phys Rev E 64: 026304
3. Maxey MR (1987) J Fluid Mech 174: 441-465
4. Pandya RVR, Mashayek F (2002) Phys Rev Lett 88: 044501
5. Reeks MW (2005) Int J Multiphase Flow 31: 93-114
6. Eidelman A, Elperin T, Kleeorin N, Krein A, Rogachevskii I, Buchholz J, Grünefeld G (2004) Nonlinear Processes in Geophysics 11: 343-350
7. Buchholz J, Eidelman A, Elperin T, Grünefeld G, Kleeorin N, Krein A, Rogachevskii I (2004) Experiments in Fluids 36: 879-887
8. Eidelman A, Elperin T, Kleeorin N, Rogachevskii I, Sapir-Katiraie I (2006) Experiments in Fluids, in press
9. Blackadar AK (1997) *Turbulence and Diffusion in the Atmosphere*, Springer, Berlin

Anemometry in Snow Particle Flows

M. Hölling[1], S. Barth[2], J. Peinke[2] and J.-D. Rüedi[3]

[1] Institute of Physics, University of Oldenburg, michael.hoelling@uni-oldenurg.de
[2] Institute of Physics, University of Oldenburg
[3] WSL Swiss Federal Institute for Snow and Avalanche Research (SLF), ruedi@slf.ch

Summary. We present velocity data collected with two different anemometers during measurements in a snow wind tunnel. This wind tunnel allows for experiments with wind blowing over naturally fallen snow. A hot-film anemometer and a Laser-Cantilever-Anemometer (LCA) were placed in the wind tunnel with an initial distance from the snow surface of about 5 cm. The main point of this experiment was to investigate the possibility of wind speed measurements within a flow loaded with snow particles. In particular the impact of snow particles on the sensors are evaluated.

1 Introduction

A major problem in dealing with velocity measurements in two phase flows is the interaction between the sensor and the different phases of the flow which can lead to corrupted data. In the case of snow particles interacting with a sensor, thermodynamical and mechanical effects can be expected. The interaction depends on the sensor type.

2 Used Anemometers

For the measurements in the snow wind tunnel at the Swiss Federal Institute for Snow and Avalanche Research (SLF) in Davos we used two different types of anemometers; a commercial hot-film anemometer and a new Laser-Cantilever-Anemometer. These anemometers differ in the physical principle of measuring the flow velocity. The hot-film anemometer detects the flow velocity via a change in the resistance of a heated hot-film due to different cooling rates depending on the flow velocity, cf. [2]. The LCA measures flow velocities by the optical detection of the velocity dependent deflection of a tiny cantilever brought into the flow [1]. Figure 1 left side shows a cantilever chip with flow optimized geometry. A laser is focused on the tip of the cantilever,

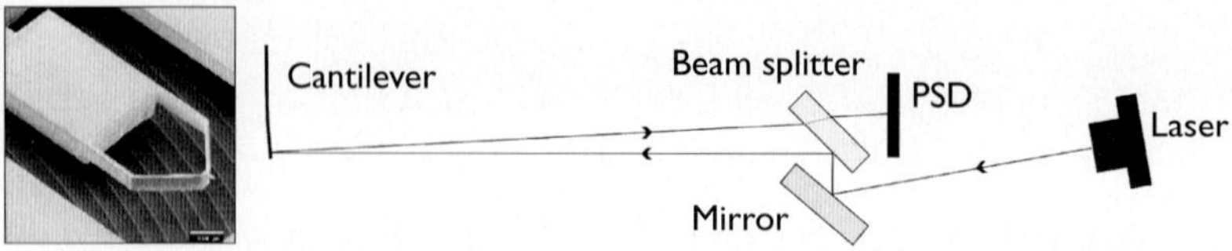

Fig. 1. Left: cantilever chip with window for flow optimization. Cantilever at the edge of triangular structure; flow direction is normal to the chip plane. Right: path of light inside the Laser-Cantilever-Anemometer

and its reflection is detected by a PSD (Position Sensitive Detector) element. Figure 1 right side shows the path of light of the laser beam inside the enclosure of the sensor. Since the deflection of the cantilever is proportional to the flow velocity squared, the measured signal allows to determine the flow velocity. Measurements under laboratory conditions showed that this new sensor provides results with a spatial-temporal resolution comparable to the ones of hot-wire measurements in air and to hot-film measurements in water [1].

3 Snow Particle - Sensor Interaction

Due to different measurement techniques the effect of a snow particle impact on the two sensors is also different. A snow particle melts after hitting a heated hot-film and leaves water that evaporates within a certain amount of time (Fig. 2). During that process, the regular cooling by the air flow is superimposed with evaporative cooling (latent heat transfer). This leads to corrupted data until all water is evaporated. From this process, we expect an exponential decaying behavior with one time constant t_0 as a response to a snow particle impact.

$$f(t) = A \cdot \exp\left(-\frac{t}{t_0}\right) + const. \tag{1}$$

Since the cantilever of the LCA is not heated, the snow particle does not melt after impact. Figure 3 shows two simple possibilities for interaction between a snow particle and the cantilever. One is a simple elastic impact where the snow particle hits the cantilever and bounces off afterwards. Another possibility arises due to the complex structure and rough surface of the snow

Fig. 2. Snow particle hits hot-film, melts and evaporates

Fig. 3. Snow particle hits cantilever - various interactions between snow particle and cantilever are possible, e.g. elastic impact or snow particle wanders along the cantilever's surface

particle. Due to this effect, a snow particle may stick to the cantilever after the impact and wander along the cantilever's surface until it finally comes off. Therefore the measured signal should show more than **just one** characteristic shape for an impact.

3.1 Results

As mentioned above, a snow particle impact on a hot-film results in melting of the snow. Figure 4(a) shows the signal (normalized by the signal mean) of three measured impacts. It shows the characteristics of an exponential decay with fluctuations. Figure 4(b) shows the average over all impact signals with a fitted function according to (1). The fit parameters deliver a time constant t_0 of around 0.72 ms for the averaged impact. After 1.67 ms the effect of the impact has dropped by 90%.

For an impact on the LCA we expect, as discussed above, a more individual response signal for each snow particle impact event. Figure 5(a) shows the measured signal, normalized by signal mean, for three impacts on the cantilever. As expected, the shapes of these impacts vary for each event. Where the curve on the far left has a clear and sharp peak, the other two curves show

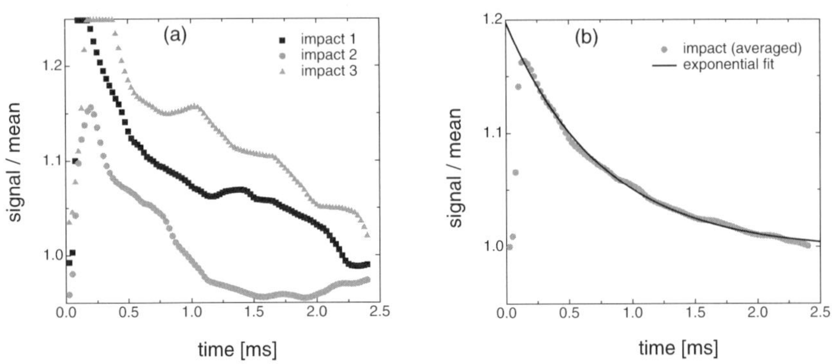

Fig. 4. (a): three impacts normalized by average. (b): signal of averaged impacts with fitted function according to (1)

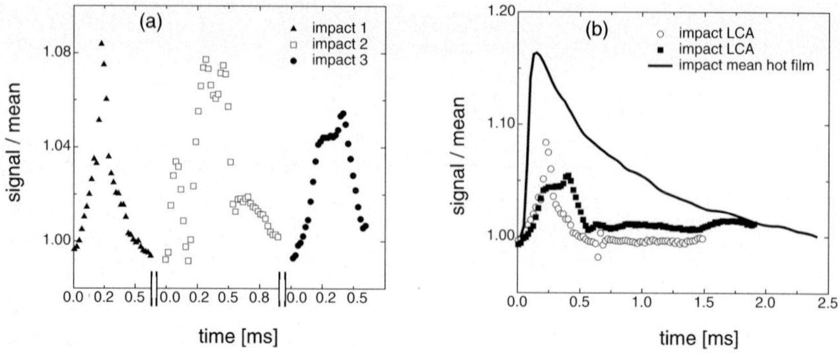

Fig. 5. Left: signals of three impacts on LCA normalized by signal mean. Right: comparison of averaged impact on hot-film with two impacts on LCA

a more or less distinct plateau before decreasing again. Figure 5(b) shows a direct comparison of impact events on the two sensors. It can clearly be seen that a snow particle impact on the LCA results in smaller signal distortion than an impact on the hot-film, both in amplitude and time. Where the hot-film is affected by the impact for 1.67 ms the LCA's relaxation time is in the order of 0.6 ms; after that time the acquired data can be related to the present flow.

4 Conclusions

Impacts of snow particles on LCA and hot-film show significant differences. Side by side measurements with both anemometers show that data collected with the hot-film is more strongly affected by a snow particle impact than data acquired with the LCA. Therefore the data collected with the LCA contains more information about the actual flow.

References

1. Barth S., et al. (2005) Laser-Cantilever-Anemometer - A new high resolution sensor for air and liquid flows, Rev. Sci. Instrum. 76 075110
2. Bruun H.H. (1995) Hot-wire Anemometry (Principles and signal analysis), Oxford Science Publications

Numerical and Experimental Investigations of the Dynamics and Structure of the Recirculation Zone in a Coaxial Jet Mixer

N. Kornev[1], S. Jahnke[2], I. Tkatchenko[1], V. Zhdanov[1] and E. Hassel[1]

[1] University of Rostock, Institute for Technical Thermodynamics, Albert-Einstein-Str. 2, 18059 Rostock, Germany
nikolai.kornev@uni-rostock.de
[2] Department of Fluid Dynamics steffen.jahnke@uni-rostock.de

1 Introduction

Mixing and particular mixing in a coaxial jet mixer has been a subject of investigations for a long time. References concerning this subject can be found in [1], [2] and [3]. The classical coaxial jet mixer consists of a pipe of diameter D with a coaxially arranged center pipe with a nozzle diameter d. Depending on the flow rate ratio there are two typical modes. Following Barchilon and Curtet [1] if $D/d < 1 + \dot{V}_D/\dot{V}_d$ the flow is similar to a free jet regime (henceforth j-mode). Otherwise we have a strong recirculation zone close the nozzle (henceforth r-mode) (Fig. 1). The recirculation enhances the mixing efficiency drastically so that the homogenous state is reached after four or five pipe diameters downstream of the nozzle what makes this mixing regime attractive for various homogenization devices. The present work is focused on investigations of flow physics within the recirculation zone in the r-mode with experimental and numerical tools.

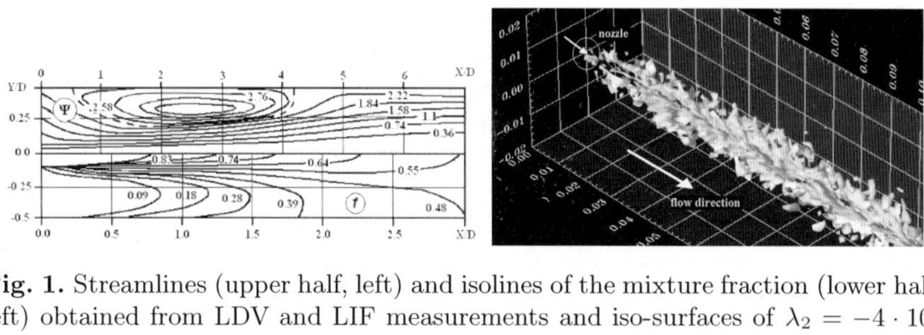

Fig. 1. Streamlines (upper half, left) and isolines of the mixture fraction (lower half, left) obtained from LDV and LIF measurements and iso-surfaces of $\lambda_2 = -4 \cdot 10^4$ (right, LES DMM)

2 Investigation Tools

The governing equations for incompressible miscible liquids consist of the filtered Navier-Stokes equations and the filtered transport equation for a normalized passive scalar, called mixture fraction $f \epsilon [0, 1]$ [2]. The unclosed terms are modelled using different subscale models including Smagorinsky, dynamic Germano, dynamic mixed (henceforth referred to as the DMM model) and vortex based models (see [5]). The dynamic mixed model with corrections of Vreman et al. [4] applied both for scalar and velocity fields was found to provide the best correlation with experiments than other subscale models [2], [3]. The wall model of Grötzbach and Piomelli [5] was used to avoid high resolution necessary for LES simulations in the wall region. Turbulent inlet conditions are calculated by superposition of fluctuations generated by a novel inflow generator with prescribed statistic properties and mean turbulent profiles. Further details of the numerical method can be found in [2]. Numerical calculations were carried out in the computational domain with the length of eight diameters of the pipe using an in-house code for LES. In terms of D the computational domain has the size $8 \times 2\pi \times 0.5$ in a cylindrical coordinate system (x, θ, r). Test calculations with grids of different resolutions showed that the grid $N_x = 256$, $N_\theta = 64$, $N_D = 32$ and $N_d = 6$ provides the necessary accuracy at the lowest consumption of computational resources. The calculations presented here have been performed for the r-mode corresponding to the flow rate ratio $\dot{V}_D/\dot{V}_d = 1.3$, the diameter ratio $D/d = 5$ and the Reynolds number $Re_d = U_d d/\nu = 10^4$. The experiments were carried out in a closed water channel [2]. The velocity field has been measured by the one component LDV system. The concentration of the passive scalar was determined by planar LIF measurements with spatial resolution 0.3 mm and repetition rate 10 Hz. All details concerning the measurements can be found in [2].

3 Results

The mixing in the r-mode is a very complicated hydrodynamic phenomenon characterized by flow separations, intermittency and unsteady motions of various time and space scales. The visualizations of instantaneous flow performed first by Barchilon et al. [1] show that the motion in the recirculation zone is highly unsteady and nearly periodic. At that time they were unable to investigate this phenomenon quantitatively due to lack of the necessary measurement technique. The instantaneous pictures of the flow revealed a consequent process of generation of regular vortices in the recirculation zone, their shedding downstream resulted in the formation of staggered rows of vortices like in the von Karman trails [1]. Unfortunately, this valuable observation has never been quantified and thoroughly analyzed in the scientific literature. The Reynolds averaged network of streamlines (Fig. 1, left) became a typical flow pattern in the jet mixer. However, this picture is not informative since the flow experiences large oscillations and the averaged pictures do not represent the flow physics. From our computational experience the URANS simulations using SST, $k - \epsilon$ and even RSM models do not reproduce the flow physics in the recirculation zone properly. The flow pattern is unrealistic smoothed, any pronounced oscillations are absent. On the contrary, the LES simulations are proved to be an appropriate numerical tool providing the detailed information on the flow physics in the r-mode. Visualization of the flow calculated using the LES DMM (Fig. 2, left) revealed the presence of a set of primary and secondary vortices induced by the primary ones at the pipe walls. They are building clusters which move downstream and thus represent the unsteady behavior of the recirculation zone. The multiple vortex structure of the recirculation zone is confirmed by flow visualizations pictures obtained from LIF (Fig. 2, right). Certainly, these pictures can be compared only qualitatively since they represent instantaneous stages of the flow. Comparison of averaged characteristics obtained by LES, URANS and measurements (a-posteriori LES assessment) is presented in our papers [2] and [3]. The LES DMM model agrees with measurements in the recirculation zone well. A very interesting result of our investigations is revealing and quantification of long period oscillations induced by the recirculation zone. The time series of the mixture fraction at two locations within the recirculation zone from both LES and measurements show that there is a long time period oscillation characterized by a dominating low frequency ω (Fig. 3, left). A Fourier analysis shows that this frequency is between 0.15 and 0.2 Hz for both experiments and LES depending on the position in the flow. The corresponding Strouhal number is about $Sh = \omega \cdot D/U_D = 0.12$. Barchilon et al. [1] observed the opposition-of-phase between the upper and the lower part of the recirculation zone. Actually, the signal history and autocorrelation function changing its sign across the pipe (Fig. 3, right) indicate that the oscillations are antisymmetric with respect to the pipe centerline. On the contrary, the oscillations in confined jets are symmetric [6]. Naturally, like in any fully developed turbulent flow, the regular long-period oscillation is superimposed on a series of short period oscillations, so that the resulting field is stochastic and irregular.

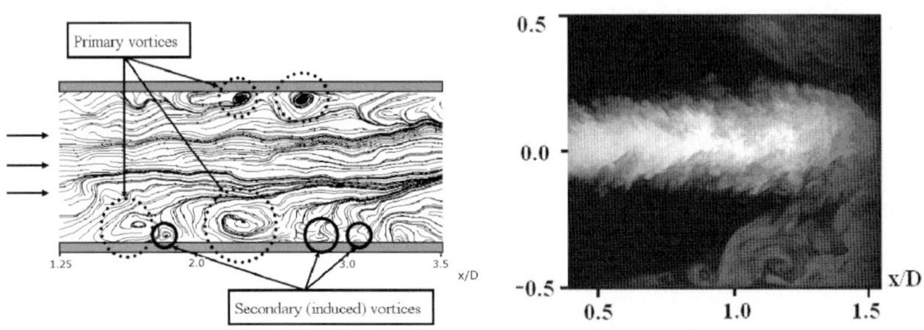

Fig. 2. Instaneous streamlines from LES-DMM (left) and a snapshot of the scalar field from LIF (right)

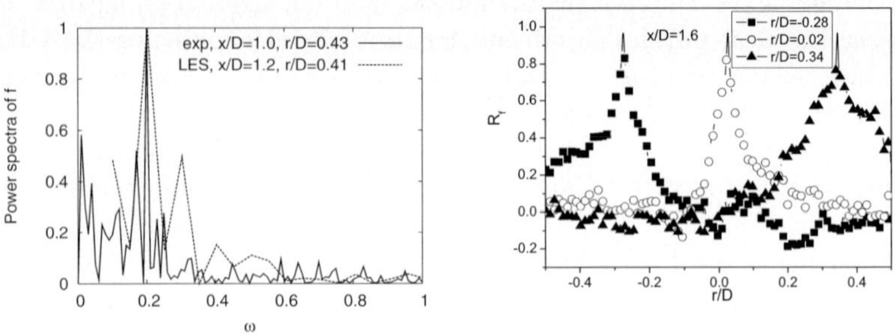

Fig. 3. Energy spectra of the mixture fraction fluctuations (left, LIF and LES) and the two-point auto correlation $R_f(r, \eta) = \overline{f'(r)f'(r+\eta)}/(\overline{f'^2(r)})$ across the mixer at three points in the cross-section $x/D = 1.6$ (right, LIF)

The presence of long period oscillations in the recirculation area has a strong impact on the methodology of LES assessment. The real time of LES simulations in this case should be 5–6 periods of long period oscillations, e.g. about 40 seconds at least to evaluate statistics properly. Assumptions that the long period oscillations are caused by large regular vortex structures like, for example, preceding vortex were not confirmed by analysis of the iso-surfaces of the λ_2-criterion (Fig. 1, right). There is no dominating vortex structures. The strongest vortex structures look like a tree with branches oriented agains the flow. Generation of large vortex clusters in the recirculation zone and their interaction leads to an intermittence of the scalar dynamics. Typical pdf distributions of the mixture fraction in the recirculation zone is of a multi-mode type what clearly shows that the the instantaneous scalar field is intermittent here.

4 Conclusion

Investigations of the r-mode of the coaxial jet mixer showed that the recirculation zone is unsteady and not dominated by a large regular vortex structure. LES and LIF analysis revealed the existence of long period oscillations of scalar in the recirculation zone. The oscillation at points symmetrical with respect to the mixer centerline are nearly antisymmetric.

References

1. Barchilon M, Curtet R (1964) J. Basic Engineering 777-787
2. Kornev N, Tkatchenko I, Zhdanov V, Hassel E, Jahnke S (2005) Proc. Turb. Shear and Flow Phenomena 4, Williamsburg VA USA: 723-728
3. Hassel E, Jahnke S, Kornev N, Tkatchenko I, Zhdanov V (2006) Large-Eddy Simulation and laser diagnostic measurements of mixing in a coaxial jet mixer. Chem. Eng. Science. 61:2908-2912

4. Vreman B, Geurts B, Kuerten H (1994) Physics Fluids 6 12:4057-4059
5. Sagaut, P (2003) Large Eddy Simulations for Turbulent Incompressible Flows. Springer
6. Villermaux E, Rehalb H (2000) J. Fluid Mech. 425:161-185

On Anisotropy of Turbulent Flows in Regions of "Negative Eddy Viscosity"

A. Liberzon[1], B. Lüthi[2], M. Guala[1], W. Kinzelbach[1] and A. Tsinober[3]

[1] Institute of Hydromechanics and Water Resources Management, ETH, Zurich, Switzerland liberzon/guala/kinzelbach@ihw.baug.ethz.ch
[2] Risø National Laboratory, Roskilde, Denmark luethi@risoe.dk
[3] Department of Fluid Mechanics and Heat Transfer, Tel Aviv University, Israel tsinober@eng.tau.ac.il

1 Introduction

"Negative eddy viscosity" is a phenomenon in which the turbulent transport of momentum occurs against the mean velocity gradient and it is related to processes in which the Reynolds stress tensor $\langle u_i u_j \rangle$ and the mean rate-of-strain tensor S_{ij} are positively correlated, so that the turbulent kinetic energy (TKE) production, $P = -\langle u_i u_j \rangle S_{ij}$ is negative [1]. This phenomenon was observed in laboratory conditions in Rayleigh-Bénard convection [2], in asymmetric channel flows, flow through annuli, wall jets and asymmetric plane jets [3]. It is known also to occur in large scale flows in geo- and astrophysics [4].

We analyze three-dimensional particle tracking velocimetry (3D-PTV) [5] measurements in a cubic Rayleigh-Bénard convection cell, reported in [6]. We repeat the experiment in which a region of negative TKE production was observed in both experiment and numerical simulations [2]. Turbulent shear flow in Rayleigh-Bénard convection is driven by buoyancy and thus is different from other shear flows in terms of the mechanisms responsible for negative TKE production, possibly related to the anisotropy and to the inhomogeneity of the flow. It is thus particularly interesting to relate the region of negative TKE production to the anisotropy and inhomogeneity of the flow and to compare it to the region of negative TKE production in a shear flow where no buoyancy is involved. It is also of fundamental interest to have an insight into the structure of the flow in which locally, in space, energy is transferred from velocity fluctuations to the mean flow, and to compare it to the structure in the region of positive turbulent energy production.

2 Experimental Setup and Method

The experiments are performed in water, in Rayleigh-Bénard convection in a cubic cell [2], and in a turbulent shear flow without buoyancy, obtained in the proximity of the wall with an array of rotating discs [6, 7]. The detailed discussion of the experiments, shown in Fig. 1, is given in [2, 6, 7] and here only a brief description is presented. The observation volumes are of the order of 2 cm^3, and shown as shaded in Fig. 1. In the first experiment the temperature difference is $\Delta T = 13K$, the mean temperature of the fluid is $T_m = 23.5°C$, the Rayleigh and Prandtl numbers are 1.6×10^9 and 6.1, respectively. In the second experiment, the turbulent flow field is maintained by the four counter-rotating disks with a constant angular speed of 250 rpm, rotating in the directions indicated by arrows in Fig 1b. We use the 3D-PTV system [5] to track the neutrally buoyant 40 μm polystyrene particles.

3 Results & Discussion

In our previous work [6] we found some correspondence between anisotropy of the Reynolds stresses and the sign of TKE production. The "tensor of anisotropy" is defined in [8], as $b_{ij} = \langle u_i u_j \rangle / \langle u^2 \rangle - 1/3\ \delta_{ij}$, where u^2 is the turbulent kinetic energy, and δ_{ij} is the Kronecker delta tensor. This tensor is dimensionless, has zero trace, vanishes identically if the turbulence is isotropic, and is described by two invariants, $II = \frac{1}{2} b_{ij} b_{ij}$, $III = \frac{1}{3} b_{ij} b_{jk} b_{ki}$. In the plane of these invariants, the universal envelope which is valid for virtually any turbulent flow, is called an anisotropy invariant map (AIM) [8].

In Fig. 2 we depict the AIM and the scatter of II, III for the flow cases with (top panel, a-b) and without buoyancy (bottom panel, c-d). The left

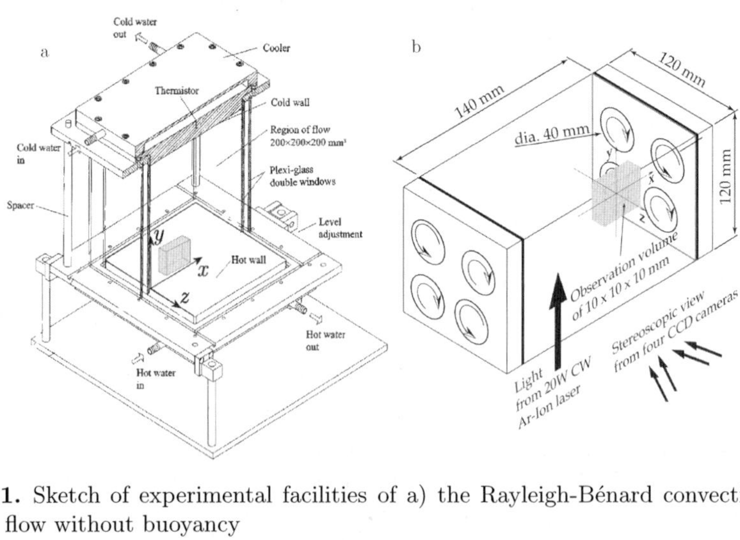

Fig. 1. Sketch of experimental facilities of a) the Rayleigh-Bénard convection b) shear flow without buoyancy

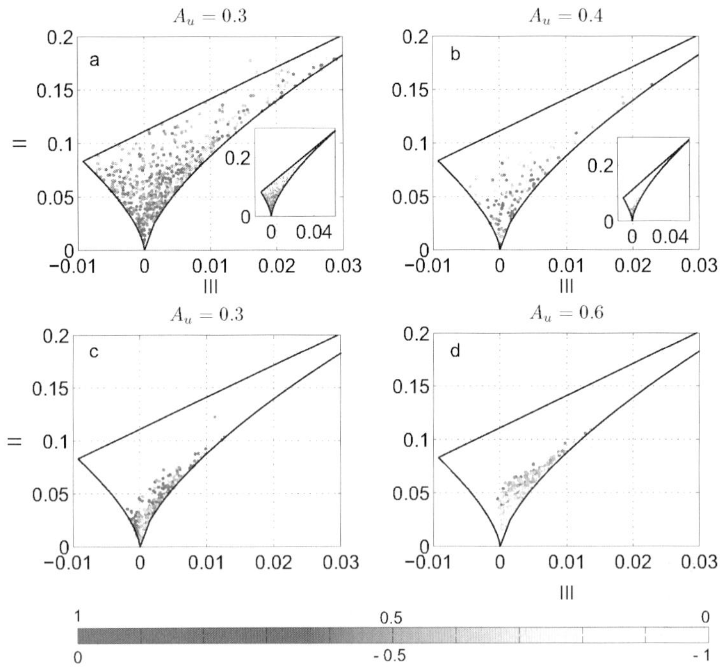

Fig. 2. AIM of the Reynolds stresses of the flow with (a,b) and without (c,d) buoyancy, in the negative (a,c) and positive (b,d) TKE production regions, respectively. The color scale corresponds to $cos(-\langle u_i u_j \rangle, S_{ij})$, and it is $0 \div 1$ / $-1 \div 0$ for the positive and negative TKE production regions, respectively

panel is devoted to the positive production regions of the convective flow (a) and of the shear flow (c). In the right panel we depict the scatter of the negative TKE production regions of the convective flow (b) and the shear flow case (d). It is noteworthy that *in both* regions of positive and negative TKE production, the Reynolds stresses are *anisotropic*, though the degree-of-anisotropy, A_u is higher in the negative TKE production regions. However, anisotropy in the regions of positive and negative TKE production is of *different* type, as it corresponds to different areas of the AIM. We emphasize the relation between the TKE production and the type of anisotropy by adding color to the scatter, according to the magnitude of normalized TKE production. Color levels vary from white (0 for positive P, -1 for negative) to black (1 for positive, 0 for negative). We observe that the scatter from the regions of positive and negative TKE production is distributed in different areas of anisotropy map. The positive production of kinetic energy is more associated with small, negative values of III (in Fig. 2a,c), and a trend is towards the two- and three-component turbulence (the two-component situation corresponds to the left-side vertex of the triangle). On the contrary, negative TKE production, shown in Fig. 2b,d, is associated more with *rod-like* turbulence

[8], in which one of the components of velocity is higher than the other two, and a trend towards the top-right vertex, which relates to a one-dimensional turbulent flow. We detect also a difference between the flows with and without buoyancy, in agreement with our results [6]. The effect of buoyancy on the type of anisotropy of the Reynolds stresses is seen in the different scatter distributions of the top panel (Fig. 2a-b) and the bottom panel (Fig. 2c-d).

4 Summary and Conclusions

Regions of negative turbulent kinetic energy (TKE) production are observed and investigated experimentally in a shear flow with buoyancy in Rayleigh-Bénard convection in a cubic cell and in a mechanically agitated shear flow without buoyancy. This work focuses on the TKE production in these regions and comprises the investigation of some aspects of the anisotropy of the Reynolds stress tensor. We utilized a non-intrusive and fully three-dimensional experimental technique, namely three-dimensional particle tracking velocimetry (3D-PTV) [5] to measure directly the velocity fields, along with the full tensor of velocity derivatives.

We found that negative TKE production regions in both flow cases are associated with an increased anisotropy as it is shown through the anisotropy invariant map (AIM) [6]. This effect is much more pronounced in the flow case with buoyancy, but in the flow case without buoyancy it is still present. The main observation in the present study is that the regions of positive and negative TKE production are connected to the different types of anisotropy: the region with $P > 0$ is more associated with two- and three-component turbulence and the $P < 0$ region is more related to axi-symmetric turbulent flow.

Acknowledgments

This research is supported by ETH research funding, under TH-18/02-4.

References

1. Monin AS, Yaglom AM (1971) Statistical fluid mechanics, MIT Press, Cambridge, MA
2. Burr U, Kinzelbach W, Tsinober A (2003) Phys Fluids 15:2313–2320
3. Hanjalic K, Launder BE (1972) J Fluid Mech 51:301–335
4. Tsinober A, (1990) in Gyr A (ed), Structure of turbuence and drag reduction, Springer, Berlin Heidelberg New York
5. Lüthi B, Tsinober A, Kinzelbach W (2005) J Fluid Mech 528:87–118
6. Liberzon A, Lüthi B, Guala M, Kinzelbach W, Tsinober A (2005) Phys Fluids 17:095110
7. Liberzon A, Guala M, Lüthi B, Kinzelbach W, Tsinober A (2005) Phys Fluids 17:031707
8. Lumley JL, Newman GR (1977) J Fluid Mech 82:161–178;
   Choi KS, Lumley J (2001) J Fluid Mech 436:59–84

Experimental Analysis on the Transition Process of Internal Gravity Waves in a Strongly Stably-Stratified Mixing Layer

Hideharu Makita[1], Katsuhisa Ohba[2] and Nobumasa Sekishita[3]

[1] Department of Mechanical Engineering, Toyohashi University of Technology, Toyohashi, Japan makita@mech.tut.ac.jp
[2] Department of Digital Engineering, Numazu College of Technology, Numazu, Japan ooba@numazu-ct.ac.jp
[3] Department of Mechanical Engineering, Toyohashi University of Technology, Toyohashi, Japan seki@mech.tut.ac.jp

1 Introduction

Internal gravity waves are generated in geophysical flow fields and their collapse produces strong turbulence in them [1]. In the present study, a strongly stably-stratified mixing layer with a large positive temperature gradient was generated in a low turbulence wind tunnel equipped with a thermal stratification generator. Simultaneous measurements were made on temperature and velocity fluctuations for spontaneously induced waves and the waves excited by a thermal fluctuation with three components satisfying Thorpe's condition [2]. Higher order correlation terms were calculated to clarify the transition process of the internal gravity waves, especially the occurrence of counter-gradient heat flux at their collapse.

2 Experiment Setup

Figure 1 shows the experimental setup. The wind tunnel has a test section of $0.42 \times 0.42m^2$ in cross section and 8.0m in length [3], [4]. A thermal stratification generator composed of sixty arrays of coil heaters was installed upstream of its contraction nozzle to reduce background turbulence to be less than 0.1%. A temperature-controlled ceiling and vacuum glass sidewalls almost completely removed thermal contamination from outside of the test section. The generator produced a mixing layer with stepwise temperature and velocity distributions of 17.8mm in initial thickness in the test section. Mean velocity was U_0 = 3.0m/s in the non-heated region and maximum temperature and velocity difference were $\Delta\Theta_{max}$=22.9K and ΔU_{max} = 0.18m/s,

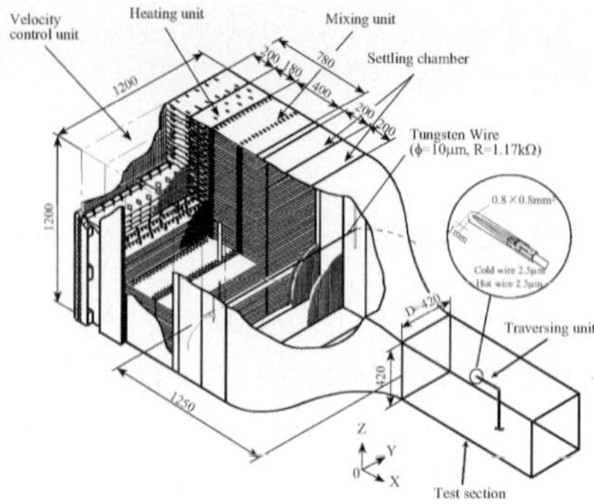

Fig. 1. Experimental setup. (unit:mm)

respectively. Maximum temperature gradient reached about 1.5×10^3K/m around the center of the mixing layer at X/D=0, as shown in fig. 2. Local Richardson number exceeded R_{ic} =0.25 [5] across the stratified mixing layer, which means that the internal gravity waves can be self-excited by buoyancy working as restoring force on the fluid motion. Small sinusoidal thermal fluctuations having three resonant components for the natural gravity waves were given to regulate the transition process of the internal gravity waves from a thin wire of $10\mu m$ in diameter horizontally settled at the center of the mixing layer just behind the generator.

Simultaneous measurement was made on temperature fluctuation, θ , and velocity fluctuations, u and w, by a high precision thermo-anemometer system (S/N ratio 60dB, DC~5kHz) [3], [4]. Its digital delay circuit instantaneously compensated the error by the gap between an upstream I cold-wire and a downstream X hot-wire based on the frozen pattern hypothesis and higher order correlation terms were calculated from the well-compensated signals. Then, the streamwise development of the natural and the excited internal

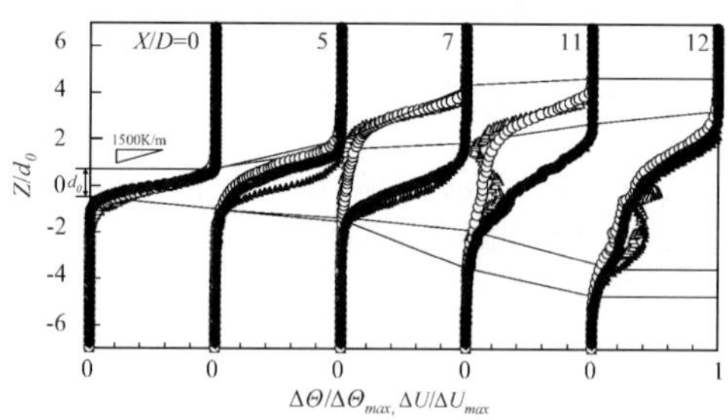

Fig. 2. Mean temperature and velocity distribution.

$\Delta\Theta$: natural ; ●, excited ; ○

U : natural ; ▲, excited ; △

gravity waves and the occurrence of counter-gradient heat flux were analyzed in detail through spectral and phase analyses.

3 Experimental Results and Consideration

Smooth stepwise mean temperature distribution at $X/D=0$ in Fig. 2 confirms that the thermal excitation scarcely affected the initial conditions of the mean flow field. The mixing layer thickness grows downstream through the usual down-gradient heat and momentum diffusion. When excited, the mean temperature and velocity distributions shift upwards from the natural case for $X/D \approx 7 \sim 11$. Rapid growth of the excited internal gravity waves may cause the change in the distributions of mean fields for the excited case. At $X/D=12$, smooth temperature distribution is suddenly broken to have complex profiles, suggesting that the transfer mechanism has changed to be complicated one. In the temperature spectrum in Fig. 3, the energy density level of temperature fluctuation apparently increases until $X/D=7$, below the Brunt-Väisälä frequency, N_B. For the natural case, three frequency components of 1.1, 2.1 and 3.2Hz satisfying Thorpe's three-wave resonance condition have developed in the mixing layer at $X/D=7$. The buoyancy caused by the large negative density gradient is known to work actively on fluid motion and grows the internal gravity waves from the smallest perturbations even in the natural unexcited mixing layer [6], [7]. When excited, the resonant frequency components are more clearly recognized already at $X/D=5$ and the occurrence of the internal gravity waves is well stabilized at the same time, suggesting that the energy becomes more efficiently transported from the mean temperature gradient to the wave components. Higher harmonics are generated in the downstream region and their nonlinear wave-wave interaction causes the breakdown of the internal gravity waves.

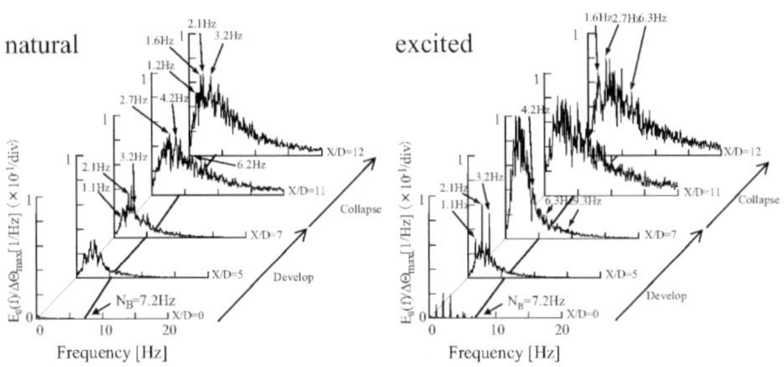

Fig. 3. Streamwise change in spectra of temperature fluctuation

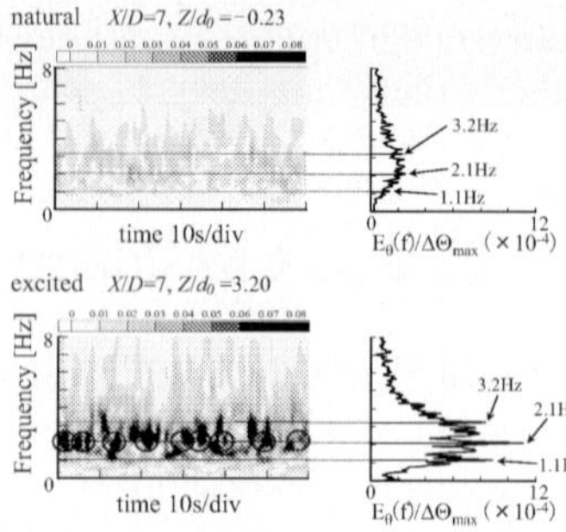

Fig. 4. Wavelet coefficient of temperature fluctuation at X/D=7

The results of the frequency and the wavelet analyses in Fig. 4 illustrate the instantaneous structure of the internal gravity waves at X/D=7. In this figure, solid lines denote the resonant frequencies of the internal gravity waves. For the natural waves, the high-energy regions appear only unsteadily on the resonant frequencies, showing that the resonant frequency components in the spectrum do not continuously exist at a fixed point of the mixing layer. For the excited case, the high-energy regions enclosed by circles are clearly observed on each excited resonant frequencies. It is known that the occurrence of the internal gravity waves is effectively stabilized and the internal gravity waves become to grow stronger in the stratified mixing layer.

Figure 5(a) compares the distributions of a time-averaged vertical heat flux, $-\overline{w\theta}/U_0\Delta\Theta_{max}$, and a local temperature gradient, dΘ/dZ, between the natural and the excited cases, where down- and counter-gradient heat fluxes are denoted by positive and negative values, respectively. And, Figs. 5(b), (c) show the production term and the turbulence intensity of w-components at X/D=12. The down-gradient heat flux proportional to the local temperature gradient dominates till X/D=7. The temperature gradient is still positive at X/D=12, however, the counter-gradient heat flux is generated in the lower

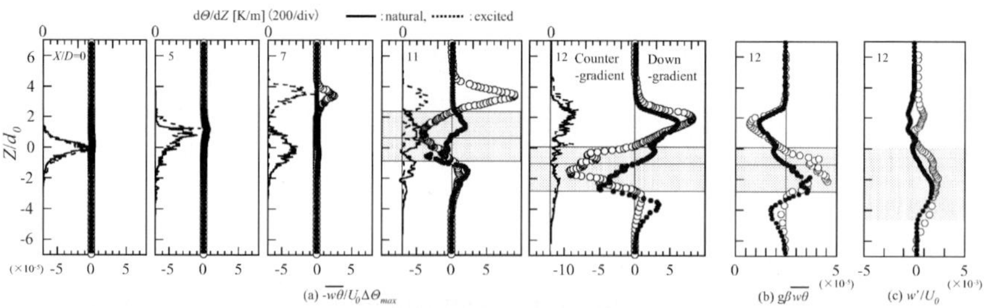

Fig. 5. Comparision among (a) vertical heat flux, (b) production term and (c) turbulence intensity.

•, —: natural, ○, ----:excited

side of the mixing layer. At the same time, the negative turbulence production of the w-components occurs in the same region of the mixing layer, as the hatched in Figs.5 (b), (c). There, the random component of the velocity fluctuation was produced by the collapse of the internal gravity waves as the results of their strong nonlinear interaction. Conclusively, the collapse of the internal gravity waves induces the negative turbulence production and generates the counter-gradient heat flux against the suppression effect by the buoyancy.

Acknowledgement

This work was partially supported by the Japanese Ministry of Education through grants-in-aid (No.15560138).

References

1. Turner,J.S.,(1973),Buoyancy effects in fluids, Cambridge University Press.
2. Thorpe,S.A.,(1966), On wave interactions in a stratified fluid, J. Fluid Mech., 24:737-751.
3. Makita, H., Mori, S. and Yahagi, A., (1994), Spontaneous generation of internal gravity waves in a wind tunnel, Proc., 4th IMA Conf. on Stratified Flow:81-91.
4. Makita, H., Ohba, K., Sekishita, N., (2002), Occurrence of internal gravity waves and counter-gradient heat-flux in a strongly stably-stratified mixing layer, Advances in Turbulence IX:605-608.
5. Hazel, P., (1967), Numerical studies of the stability of inviscid stratified shear flows, J. Fluid Mech., 51:39-61.
6. Stewart, R. W., (1969), Turbulence and waves in a stratified atmosphere, Radio Science, 4:1269-1278.
7. McBean, G. A., Miyake, M., (1972), Turbulent transfer mechanisms in the atmospheric surface layer, Qurt. J. Met. Soc., 98:383.

The Correlation Between Velocity and Acceleration in Turbulence

Jakob Mann, Søren Ott, Jacob Berg and Beat Lüthi

Wind Energy Dept., Ris National Laboratory, Roskilde, Denmark
jakob.mann@risoe.dk

1 Introduction

It can be proven that the single point correlation of velocity and acceleration $\langle \mathbf{v} \cdot \mathbf{a} \rangle$ in homogeneous and stationary turbulence is zero. However, in locally homogeneous flow without body forces the correlation is minus the energy dissipation, i.e.

$$\langle \mathbf{v} \cdot \mathbf{a} \rangle = -\varepsilon. \tag{1}$$

This relation seems also fulfilled in real turbulent flows [3, 2, 4]. The Eulerian velocity–acceleration structure function $\langle \delta\mathbf{v} \cdot \boldsymbol{\delta}\mathbf{a} \rangle$, where $\delta\mathbf{v} = \mathbf{v}(\mathbf{x}+\mathbf{r}) - \mathbf{v}(\mathbf{x})$ and likewise for $\mathbf{a}$, can be shown theoretically to be -2ε independent of $r = |\mathbf{r}|$ in the inertial subrange. This is true for homogeneous, stationary turbulence, where the body forces act on larger scales, as well as for locally homogeneous flows.

For real flows and for Direct Numerical Simulation (DNS) the relation

$$\langle \delta\mathbf{v} \cdot \delta\mathbf{a} \rangle = -2\varepsilon \tag{2}$$

seems to be fulfilled approximately, often better than the other "exact" relation $\left\langle \delta v_{\parallel}^3(r) \right\rangle = -\frac{4}{5}\varepsilon r$. Here we investigate whether (1) and (2) can be used in Lagrangian experiments to determine the energy dissipation. We present new particle tracking data where the energy dissipation has been measured directly through the velocity differences in the viscous subrange simultaneously with the accelerations.

2 Experiment

Four cameras from Basler record 50 frames per second with a resolution of 1000×1000 pixels. A typical run lasts 200 s corresponding to 10^4 frames. A tank with water is illuminated by a stroboscope (Fig. 1) capable of 250

Fig. 1. Experimental $500 \times 320 \times 320$ mm^3 tank with four cameras, stroboscope and 50–100 μ cenospheres used for micro-tracking

flashes per second with a maximum average power of 200 W. We run the experiment in two modes of operation: *micro- and macro-tracking.* Micro-tracking covers (15 mm)^3. Macro-tracking covers (200 mm)^3, i.e. many integral scales.

For micro-tracking we used cenospheres, hollow glass spheres which are a byproduct of power plant coal combustion. They are perfectly round and reflect the light well, but their density is ≈ 0.7 (water = 1). For macro-tracking we use 500 μm gas saturated polystyrene particles that have been "puffed" slightly to achieve a density of 1 within 1:10000. Tracking gets difficult when the number of particles goes beyond 1000 and when the velocity is too large.

3 Results

Both results from micro- and macro-tracking is reported.

3.1 Micro-tracking

The simplest way to determine the energy dissipation is from the second order structure function, given in the *viscous* subrange by

$$\left\langle |\delta \mathbf{v}(\mathbf{r})|^2 \right\rangle = \frac{\varepsilon r^2}{3\nu}, \tag{3}$$

where ν is the kinematic viscosity. In presence of noise the structure function will have a constant offset from the parabolic law, given in terms of the uncertainty of the particle position determination σ_x. Examples of viscous range structure functions and the derived parameters are shown in Fig. 2.

Thus determined, the energy dissipation and the velocity–acceleration covariance is compared according to (1). Beyond $\varepsilon \approx 100$ $mm^2 s^{-3}$ the good agreement deteriorates (Fig. 3). At present we believe this is due to problems with the tracking algorithm for the fastest runs.

Equation (2) also compares well for the slower runs as seen from the micro-tracking runs in Fig. 4.

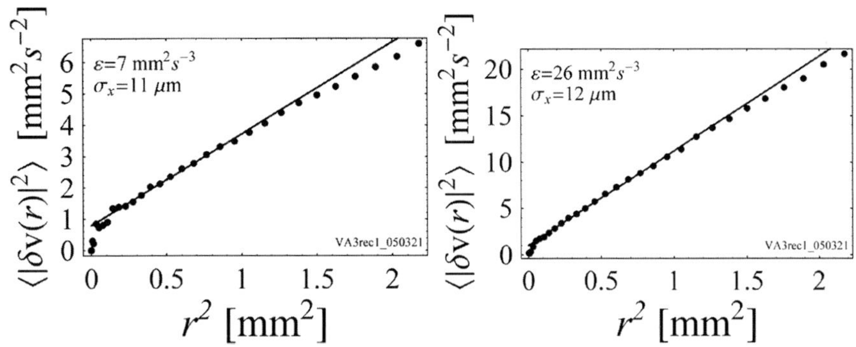

Fig. 2. The second order structure function versus distance squared for two different runs. The slope of the straight line gives ε, the offset the position determination noise

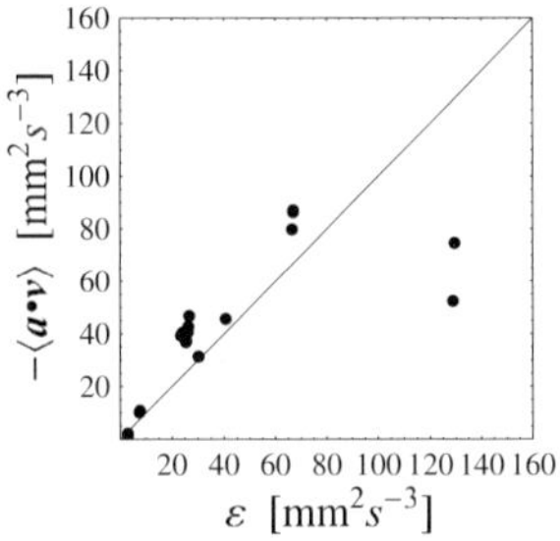

Fig. 3. The energy dissipation versus the velocity–acceleration covariance

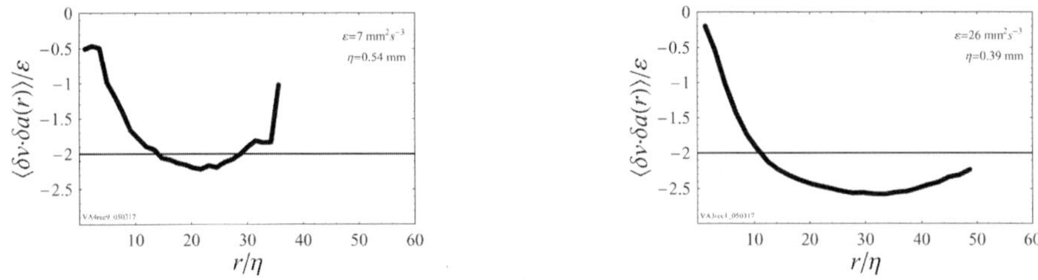

Fig. 4. The velocity–acceleration structure function for two different values of ε

3.2 Macro-tracking

Comparison with the DNS of [1] shows that $\langle \mathbf{v} \cdot \mathbf{a} \rangle = 0$ within 1% of ε as expected. The relation (2) seems to be fulfilled over a wider range in the experiment than in DNS for a comparable Reynolds number.

Conversely, the third order structure function matches the theoretical inertial range expectation $-4/5\varepsilon r$ in a narrower range in the experiment than in DNS.

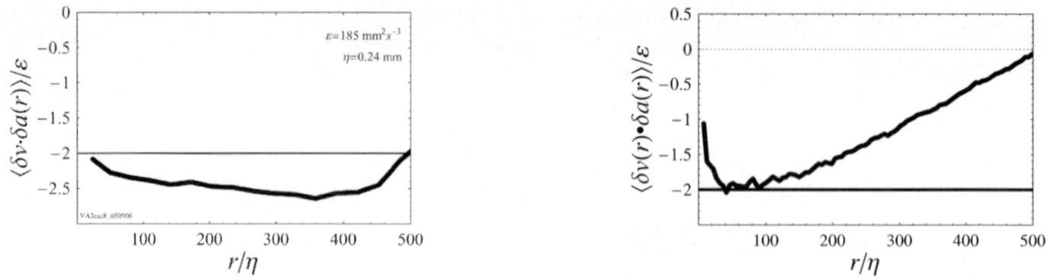

Fig. 5. The velocity acceleration structure function (2) from particle tracking data (left) and DNS (right)

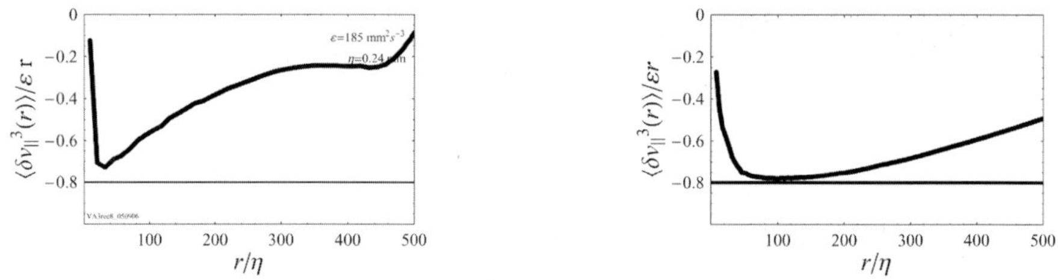

Fig. 6. The third order velocity structure function from particle tracking data (left) and DNS (right)

4 Conclusion

The velocity–acceleration correlations (1) and (2) estimate the energy dissipation in Lagragian experiments provided the tracking efficiency is good. Some of the data presented here are rather new leaving much analysis to do.

We acknowledge G. Boffetta for access to the Lagrangian DNS database. B. L. has been supported by the EU Network "Fluid Mechanical Stirring and Mixing: the Lagrangian Approach" funded by the European Commission's 5th Framework Improving Human Potential programme.

References

1. L. Biferale, G. Boffetta, A. Celani, B. J. Devenish, A. Lanotte, and F. Toschi. Multifractal statistics of lagrangian velocity and acceleration in turbulence. *Phys. Rev. Lett.*, 93(6):064502, 2005.
2. B. Lüthi, A. Tsinober, and W. Kinzelbach. Lagrangian measurement of vorticity dynamics in turbulent flow. *J. Fluid Mech.*, 528:87–118, 2005.
3. S. Ott and J. Mann. An experimental investigation of the relative diffusion of particle pairs in three dimensional turbulent flow. *J. Fluid Mech.*, 422:207–223, 2000.
4. S. Ott and J. Mann. An experimental test of Corrsin's conjecture and some related ideas. *New Journal of Physics*, 7, 2005.

PIV Measurement of Coherent Structures and Turbulence Created by an Oscillating Flow at the End of a Thermoacoustic Stack

Xiaoan Mao, David Marx and Artur J. Jaworski

University of Manchester, MACE, Sackville St., Manchester M60 1QD, UK
x.mao@student.manchester.ac.uk, david.marx@manchester.ac.uk, artur.jaworski@manchester.ac.uk

1 Introduction and Background

In thermoacoustic devices, an acoustic wave interacts with a stack of tightly spaced plates either to produce acoustic power, induced by a temperature gradient, or to obtain a refrigeration effect, induced by an acoustic wave. This is based on the thermoacoustic effect whereby appropriately phased pressure and velocity oscillations enable the compressible fluid to undergo a thermodynamic cycle in the vicinity of a solid body. The interaction of the acoustic field with the solid boundary is governed by complex thermo-fluid processes, which are not fully understood, especially for large acoustic excitations (drive ratios D_r>1%, where D_r is defined as the max acoustic pressure divided by the mean pressure) with large fluid displacements [1]. This work is driven by research into the fluid-mechanical aspects of thermoacoustic systems, in particular the identification of the flow morphology and turbulence characteristics in the vicinity of the parallel-plate thermoacoustic stack.

2 Experimental Apparatus

The experimental apparatus shown in Fig. 1(a) is fully described in [2]. It is an 8.3 m long resonator made of: a 0.6 m long "loudspeaker box", a 0.3 m long pyramidal contraction, and a 7.4 m long square pipe (136x136 mm inside). This has transparent walls for PIV measurements (TSI system with a 4 MP camera). The resonator is filled with atmospheric air. The first quarter-wavelength mode is used, of frequency 13.1 Hz, corresponding to the viscous boundary layer thickness [1] $\sim$ 0.6 mm. Two stack models were tested: a "thin" stack with plate thickness 1 mm and plate spacing 5 mm, and a "thick" stack with plate thickness 5 mm and plate spacing 10 mm. Both were 200 mm long and 136 mm wide, and located at a distance Δ_{st} = 4.11 m from the resonator

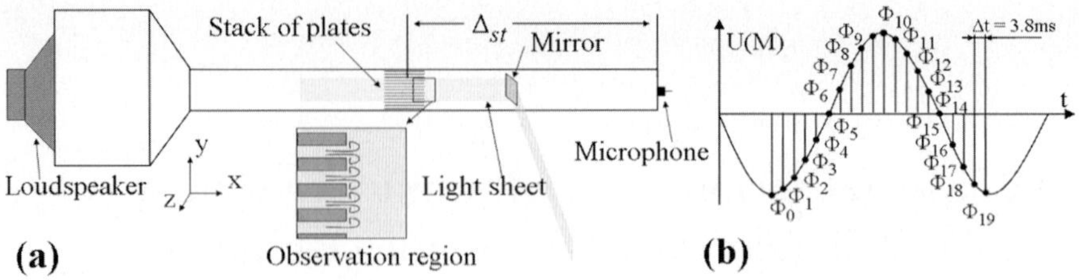

Fig. 1. (a) Schematic of the experimental apparatus and the PIV setup; (b) the phases for PIV measurements during the acoustic cycle relative to U(M)

end. The experiments aimed to identify the flow structures generated at the end of the stack, at 20 phases (Φ_i, i=0, 1, 2, ..., 18, 19) within one cycle, as a function of the acoustic excitation (D_r = 0.1, 0.3, 0.6, 1.0, 1.5, 2.0 and 3.0%) for each of the two stacks. Fig. 1(b) illustrates the selected phases relative to velocity U(M), measured in the middle of the channel entry. A 3% drive ratio corresponds to the fluid displacement amplitudes ~ 11 cm and the velocity amplitudes ~ 10 ms^{-1}. In a typical experimental run, 100 PIV images were taken for each of the 20 phases, which produced samples 2000 images long. The results were processed to obtain the average phase-locked velocity and vorticity fields as well as the fluctuating component of velocity (turbulence intensity). It was shown that 100 frames compared relatively well with a widely accepted figure of 1000 frames for calculating turbulence statistics [3]; the choice dictated by the limited data storage capacity and post-processing time. The "viewing field" of the PIV images was either 25x25 mm or 60x60 mm. The maximum spatial resolution of the former was 0.1 mm (6 measurement points in the laminar viscous layer).

3 Sample Results and Discussion

In general the oscillating flow has two main phases: the ejection phase when the velocity is towards x>0 (Φ_6-Φ_{14}), and the suction phase when the velocity is towards x<0 (Φ_0-Φ_4 and Φ_{16}-Φ_{19}). However, there are significant differences between the low drive ratios (D_r<1%) and high drive ratios (D_r>1%). In the former case the flow is dominated by laminar-like features, in particular the vortices formed at the edges of the plates in the ejection phase remain attached and subsequently their remains are sucked back between the plates when the velocity changes its direction. This is illustrated in Fig. 2 where phase Φ_9 (ejection) and phases Φ_{17} and Φ_{19} (suction) are shown for the thick stack at D_r=0.3%. For the thin stack the high vorticity regions are much more elongated; this is in agreement with [4]. At high drive ratios the flow morphology changes markedly. During the ejection phase coherent vortex structures are being shed from the edges of the plates as illustrated in Fig. 3 for the thick stack (left) and for the thin stack (right) at a D_r of 1.0 and 1.5%, respectively,

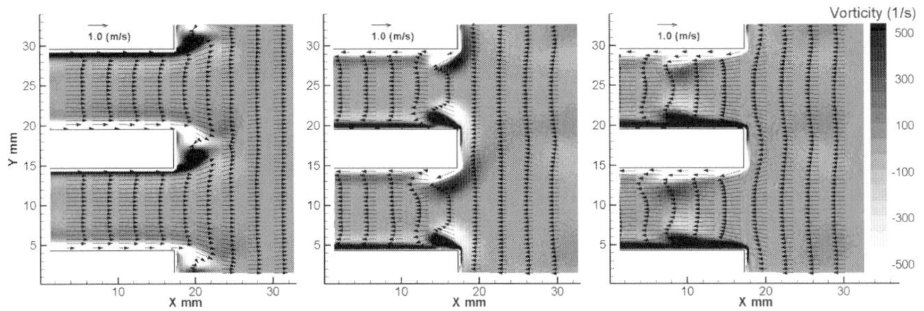

Fig. 2. Mean x-wise velocity field and the vorticity field for phases Φ_9 (left), Φ_{17} (middle) and Φ_{19}(right) obtained for the "thick" stack and D_r=0.3%

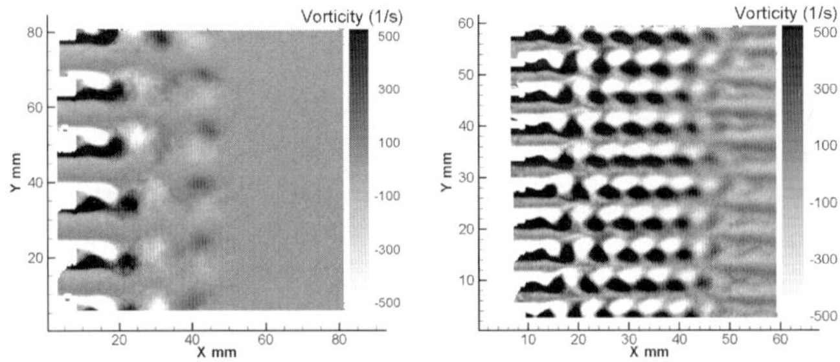

Fig. 3. Vortex shedding from the edges of the plates during the ejection phase (Φ_{11}) from the thick stack at D_r=1% (left) and the thin stack at D_r=1.5% (right)

both measurements obtained at Φ_{11}. For the thin stack, however, the formation of the train of vortices is preceded by the formation of a thin elongated shear layer which destabilises into a wavy pattern and then breaks up into individual vortices. For the thick stack the vortices seem to be shed without this intermediate phase. The unsteady component of the velocity field (corresponding to the turbulence intensity) is illustrated as the geometrical sum of the fluctuating velocity components in x and y directions, normalised by the amplitude of velocity: $[(u'^2+v'^2)]^{0.5}/U(M)$. Figure 4 compares the turbulence intensity for the thick stack at D_r=0.3% (Φ_{12}) with that obtained for D_r=2% (Φ_{10}). The field of view for the latter had to be enlarged due to the higher fluid displacement. When looking at the time history of turbulence intensity it is clear that turbulence generated during the ejection phase dissipates by the end of this phase and very little is transported back into the stack during the suction phase. This is shown in Fig. 5 (left) obtained for the thick stack at D_r=2% but is equally true for the thin stack. It is also observed that most of the turbulence in the suction phase is generated at the edges of the plates - see the middle and right graphs of Fig. 5.

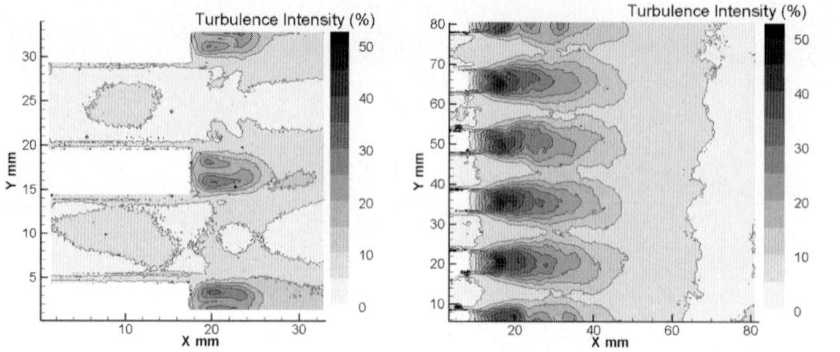

Fig. 4. Turbulence intensity distribution for D_r=0.3% and phase Φ_{12} (left); and D_r=2% and phase Φ_{10} (right) for the "thick" stack during the ejection phase

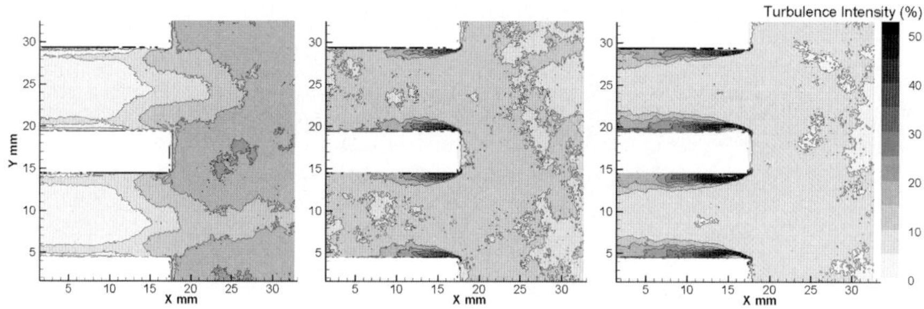

Fig. 5. Turbulence intensity distribution for phases Φ_{16} (left), Φ_{18} (middle) and Φ_0 (right) obtained for the "thick" stack at D_r = 2%, during the suction phase

4 Future work

Current work covers a preliminary mapping of the existing flow regimes. LDA and HWA will be used in future to obtain better turbulence statistics. A wider range of geometrical parameters will be covered and the dimensional analysis will be used to predict the flow nature as a function of normalized parameters: the Reynolds number, the stack porosity or the normalized fluid displacement.

Acknowledgments

Support from EPSRC (UK) is gratefully acknowledged. X. Mao thanks Universities UK and the University of Manchester for their funding.

References

1. Swift GW (1988) J Acoust Soc Am 84: 1145-1180
2. Marx D, Mao X and Jaworski AJ (2006) Appl Acoust 67: 402-419
3. Hyun BS, Balanchadar R, Yu K and Patel VC (2003) Exp Fluids 35: 262-267
4. Blanc-Benon P, Besnoin E and Knio O (2003) CR Mecanique 331:17-24

Statistics of the Temperature Fluctuations as a Passive Scalar in a Freejet Experiment

Marco Munzel[1], Joachim Peinke[2], and Achim Kittel[3]

[1] University of Oldenburg marco@ehf.uni-oldenburg.de
[2] University of Oldenburg joachim.peinke@uni-oldenburg.de
[3] University of Oldenburg achim.kittel@uni-oldenburg.de

The measurement of the fluctuations of a passive scalar in fluids is important for the investigation of the transport features in turbulent flows [1]. A detection of the passive scalar is required with high temporal and spatial resolution in order to gain information over the entire turbulent cascade. We present data of temperature fluctuations taken in a heated free-jet experiment with a newly developed miniaturized thermocouple sensor. This Au/Pt-thermocouple possesses an enhanced temporal resolution extending to the range of microseconds. The measured power spectra show a slight shift of the Reynolds number if the position of the sensor is varied along symmetry axis of the jet. Furthermore, a shift of the inertial range can be observed while approaching the nozzle. The passive scalar spectrum is less steep than anticipated by the theory of ideal turbulence. An -4/3 scaling is observed due to the flow situation at still modest Re numbers [2].

1 The Temperature Sensor

The temperature measurement system is depicted in Fig. 1. It consists of a conical measurement head supporting a coaxial thermocouple made of a micro pipette and a watertight enclosure for a low-noise preamplifier. The enclosure is made of stainless steel with capton sealings to withstand water or even aggressive fluids. The dimensions of the pen-size body are 14 mm in diameter and 15 cm in length. The standard mounting of the measuring head provides an easy replacement and, therefore, an adjustability to various measurement situations, like through hole and near wall measurements. The design of the sensor is a coaxial thermocouple which consists of a platinum core, an insulating layer of glass, and an outer coating of gold, which serves as the second contact of the thermocouple. The two metals form the thermocouple only at the very end of the tip (about 200 nm in diameter and 200 nm in length). The Au/Pt-thermocouple head is connected to the preamplifier circuit board via two gold plated spring-loaded contacts to which a Pt-100 sensor is attached

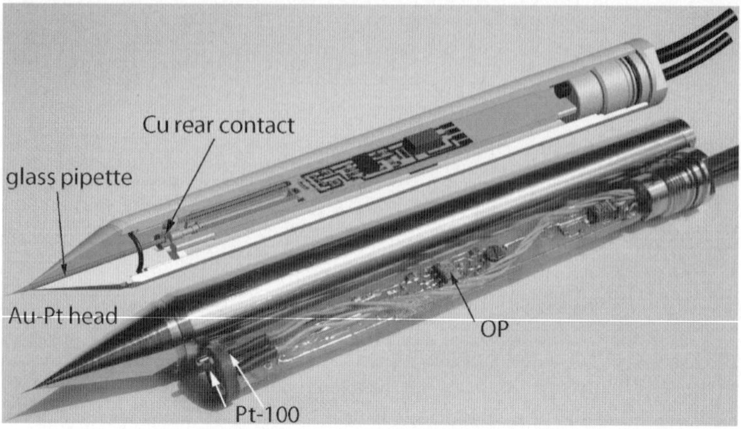

Fig. 1. The temperature measurement system consists of the detachable measuring head with the embedded thermocouple pipette and a waterproof stainless steel enclosure housing the amplifier module

with thermally conducting cement. The Pt-100 sensor monitors the temperature of the thermocouple reference junction. This temperature information is needed in order to calculate absolute temperature values from the thermocouple data. The thermoelectric voltage is amplified by a factor of 1000 with a single stage operational amplifier (OPA) sitting inside the the watertight enclosure of the sensor. The amplifier provides a bandwidth of 450 kHz at this level of gain. The AD797 from Analog Device was chosen for the OPA because of its outstanding low spectral noise density of $0.9\text{nV}/\sqrt{\text{Hz}}$ and harmonic distortion for source impedances less than 1 kΩ as present in our case. This results in a temperature resolution of the entire measurement system of about 0.08°C with our Au/Pt-thermocouple.

2 Measurement Setup of the Free-jet

The measurement setup is a heated free-jet of water in water. A water tank 50 cm in diameter and 80cm in height is held at 16°C. Hot water is injected into the tank from the bottom by a specially formed nozzle that provides a uniform velocity profile at the nozzle outlet with diameter D = 2mm. In order to keep the temperature of the free-jet surrounding water constant, cold water is supplied over the entire bottom area with a slow flow velocity (below 1 mm/s). The highest achievable Re number in this setup is equal to 80000 due to an onset of cavitation at the nozzle outlet at velocities above 38 m/s. The water supplied at the bottom can drain off the tank via a circumferential drainage system at the free surface.

3 Results

In the following two series of measurements are discussed. In the first one the temperature fluctuations are measured for different Re numbers. In the second one the temperature fluctuations are measured for different positions with respect to the nozzle outlet on the symmetry axis of the free-jet. The time series are 10MSamples in length and taken with a 16-bit A/D-converter at a sampling rate of 60 kHz after filtering with a 30 kHz low-pass filter to avoid aliasing effects. From these time series the power spectral density $P(f)$ is calculated. The power spectra shown in Fig. 2 display the results of two experimental series which were taken on the symmetry axis of a free-jet. In both cases the free-jet had a temperature of 35°C at the nozzle and was injected into a water tank of 16°C average temperature. In the first series of measurements shown in Fig. 2(a) the thermometer was kept at a distance of 290 mm, which corresponds to 145 diameters D of the used nozzle, while the Re number was varied ranging from 6000 to 80000. In the second measurement series the separation of the thermometer to the nozzle outlet was varied Fig. 2(b) from 8 cm to 28 cm at a Re number of 36000. In both cases the observed inertial range spans two orders of magnitude for all Re numbers and each distance to the nozzle. The slope of -4/3 is in contradiction to the KOC-theorem for ideal turbulence at high Re numbers [3]. Such a low value was also reported by Sreenivasan et al. and Zhou et al. as a typical feature of passive scalars in low Re number shear flows [4, 5, 6]. This feature may also occur in grid generated turbulence [7]. For higher Re numbers the inertial range extends towards higher frequencies. At the high frequency end in all cases the slopes increase indicating the end of the inertial range and the increasing dominance of the dissipation.

The spectra shown in Fig. 2(b) were fitted with the empirical formula

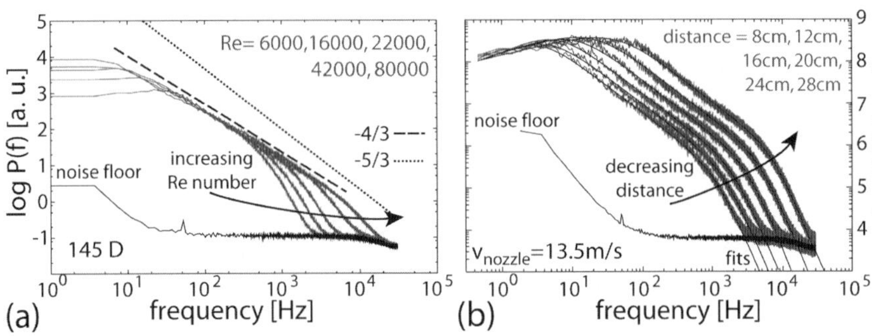

Fig. 2. The frequency power spectra of temperature fluctuations show a constant slope of -4/3 in the inertial range for various Re numbers at a constant distance to the nozzle (a) and for various distances to the nozzle at a constant Re number (b)

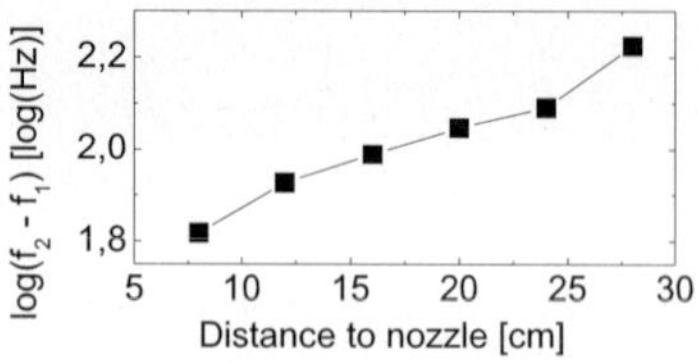

Fig. 3. The logarithm of the length of the inertial range is plotted versus the distance of the thermometer to the nozzle outlet. f_1 and f_2 mark the beginning and the end of the inertial range, respectively, determined by means of (1)

$$p = \frac{1}{af^{\alpha} + bf^{\beta} + cf^{\gamma}}, \tag{1}$$

where a, b, c, γ are fit parameters and α and β are fixed at 1.3 and 4 respectively. The fitted curves coincide quite well with the measured data. The key feature of a free-jet, a constant Re number along the jet's axis, can clearly be seen by the shift of the inertial range towards higher frequencies while approaching the nozzle. The length of the inertial range increases from a value slightly below two frequency decades to a value slightly above. This might be due to a thermal effect of the heated free-jet. Only for the closest positions to the nozzle of the thermometer the values are smaller indicating, that the turbulence may not fully developed in these cases (see Fig. 3).

Our new developed coaxial thermocouple sensor has proven to be a valuable addition and enhancement to the established measurement techniques for passive scalars in turbulent flows. The findings of Sreenivasan and Warhaft in gaseous flows have been confirmed and extended to liquid flows with our results. In the nearest future the flows will be characterized in more detail by scanning the free-jet on a three-dimensional grid. We want measure at each position the temperature fluctuations and analyze them concerning their statistics by means of quantities like probability density of the temperatures increments, the skewness, and the kurtosis of the increment distribution. Furthermore correlations will be determined at the different positions.

References

1. Falkovich G, Gawędzki K, Vergassola M (2001) Reviews of Modern Physics 73:913–975
2. Warhaft Z (2000) Annual Review of Fluid Mechanics 32:203–240
3. Corrsin S (1951) Journal of Applied Physics 22:469–473
4. Sreenivasan K R (1991) Proceedings of the Royal Society A 434:165–182
5. Sreenivasan K R (1996) Physics of Fluids 6:189–196
6. Zhou T (2003) Experiments in Fluids 34:449–459
7. Jayesh (1994) Physics of Fluids 6:306–312

Part III

Coherent Structure

Coherent Structures Identification in 2D Turbulence

Ch.-H. Bruneau[1], P. Fischer[1] and H. Kellay[2]

[1] MAB, Université Bordeaux 1, 351 Cours de la Libération, 33405 Talence, France CNRS UMR 5466 - INRIA FUTURS Team MC^2
bruneau@math.u-bordeaux1.fr, fischer@math.u-bordeaux1.fr

[2] CPMOH, Université Bordeaux 1, 351 Cours de la Libération, 33405 Talence, France hamid.kellay@u-bordeaux1.fr

Experimental results and direct numerical simulations show the coexistence of both enstrophy cascade and inverse energy cascade in two dimensional turbulence. The computation of the energy flux shows clearly the inverse energy cascade from the injection scale to the large scales and the computation of the enstrophy flux shows clearly the direct enstrophy cascade from the injection scale to the small scales. A wavelet packet filtering allows to split the whole flow into two sub parts, namely the vortices and the vorticity filaments that appear to be both coherent structures with their own dynamics.

1 Coexistence of Both Enstrophy Cascade and Inverse Energy Cascade

A direct numerical simulation of two dimensional turbulent flows behind arrays of cylinders in a soap film channel experiment as described in [6] is achieved solving the 2D incompressible Navier-Stokes equations. There is no force added as the injection scale is given by the diameter of the cylinders (see [2] for more details). The vorticity fields around the cylinders for both the numerical simulation and the soap film experiment are represented in Fig. 1. It is obvious that the addition of cylinders along the wall of the channel reinforces the injection of small scales into the main flow. The numerical simulation is performed on a 2560×640 cells grid and a square box of the flow of size 640×640 located after the arrays of cylinders is selected to analyse the flow. A snapshot at a given time is represented on Fig. 3 at the top left. A collection of 25 windowed snapshots thanks to Tukey window with a small parameter (see [5]) is used to compute the energy spectrum and the fluxes. The time between the snapshots is long enough for having a significant evolution of the flow. Indeed between the 1st and the 25th snapshots, the vortices have crossed several times the observation box. The results are plotted

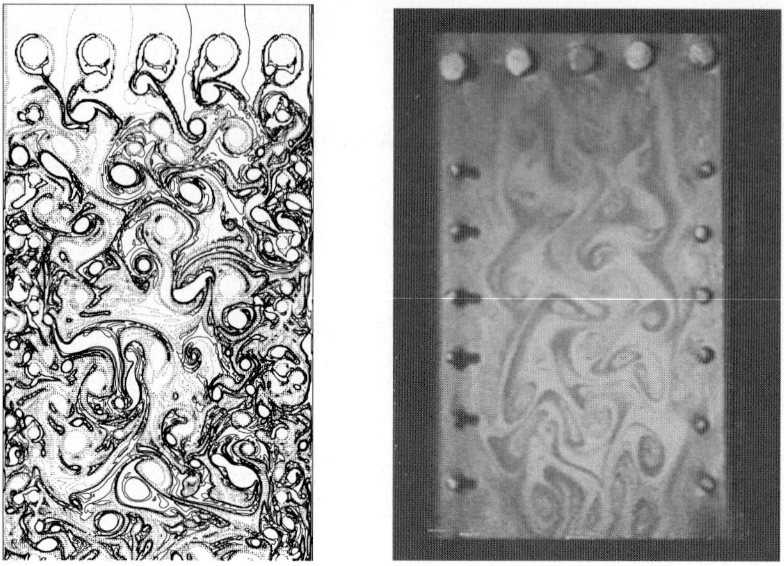

Fig. 1. Vorticity field of the numerical simulation (left) and the experiment (right)

on Fig. 2. The spectrum shows clearly that two ranges coexist on both sides of the injection scale 20 as the diameter of the cylinders along the wall is equal to $W/20$ where W is the width of the channel. There is no scaling before 4 as the observation box is closed to the cylinders and so the size of the biggest structures is about one fourth of the channel width. The spectrum is not significant beyond 120 as the mesh size is not fine enough to capture the small scales beyond this value. In between two distinct straight lines are found with slopes steeper that those given by the KLB theory. However this simulation corresponds to a real computation in a bounded domain with no-slip boundary conditions on the walls of the channel, a Poiseuille flow at the entrance section and a non-reflecting boundary condition at the exit section of the computational domain [1]. So the KLB theory does not apply in this configuration.

To confirm the coexistence of both cascades the energy flux and the enstrophy flux are plotted on Fig. 2. A negative energy flux is obtained on the left hand side of the injection scale showing an inverse energy cascade from this injection scale to larger scales and the flux vanishes for larger scales. A main positive part of the enstrophy flux is obtained on the right hand side of the injection scale showing a direct enstrophy cascade from this scale to the smaller scales. Let us note also a negative part of the flux before the injection scale showing that some vorticity is brought by the inverse energy cascade.

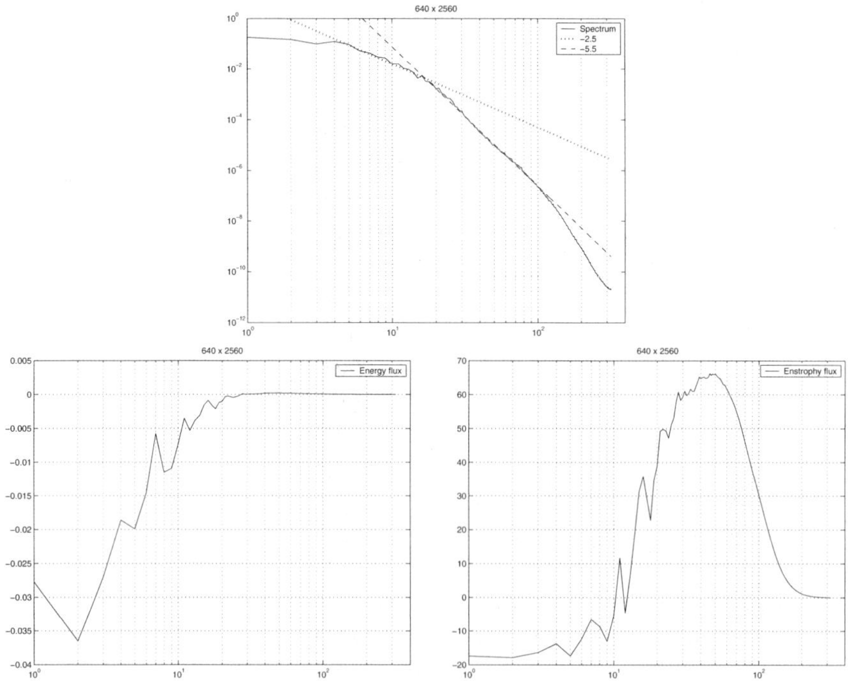

Fig. 2. Analysis of 2D flow on grid 2560 × 640: energy spectrum (top), energy flux (middle) and enstrophy flux computed on a set of 25 windowed snapshots

2 Wavelet Packets Filtering

To perform a coherent structures identification, a wavelet packets decomposition is performed as in [3]. In that paper the decomposition was used on an academic case made of three vortices, here it is used on the observation box in the way described in [4]. Using a threshold on the wavelet packets coefficient it is possible to split the flow into two orthogonal sub flows, isolating on the one hand the solid rotation vortices and on the other hand the vorticity filaments that roll up in spirals inside the vortices as illustrated in Fig. 3.
An interesting feature of such an orthogonal decomposition is that the global energy spectrum is the sum of the two energy spectra corresponding to each sub flows. The spectrum of the vortices is almost superimposed with the global spectrum on the left hand side of the injection scale and very small on the right hand side while the spectrum of the filaments is small on the left hand side and almost superimposed on the right hand side. This shows clearly that each part of the orthogonal decomposition is coherent and has its own dynamics. In addition the inverse energy cascade is mainly due to the vortices and the direct enstrophy cascade is mainly due to the filaments.

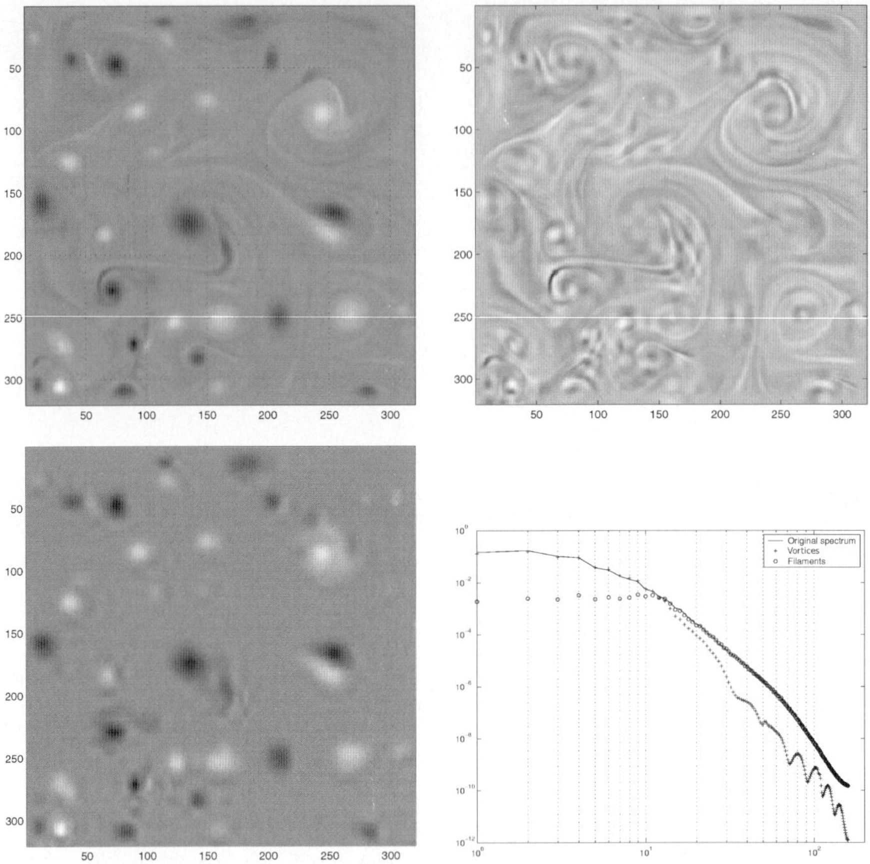

Fig. 3. Filtering of a part of the flow located after the arrays of cylinders. The whole vorticity field (top left), the vortices corresponding to the higher wavelet packets coefficients (bottom left) and the vorticity filaments corresponding to the lower coefficients (top right). The spectra of the three different parts (bottom right) show that the sum of the spectra of the two sub fields corresponds to the spectrum of the whole flow

References

1. Bruneau CH (2000) M2AN 34:303–314
2. Bruneau CH, Kellay H (2005) Phys. Rev. E 71:046305
3. Farge M, Kevlahan N (1997) J. Fluid Mech. 346:49–76
4. Fischer P (2005) Discr. Cont. Dyn. Syst. B 5:659–686
5. Fischer P, Bruneau CH (submitted)
6. Kellay H, Goldburg WI (2002) Rep. Prog. Physics 65:845–894

Multi-resolution Analysis of the Large-scale Coherent Structure in a Turbulent Separation Bubble Affected by Unsteady Wake

Sejong Chun[1], Ying Zheng Liu[2], and Hyung Jin Sung[3]

[1] Division of Physical Metrology, Korea Research Institute of Standards and Science, Daejeon, Korea sjchun@kriss.re.kr
[2] School of Mechanical Engineering, Shanghai Jiao Tong University, Shanghai, China yzliu@sjtu.edu.cn
[3] Department of Mechanical Engineering, Korea Advanced Institute of Science and Technology, Daejeon, Korea hjsung@kaist.ac.kr

1 Introduction

Multi-resolution analysis (MRA) was made of the large-scale coherent structure in a turbulent separation bubble affected by unsteady wake. The unsteady wake was generated by using a spoked-wheel type wake generator. The wake generator was installed in front of the separation bubble. The wake generator was rotated either clockwise or counter-clockwise directions (CW and CCW) with the normalized passing frequency ($St_H = 0.2$). The Reynolds number was $Re_H = 6,000$ based on the half-thickness of the blunt body. To show the unsteady dynamic flow structure between 'cutting' and 'wrapping' regimes, a multi-resolution analysis using the maximal overlap discrete wavelet transform (MODWT) was performed. This method enabled scale-resolved analysis to delineate the coherent structure of turbulent separation bubble. The reconstruction of flow field in combination with conditional averaging was attempted.

2 Results and Discussion

Prior to showing unsteady patterns of flow structures, the autocorrelation of the continuous wavelet transformed wall pressure was made to see footprints of vortical structures [1, 2]. Based on the Mexican hat wavelet, the continuous wavelet transform coefficient of a real-time series of data is defined as

$$wf(b,a) = \frac{1}{a}\int_{-\infty}^{+\infty} f(t)\Psi(\frac{t-b}{a})dt, \tag{1}$$

where a is the timescale dilation parameter, b is the time transition parameter and Ψ is the Mexican hat wavelet, $\Psi(t) = (1 - t^2)\exp(-0.5t^2)$. The wavelet autocorrelation function is given as

$$wc(a,\tau) = \lim_{T\to\infty} \frac{1}{T} \int_{-T/2}^{T/2} wf(b,a)wf(b+\tau,a)db \tag{2}$$

where τ is the time delay of wavelet coefficients in the wavelet space. The autocorrelation of the continuous wavelet transformed wall pressure at $x/H = 0.8$, which is immediately downstream of the separation edge, is displayed in Fig. 1. Periodic wake signatures are subsequently generated with 1 period in $5\tau U_\infty/H$ at $\log(aU_\infty/H) = 0.7$ ($St = 0.2$) for both CW and CCW conditions. When the wake generator is in CW rotation, the wakes are very strong; otherwise, the wakes are significantly attenuated, suggesting the enhanced mixing between unsteady wakes and vortical structures persistent in the separation bubble. Recall that the shedding frequency of large-scale vortical structures is $St = 0.09$ [3]. In CW rotation, inspection of the autocorrelation along $\tau U_\infty/H = 0$ shows that the large-amplitude correlation region is extended down to low frequency and two other discrete areas are detected near the denoted $St = 0.09$ at $\tau U_\infty/H = \pm 8$. In CCW rotation, the low-frequency region is not fully separated from footprints of the unsteady wakes, showing the stronger interaction between unsteady wakes and large-scale vortical structures.

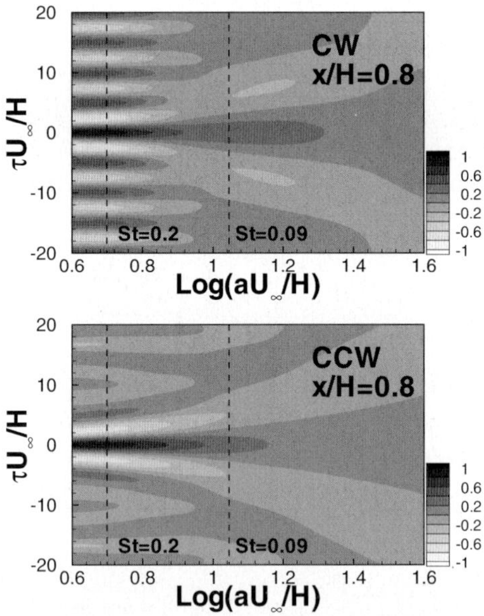

Fig. 1. Autocorrelation of continuous wavelet transformed wall pressure

To resolve the influence of vortices at successively increasing frequencies on the flow field, which is distributed by the CW rotation of the wake generator, the flow field is reconstructed by serially adding higher frequency signals as shown in Fig. 2 (left). At $St = 0.03$, a recirculation region is observed in $0 \leq x/H \leq 5.6$. Superposition of $St = 0.08$ divides the recirculation region into two parts, e.g. the recirculation region confined within $0 \leq x/H \leq 2$ and a large flow structure within $2 \leq x/H \leq 5.6$. The center of flow structure corresponds with the location of the maximum turbulent kinetic energy, indicating the presence of a turbulent vortex [4]. Another maximum of the turbulent kinetic energy is found at $(x/H, y/H) = (2.4, 0.7)$. This may suggests the turbulent engulfment between two consecutive vortices. The location of maximum turbulent kinetic energy at $(x/H, y/H) = (2.4, 0.7)$ also indicates that small scale turbulent motions reside in the separation bubble when the large-scale vortex sheds from the recirculation region. As shown in Fig. 2e (left), the concentration of the turbulent kinetic energy along the initial phase of the separated shear layer is primarily contributed by the shedding vortices centered at $St = 0.53$.

The reconstruction of multi-scale resolved flow velocity in case of the CCW rotation is displayed in Fig. 2 (right). The recirculation region is found in $0 \leq x/H \leq 2.8$ at $St = 0.03$. Superposition of the flow at $St = 0.08$ divides the recirculation region into two parts, i.e. $0 \leq x/H \leq 1.2$ and $1.2 \leq x/H \leq 2.8$. At $St = 0.13$, the maximum turbulent kinetic energy is found at $(x/H, y/H) = (2.2, 0.6)$, in good conformity with the position of the turbulent vortical structure. Another maximum of turbulent kinetic energy is found at $(x/H, y/H) = (1.2, 0.4)$, being a saddle point located between the

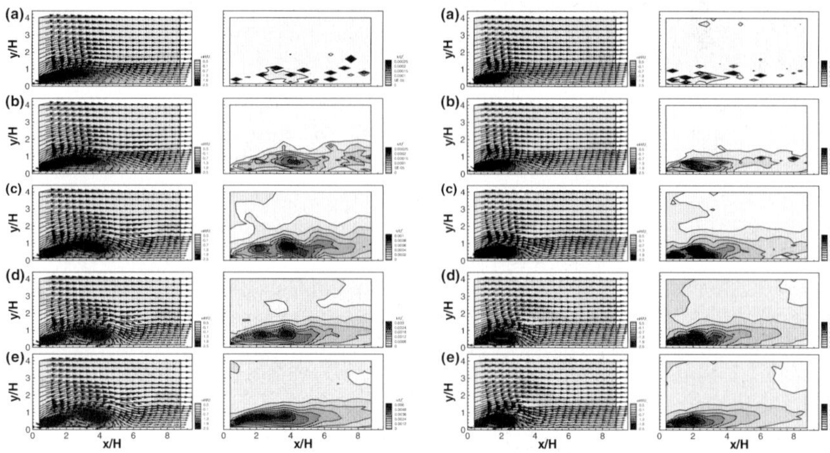

Fig. 2. Multi-resolution analysis of velocity field at CW (left) and CCW (right) rotation of wake generator for $St_H = 0.2$. a) $St = 0.03$ b) $St < 0.08$ c) $St < 0.13$ d) $St < 0.26$ e) $St < 0.53$

recirculation region and the large turbulent vortex. No clear difference of the mean flow field with increasing frequency region is found in Fig. 2 (right). This means that the small-scale flow structure gives little contribution to the time-averaged flow pattern, although it induces large fluctuation of the velocity. The overall glance of the intensified turbulent kinetic energy with increasing frequency level in Figs. 2a-e (right) demonstrates the large velocity fluctuations induced from the small-scale vortices.

3 Conclusions

The multi-resolution analysis (MRA) based on the maximal overlap discrete wavelet transform (MODWT) was extensively employed to delineate the unsteady events buried in the turbulent separation bubble which was subjected to disturbance of unsteady wakes. The major concern was directed toward the intermittent interaction between the periodic wakes and the scale-variant vortical structures at $St_H = 0$ and 0.2 (CW, CCW). Toward this end, the simultaneous measurements of wall pressure fluctuations and velocity fluctuations were accomplished by using microphone array and split-film, respectively. Conditional average was applied to capture the events associated with shedding of large-scale vortical structures. Preliminary analysis based on the continuous wavelet transform (CWT) demonstrates that the unsteady wakes in the CCW configuration mixed well with the large-scale vortical structures buried in the separation bubble, resulting faster decay than that in CW configuration. By using the MRA, it was found that most of energy content of coherent structures was concentrated on the large-scale vortices and the initial phase of flow separation. Addition of three velocity fields from the lowest scale, i.e. $St = 0.03$, 0.08 and 0.13 was enough to reconstruct a converged time-averaged coherent structure.

References

1. Chun S, Liu YZ, Sung HJ (2004) Wall-pressure fluctuations of turbulent separated and reattaching flow affected by unsteady wake. Exp Fluids 37: 531–546
2. Liu YZ, Kang W, Sung HJ (2005) Assessment of the organization of a turbulent separated and reattaching flow by measuring wall pressure fluctuations. Exp Fluids 38: 485–493
3. Chun S, Sung HJ (2002) Influence of unsteady wake on a turbulent separation bubble. Exp Fluids 32: 269–279
4. Kiya M, Sasaki K (1985) Structure of large-scale vortices and unsteady reverse flow in the reattaching zone of a turbulent separation bubble. J Fluid Mech 154: 463–491

Effect of Spanwise System Rotation on Longitudinal Vortical Structure of Homogeneous Shear Flow

Oaki Iida and Yasutaka Nagano

Dept. of Mech. Eng., Nagoya Institute of Technology, Gokiso-cho, Showa-ku, Nagoya 466-8555, Japan
Email: iida.oaki@nitech.ac.jp

1 Introduction

In a homogeneous shear flow, the longitudinal streamwise (x-direction) vortical structure, which is similar to that in the near-wall turbulence, is found to be generated. The elongated vortical structure tends to tilt into the spanwise direction (z-direction), and the sense of the tilting is different between the vortices with the positive and negative streamwise vorticity [1].

This asymmetry must be due to the mean spanwise rotation associated with mean shear. Hence, imposing system rotation in the spanwise direction must change the mechanism of the tilting, which might give us the clue for it.

The objective of this study is to learn the rotational effect on the typical longitudinal vortical structure, and investigate the mechanism of the tilting.

2 Numerical Methods

We performed DNS of a homogeneous shear flow mimicking the near-wall turbulence at a low Reynolds number. The governing equations are the Boussinesq form of the Navier-Stokes equations. The same numerical algorithm is used as Rogallo [2]. As shown in Fig.1, x and z are assumed to be streamwise and spanwise direction, respectively. The constant mean shear $S = dU/dy$ is imposed on the flow. The system spanwise rotation with the rotation frequency f is also imposed by including the Colioris term in the Navier-Stokes equation.

The numerical conditions are similar to those used in Iida et al, except 128^3 grid resolution used for this study. The sides of the computational region L_1, L_2, and L_3 are $4\pi\sqrt{S/\nu}(= 380)$, $2\pi\sqrt{S/\nu}$ and $2\pi\sqrt{S/\nu}$ in the x_1, x_2 and x_3 directions, respectively. The Reynolds number $Re_t = q^4/\epsilon\nu$ is 66 at $St = 0$, while shear rate parameter $S^* = Sq^2/\epsilon$ is 14, where q and ϵ are the

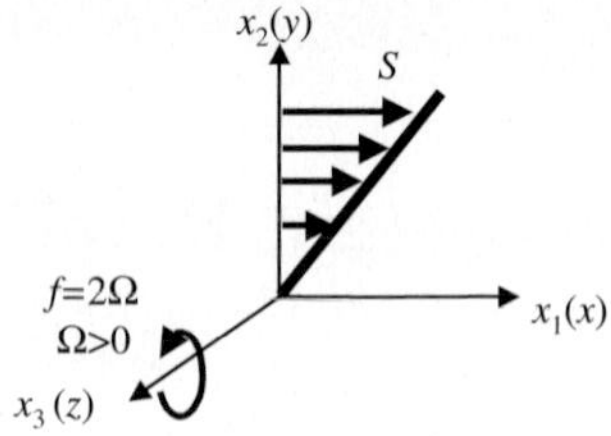

Fig. 1. Coordinate system and flow geometry representing anticyclonic case. Ω is angular velocity

turbulent velocity $\sqrt{u_j u_j}$ and the dissipation rate of turbulent kinetic energy $q^2/2$, respectively.

The several numerical simulations are performed by changing the rotation number $Ro = -f/S$. The positive rotation number represents the cyclonic cases (Case C) where the vorticity of the imposed system rotation is in the same sense as that of the mean vorticity, while the negative rotation number represents the anticyclonic cases (Case A). In Cases C1 and C2, *Ro* is 0.25 and 0.5, respectively, while in the case A1, *Ro* is -0.5. Case N represents the flow without a system rotation.

3 Results and Discussion

To investigate how the system spanwise rotation affects the vortical structure, we conditionally averaged it from the instantaneous flow field, and the averaged velocities and the Reynolds shear stress are investigated in detail. Vortical structure is identified by the second invariant of the deformation tensor defined as $II = -\partial u_i/\partial x_j \partial u_j/\partial x_i$. The details of the numerical procedure are same as used in Iida et al. [1] and Jeong et al. [3], and summarized as follows:

(1) Detection of the local maximum of II in cross streamwise planes (positive II is used to identify the vortical structure, while the streamwise vorticity ω_1 is used to distinguish between structures with a different sense of rotation); (2) linking the local maximum of II as the core of the longitudinal vortex; (3) when the local maximum point of II in the cross-streamwise plane shifts by more than 2 grids points from the cross-streamwise plane of one step before, the structure is considered to be terminated, and detection of a new vortical structure begins; (4) each vortex must have a streamwise length greater than $1/13L_1$, which is approximated as the streamwise longitudinal integral scale $L_{11,1}$ at $St = 6$ in case N (the condition required to capture the fully developed longitudinal vortices); and (5) the educed vortical structures are averaged with the sense of rotation by aligning the mid-point of their streamwise extent.

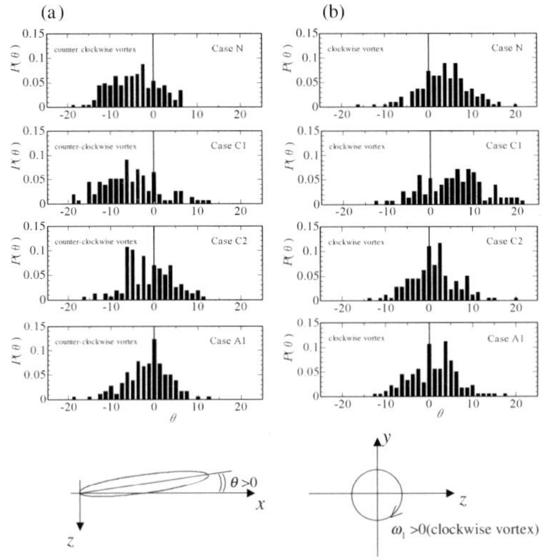

Fig. 2. P.d.f. of tilting angle at $St = 6$ in the educed longitudinal vortices. (a) is for the vortices with a negative streamwise vorticity, while (b) is for those with a positive streamwise vorticity

The number of vortical structures educed is more than two hundreds in all cases. Figure 2 shows the probability distribution function, p.d.f. of the tilting angle of each educed vortical structure. The tilting angle θ is determined as the angle in the x-z plane between the x axis and the segment connecting the cores of the vortices at both edges. In the case without system rotation, the vortices with the positive and negative streamwise vorticity tend to have positive and negative tilting angles, respectively. Hence, the skewness is definitely observed in the p.d.f. of θ. It is also noted that while the weak cyclonic system rotation enhances the tilting, the anticyclonic and strong cyclonic rotation reduces it.

Figure 3 shows the distribution of the spanwise velocity $< w >$ and the Reynolds shear stress $< v >< w >$ in the different cross streamwise plane, while $<>$ represents the value after the averaging procedure mentioned above. At the edge of the longitudinal vortex, there is the definite asymmetry in the intensity of the spanwise velocity. At the upwind edge of the longitudinal vortex, the spanwise velocity at the upper side is larger than that at the lower side; at the downwind edge, the spanwise velocity at the lower side is larger than that at the upper side. These spanwise-velocity pools could tilt the longitudinal vortex in the spanwise direction as long as the spanwise momentum is transferred into the longitudinal vortex.

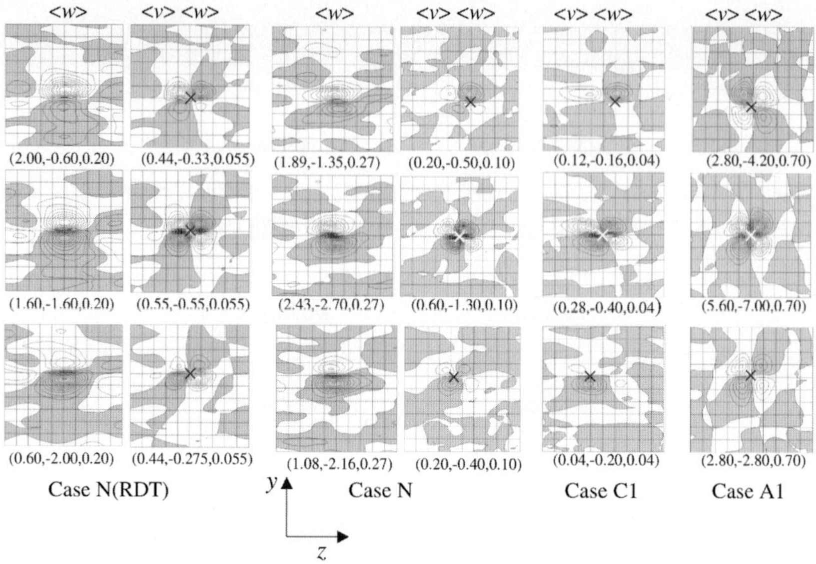

Fig. 3. Distribution of w and vw in Cases N. The first and third rows represent the upwind and downwind cross streamwise sections, respectively. The second row represents the cross section in the middle of the longitudinal vortex. The symbol cross represents the core of vortex. The numbers in the blacket represent the maximum, minimum and interval of the contours, respectively. The shaded regions represent the negative value. Each side of region is $31\sqrt{\nu/S}$

The turbulent-momentum transfer is verified by the Reynolds shear stress $< v >< w >$ across the fluctuating shear $\partial < w > /\partial y$. In Case N, $\partial < w > /\partial y$ shows the marked asymmetry between the positive and negative regions. In the RDT, however, the positive and negative $< v >< w >$ are balanced with each other, and hence the spanwise momentum is not virtually transferred into the longitudinal vortex. In Case C1, the asymmetry is more marked, while it almost disappears in the cases (Cases C2 and A1) where the longitudinal vortex becomes parallel to the streamwise direction as in the RDT, and the averaged tilting angle is almost zero.

4 Conclusions

Direct numerical simulations of a homogeneous shear flow under both cyclonic and anticyclonic system rotation are performed to study the mechanism of the tilting of the longitudinal vortical structure.

Weak cyclonic system rotation increases the tilting in comparison to the case without a system rotation, while the anticyclonic and strong cyclonic system rotation decrease the tilting. The spanwise-velocity pools $< w >$ associated with the longitudinal vortical structure become the driving force for tilting in the spanwise direction, which is strongly related to the down gradient flux for the local shear $\partial < w > /\partial y$. When tilting is not observed, however, the down gradient flux is balanced by the counter gradient flux for $\partial < w > /\partial y$.

Acknowledgement

This work was supported by the Ministry of Education, Culture, Sports, Science, and Technology through a Grant-in-Aid for Scientific Research (No. 16760153), and by the Toyota Physical and Chemical Research Institute.

References

1. O. Iida, M. Iwatsuki & Y. Nagano, 2000, Phys. Fluids 12, 2895-2905.
2. S. Rogallo, Numerical experiments in homogeneous turbulence, NASA TM-81315, 1981.
3. J. Jeong, F. Hussain, W. Schoppa, & J. Kim, 1997, J. Fluid Mech. 332, 185-214.

Generation of Large-scale Vorticity in Sheared Homogeneous and Rotating Inhomogeneous Turbulence

T. Elperin, I. Golubev, N. Kleeorin, I. Rogachevskii

Department of Mechanical Engineering, Ben-Gurion University of the Negev, POB 653, 84105 Beer-Sheva, Israel; elperin@menix.bgu.ac.il

The main goal of this communication is to review our recent theoretical studies on the generation of large-scale vorticity in sheared homogeneous turbulence and in rotating inhomogeneous turbulence.

1 Sheared Non-rotating Homogeneous Turbulence

Let us first discuss the generation of large-scale vorticity in sheared homogeneous turbulence which is related to the effects of mean velocity shear on turbulence and the effective force determined by the divergence of Reynolds stresses [1]. We consider homogeneous divergence-free turbulence with a mean velocity shear of the form $\mathbf{U}^{(s)} = (0, Sx, 0)$ and hence the mean vorticity $\mathbf{W}^{(s)} = (0, 0, S)$. For simplicity we use perturbations of the mean vorticity in the form $\mathbf{W} = (W_x(z), W_y(z), 0)$. We derived the following equations for the perturbations of the mean vorticity:

$$\frac{\partial W_x}{\partial t} = S W_y + \nu_{_T} W_x'' \,, \tag{1}$$

$$\frac{\partial W_y}{\partial t} = -\beta_* \, S \, l_0^2 \, W_x'' + \nu_{_T} W_y'' \,, \tag{2}$$

where $W'' = \partial^2 W / \partial z^2$, $\nu_{_T}$ is the turbulent viscosity, l_0 is the maximum scale of turbulent motions and the parameter $\beta_* = 3.5$. Details of the derivations of (1) and (2) can be found in [1]. In (1) we took into account that the characteristic scale L_W of the mean vorticity variations is much larger than the maximum scale of turbulent motions l_0. This assumption corresponds to the mean-field approach. We seek for a solution of (1) and (2) in the form $\propto \exp(\gamma t - iKz)$. Thus, when $\beta_* > 0$ perturbations of the mean vorticity can grow in time and the growth rate of the instability is given by $\gamma = \sqrt{\beta_*} \, S \, l_0 \, K - \nu_{_T} \, K^2$, where K is the wave number. The maximum growth rate of perturbations of the mean vorticity, $\gamma_{\rm max} = \beta_* \, (S \, l_0)^2 / 4\nu_{_T}$, is attained at $K = K_m = \sqrt{\beta_*}\, S\, l_0 / 2\nu_T$. The sufficient condition $\gamma > 0$ for the excitation of the instability is $L_W / l_0 > 2\pi / (3\sqrt{\beta_*}\, \tau_0 S)$, where $L_W \equiv 2\pi / K$ and we consider a weak velocity shear $\tau_0 S \ll 1$.

Now let us discuss the mechanism of this instability. The first term, $S W_y = (\mathbf{W}^{(s)} \cdot \boldsymbol{\nabla})\, U_x$, in (1) describes a "skew-induced" generation of perturbations of the mean vorticity W_x by quasi-inviscid deflection of the equilibrium mean vorticity $\mathbf{W}^{(s)}$. Here $\mathbf{U}$ are the perturbations of the mean velocity. The mean vorticity $W_x \mathbf{e}_x$ is generated from $W_y \mathbf{e}_y$ by equilibrium shear motions with the mean vorticity $\mathbf{W}^{(s)} = S\, \mathbf{e}_z$, i.e. $W_x \mathbf{e}_x \propto (\mathbf{W}^{(s)} \cdot \boldsymbol{\nabla}) U_x \mathbf{e}_x \propto W_y \mathbf{e}_y \times \mathbf{W}^{(s)}$. Here $\mathbf{e}_x$, $\mathbf{e}_y$ and $\mathbf{e}_z$ are the unit vectors along x, y and z axes, respectively. On the other hand, the first term, $-\beta_*\, S\, l_0^2\, W_x''$, in (2) determines a "Reynolds stress-induced" generation of perturbations of the mean vorticity W_y by the Reynolds stresses. In particular, this term is determined by $(\boldsymbol{\nabla} \times \boldsymbol{\mathcal{F}})_y$, where $\boldsymbol{\mathcal{F}}$ is a divergence of Reynolds stresses (the effective force). This implies that the mean vorticity $W_y \mathbf{e}_y$ is generated by an effective anisotropic viscous term $\propto -l_0^2\, \Delta\, (W_x \mathbf{e}_x \cdot \boldsymbol{\nabla})\, U^{(s)}(x) \mathbf{e}_y \propto -l_0^2\, S\, W_x'' \mathbf{e}_y$. This mechanism of the generation of perturbations of the mean vorticity $W_y \mathbf{e}_y$ can be interpreted as a stretching of the perturbations of the mean vorticity $W_x \mathbf{e}_x$ by the equilibrium shear motions $\mathbf{U}^{(s)} = S\, x\, \mathbf{e}_y$ during the turnover time of turbulent eddies. The growth rate of this instability is caused by a combined effect of the sheared motions ("skew-induced" generation) and the "Reynolds stress-induced" generation of perturbations of the mean vorticity.

The results obtained in this study are different from [2, 3] whereby the generation of large-scale vorticity in the helical turbulence occurs due to the hydrodynamic α effect. The latter effect is associated with the term $\alpha \mathbf{W}$ in the equation for the mean vorticity, where α is determined by the hydrodynamic helicity of turbulent flow. We considered a nonhelical homogeneous turbulence which implies that (1) and (2) do not have the term $\alpha \mathbf{W}$.

The analyzed effect of the mean vorticity production may be of relevance in different turbulent industrial, environmental and astrophysical flows. Thus, e.g., the suggested mechanism can be used in the analysis of the flows associated with Prandtl's turbulent secondary flows. The obtained results may be also important in astrophysics, e.g. in extragalactic clusters, merged protostellar and protogalactic clouds.

2 Rotating Inhomogeneous Turbulence

Let us discuss the formation of large-scale structures in a rotating inhomogeneous non-helical turbulence [4]. In particular, we investigate the excitation of large-scale inertial waves. A mechanism for the excitation of the large-scale inertial waves is associated with a generation of a large-scale vorticity due to the excitation of the large-scale instability in a uniformly rotating inhomogeneous turbulence. The excitation of the mean vorticity in this system requires inhomogeneity of turbulence.

We derived equation for the effective force in rotating inhomogeneous turbulence. This allowed us to obtain the mean-filed equations and to study the excitation of large-scale inertial waves. In particular, we derived the following non-dimensional equations for the perturbations of the mean vertical velocity and the mean vertical vorticity

$$\Delta \frac{\partial U_z}{\partial t} = -\hat{G}\, W_z + [\nu_{_T}\, \Delta^2 + \nu_{_T}\, \Lambda\, \Delta \nabla_z + \nu_{_U}\, \Lambda\, \Delta_\perp \nabla_z + \eta_{_U}\, \Delta_\perp \nabla_z^2] U_z \,, \quad (3)$$

$$\frac{\partial W_z}{\partial t} = (\hat{G} - \beta\, \Lambda\, \Delta_\perp) U_z + [\nu_{_T}\, \nabla_z^2 + \nu_{_T}\, \Lambda \nabla_z + \eta_{_W}\, \Delta_\perp] W_z \,, \quad (4)$$

Details of the derivations of (3) and (4) can be found in [4]. For simplicity we considered the case when turbulence is inhomogeneous along the rotation axis, i.e. $\mathbf{\Lambda} = \Lambda(z) \mathbf{e}_z$ and $\mathbf{\Omega} = \Omega \mathbf{e}_z$. Here $\Delta = \Delta_\perp + \nabla_z^2$, $\nabla_z = \partial / \partial z$, $\mathbf{\Lambda} = (\mathbf{\nabla} \nu_{_T}^{(0)}) / \nu_{_T}^{(0)}$, $\hat{G} = 2\, a_*\, \nabla_z + \beta[\Lambda\, \nabla_z^2 + \Delta \nabla_z] + 2\beta_{_U}\, \Delta_\perp \nabla_z$, $a_* = \Omega L^2 / \nu_{_T}^* \gg 1$ and $\nu_{_T}^*$ is the characteristic value of turbulent viscosity. The parameters β, $\beta_{_U}$, $\nu_{_T}$, $\nu_{_U}$, $\eta_{_U}$ and $\eta_{_W}$ in (3) and (4) are proportional to the function $\nu_{_T}^{(0)}(z)$ and depend on the angular velocity Ω. We determined these dependencies on the rotation rate. In (3) and (4) we use the following dimensionless variables: length is measured in units of L, time in units of $L^2 / \nu_{_T}^*$, the function $\Lambda(z)$ is measured in the units of L^{-1}, the function $\nu_{_T}^{(0)}(z)$ is measured in the units of $\nu_{_T}^*$, the perturbations of velocity U_z and vorticity W_z are measured in units of U_* and U_* / L, respectively.

Assume now that functions $\nu_{_T}^{(0)}(z)$ and $\Lambda(z)$ vary slowly with z in comparison with the variations of the mean velocity $U_z(z)$ and mean vorticity $W_z(z)$. Let us seek for the solution of (3) and (4) in the form $\propto \exp(\gamma t - i \mathbf{K} \cdot \mathbf{R})$. Let us first consider perturbations with the wave numbers $K_\perp^2 \ll K_z^2$. The growth rate of the inertial waves with the frequency $\omega_w = -2\, a_*\, (K_z / K)\, \mathrm{sgn}(\beta\, \Lambda)$ is given by $\gamma_w = |\beta(\Omega)\, \Lambda\, K_z| - \nu_{_T}(\Omega) K^2$, where $\gamma = \gamma_w + i \omega_w$, the wave number K is measured in the units of L^{-1} and γ is measured in the units of $\nu_{_T}^* / L^2$. The inertial waves are helical, i.e. the large-scale hydrodynamic helicity of the motions in the inertial waves is not zero. Let us consider the opposite case, i.e. the perturbations with the wave numbers $K_\perp^2 \gg K_z^2$. The growth rate of perturbations with the frequency $\omega_w = 2\, a_*\, (K_z / K)\, \mathrm{sgn}(\beta\, \Lambda)$ is given by $\gamma_w = (1/2) \{ |\beta(\Omega)\, \Lambda|\, K - [\nu_{_T}(\Omega) + \eta_{_W}(\Omega)]\, K^2 \}$.

Now we take into account the inhomogeneity of the functions $\nu_{_T}^{(0)}(z)$ and $\Lambda(z)$. Let us consider an eigenvalue problem for a system of (3) and (4). We seek for a solution of (3) and (4) in the form $\propto \Psi(z) \exp(\gamma t) J_0(K_\perp r)$, where $J_0(x)$ is the Bessel function of the first kind. After the substitution of this solution into (3) and (4) we obtained the system of the ordinary differential equations which was solved numerically.

We determined numerically the range of parameters at which the large-scale instability occurs. We found the rotation rate dependencies of the growth rate γ_w of the large-scale instability and the frequency ω_w of the generated inertial waves due to the large-scale instability. There is a threshold in the rotation rate for the large-scale instability: $\Omega_* \tau_0 \approx 0.025$, and when $\Omega > \Omega_*$, the instability is excited. The instability threshold in the parameter L is $L > 10 l_0$. Note that the characteristic time ($\sim 2\pi/\gamma_w$) of the growth of perturbations of the mean fields **U** and **W** is by 5 orders of magnitudes larger than the turbulent correlation time τ_0. The period of oscillations $T = 2\pi/\omega_w$ of inertial waves is at least 10 times larger than the turbulent correlation time τ_0. The minimum value of the period of rotation $T_R = 2\pi/\Omega$ is at least 20 times larger than the turbulent correlation time τ_0. The source of the large-scale instability is the energy of the small-scale turbulence. Therefore, the rotation and inhomogeneity of turbulence provide a mechanism for transport of energy from small-scale turbulence to large-scale motions.

3 Conclusions

We studied an effect of a mean velocity shear on a turbulence and on the effective force determined by the divergence of the Reynolds stresses. The mean vorticity in a homogeneous nonhelical turbulent flow with an imposed mean velocity shear is generated due to an excitation of a large-scale instability. The instability is caused by a combined effect of the large-scale shear motions and "Reynolds stress-induced" generation of perturbations of mean vorticity. Spatial characteristics of the instability, such as the minimum size of the growing perturbations and the size of perturbations with the maximum growth rate, are determined.

We also studied the generation of large-scale vorticity and excitation of the large-scale inertial waves in rotating inhomogeneous turbulence caused by an excitation of a large-scale instability. The mechanism of the instability is associated with the inhomogeneity of the turbulence and the uniform mean rotation. We determined the range of parameters whereby the large-scale instability is excited, the growth rate of the instability and the frequency of the generated large-scale inertial waves. The source of these large-scale instabilities is the energy of the small-scale turbulence.

References

1. Elperin T, Kleeorin N, Rogachevskii I (2003) Phys Rev E 68: 016311
2. Khomenko GA, Moiseev SS, Tur AV (1991) J Fluid Mech 225: 355-364
3. Chkhetiany OG, Moiseev SS, Petrosyan AS, Sagdeev RZ (1994) Physica Scripta 49: 214-221, and references therein
4. Elperin T, Golubev I, Kleeorin N, Rogachevskii I (2005) Phys Rev E 71: 036302

Mean Modal Analysis of the Coherent Structures in a Two-dimensional Vortex Merger

Fereidoun Sabetghadam[1] and Shervin Sharafatmandjoor[2]

[1] Assistant Professor, Science and Research Branch, Azad University (IAU), Tehran, Iran. fsabet@sr.iau.ac.ir

[2] Ph.D. Student, Science and Research Branch, Azad University (IAU), Tehran, Iran. Sharafatmand@sr.iau.ac.ir

Summary. Time evolution of mean behavior of the coherent structures in two-dimensional vortex merger process is analyzed via solution of the mean modal form of the Navier-Stokes equations. A ten member ensemble has been constructed and used to obtain appropriate modal viscosity which has been used to close the mean modal equations. Results show very good ability of the method to capturing the large scale motions and to prediction the energy and enstrophy spectrum at least for the low wave numbers.

1 Introduction

Turbulent flow computations, allowing for the simulation of the time evolution of the Coherent Structures (CS), have been in the center of attention for their theoretical as well as for their applied importance. On the other hand, the fact that any CS covers almost the whole spectrum, and capability of the ensemble averaging to structural filtering, make this averaging process as a suitable filtering tool when capturing the CS is desired.

Recently the Mean Modal Decomposition (MMD) method has been suggested as a new tool for analysis of the turbulent flows [1]. In its essentials, the method consists of projection of the field onto an orthogonal function space, which could be for example any of the classical Fourier modes, the Proper Orthogonal Decomposition (POD) modes or others, and then ensemble averaging on each mode in that function space. Because of using the ensemble averaging as the filtering process, the method is expected to have the ability of capturing the CS from the field. In fact, this ability has been shown completely for a spatio-temporal one-dimensional chaotic field with both Fourier and POD modes [2], and partially for the two-dimensional homogenous turbulence [3].

As a continuation of our previous works, in this paper, the mean behavior of the CS during two-dimensional vortex merger process, which causes an inverse energy cascade and direct enstrophy cascade [4], is simulated and observed via solution of the MMD form of the Navier-Stokes equations in the vorticity formulation.

2 Formulation

As mentioned earlier, in this paper, the mean modal formulation is applied on the vorticity form of two-dimensional Navier-Stokes equations. Consider an 'M' member ensemble $\{\boldsymbol{\omega}(\mathbf{x},t)\}_{m=1}^{M}$ of a two-dimensional vorticity field $\boldsymbol{\omega}(\mathbf{x},t) = (0,0,\omega(\mathbf{x},t))$, each of them satisfies the vorticity form of the Navier-Stokes equations

$$\partial_t \omega^m + \mathcal{N}\{\omega^m\} = \nu \nabla^2 \omega^m, \tag{1}$$

on $\mathbf{x} \in \Omega$ and during $t \in [t_0, t_0 + T]$. Here the $\mathcal{N}\{\omega\} = (\mathbf{u}.\nabla)\omega$ denotes all the nonlinear terms in which the convective velocities are formally replaced with their equivalent ω via the Biot-Savart integral. The general mean modal form of the Navier-Stokes equations can be found in [1].

2.1 Mean Modal Decomposition (MMD) Procedure

Each member of the ensemble is decomposed into a mean modal part, with a preassumed maximum mode number 'N' and time duration 'T', and a fluctuation part which is assumed to be the deviation of each member from the mean modal representation of the ensemble.

$$\omega^m(\mathbf{x},t) = \langle \omega \rangle_{N,T}(\mathbf{x},t) + \omega_{N,T}^{'m}(\mathbf{x},t) \tag{2}$$

Projection of this relation onto an appropriate orthogonal function space $\{\phi_i(\mathbf{x})\}_{i=0}^{\infty}$, ensemble averaging and definition of the mean modal term as:

$$\bar{\hat{\omega}}_k = \mathcal{M}_k\{\omega^m\} = \langle (\omega^m, \phi_k) \rangle = \frac{1}{\mathrm{M}} \sum_{m=1}^{M} (\omega^m, \phi_k) \qquad \text{for} \quad |k| \leq N \tag{3}$$

yields a generalization of the Reynolds definition of the fluctuation part as:

$$\mathcal{M}_k\{\omega_{N,T}^{'m}\} = 0 \qquad \text{for} \quad |k| \leq N. \tag{4}$$

Here, (ω^m, ϕ_k) denotes the inner product in the function space and the mean modal operator $\mathcal{M}_k\{\cdot\}$ is defined just for the sake of simplicity. With the above definitions, the mean modal representation of the ensemble is:

$$\langle \omega \rangle_{N,T} = \sum_{k=-N}^{+N} \bar{\hat{\omega}}_k^m \phi_k(x). \tag{5}$$

For more details of the definitions and the properties of the mean modal operator may consult [1].

2.2 Governing Equation for the Mean Modes

By introducing the decomposition (2) into the (1) we obtain

$$\partial_t \langle\omega\rangle_{N,T} + \mathcal{N}\{\langle\omega\rangle_{N,T}\} - \nu\nabla^2\langle\omega\rangle_{N,T} = \\ -[\partial_t \omega'^{m}_{N,T} + \mathcal{N}_1\{\omega'^{m}_{N,T}, \langle\omega\rangle_{N,T}\}] - \mathcal{N}\{\omega'^{m}_{N,T}\} \quad (6)$$

in which the $\mathcal{N}_1$ contains all the nonlinear coupling of the mean modes and the fluctuations. Application of the mean modal operator (3) on the above equation yields:

$$\partial_t \bar{\hat{\omega}}_k + (\bar{\hat{\omega}} \otimes \bar{\hat{\omega}})_k = -\nu k^2 \bar{\hat{\omega}}_k - (\hat{\omega}' \otimes \hat{\omega}')_k, \quad (7)$$

where $(\cdot \otimes \cdot)_k$ formally denotes k^{th} mode of the mean modes convolution. Obviously this equation is unclosed because of the fluctuation part in the right hand side which can be viewed as a general Reynolds stress. The simplest modeling method which is used in this paper, is to define a Boussinesq viscosity type which we call it the modal viscosity ν_m. Using this modal viscosity, finally we have:

$$\partial_t(\bar{\hat{\omega}}_k) + (\bar{\hat{\omega}} \otimes \bar{\hat{\omega}})_k = -(\nu + \nu_m) k^2 \bar{\hat{\omega}}_k. \quad (8)$$

Having the solution of this equation, one can construct the mean time evolution of the field using the definition (5).

3 Results and Discussion

System of (1) is solved directly on a $(2\pi \times 2\pi)$ box using pseudospectral method with 256^2 modes until the development of a k^{-3} energy spectrum.

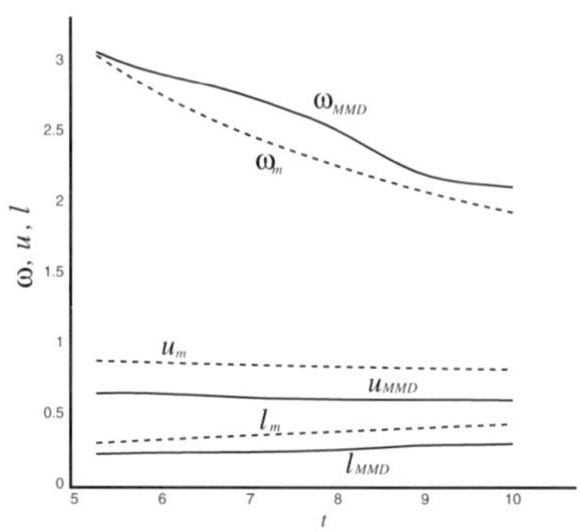

Fig. 1. Time variations of the mean vorticity 'ω', velocity 'u' and length scale 'l'. **dashed lines:** DNS results with 256^2 modes, **solid lines:** MMD with 128^2 modes

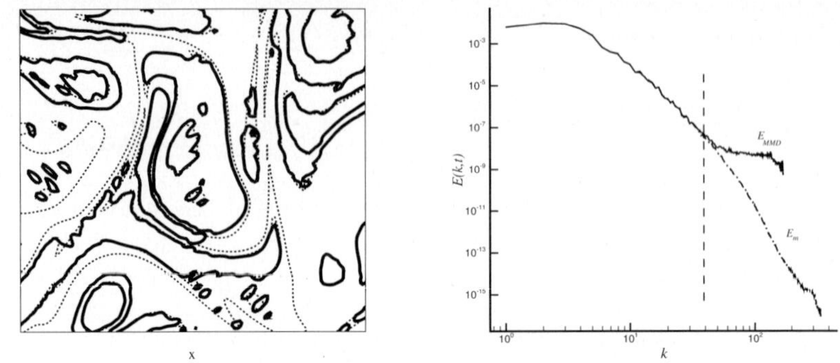

Fig. 2. Comparison of the vorticity (in physical space) and energy spectrum for MMD and DNS. Dashed and solid lines are as mentioned on Fig. 1

The solution method and the initial condition have been similar to [5]. A ten member ensemble then is constructed choosing this condition as the initial condition and a ±10 percent random disturbance on the Fourier modes. The ensemble then is used to obtain modal viscosity for solution of the MMD equations (8) with 128^2modes. Figure 1 shows the time variations of the mean vorticity, velocity and length scales obtained from DNS with 256^2 modes and the MMD computations. Obviously although there are some differences, the MMD has captured the inverse energy cascade. Vorticity, in physical space and the energy spectrum of DNS and MMD are compared in Fig. 2. As one can see, except for the small scales, after about 50th mode, where there are some differences, the vorticity field in the large scales is predicted accurately.

References

1. Sabetghadam, F. (2004). Generalization of the RANS Equations Using Mean Modal Decomposition of the Navier-Stokes Equations, Proceedings of the ASME Engineering System Design and Analysis (ESDA) Conference, Manchester, UK.
2. Sabetghadam, F. (2005). POD and MMD Comparison in Extraction of the Coherent Structures, IASME transactions, Issue 1, Vol. 2, 92-97.
3. Sabetghadam, F., Sharafatmandjoor, S. (2005). Extraction of the Coherent Structures in a Homogenous Two-dimensional Turbulent Flow Using the Mean Modal Decomposition Method, Proceedings of 13th Annular conference of Computational Fluid Dynamics (CFD 2005), St. Johns, NL, Canada.
4. Tabeling, P. (2002). Two-dimensional Turbulence: A Physicist Approach, *Phys. Reports* 362:1-62.
5. Chasnov, J. R. (1997). On the Decay of Two-dimensional Homogeneous Turbulence, *Phys. Fluids* 9:171-180.

Low-Frequency Coherent Structures in Turbulent Flows

Rüdiger Schwarze[1] and Frank Obermeier[2]

1 Institut für Mechanik und Fluiddynamik, TU Bergakademie Freiberg, Lampadiusstr. 4, 09596 Freiberg, Germany
Ruediger.Schwarze@imfd.tu-freiberg.de
2 Institut für Mechanik und Fluiddynamik, TU Bergakademie Freiberg, Lampadiusstr. 4, 09596 Freiberg, Germany
Frank.Obermeier@imfd.tu-freiberg.de

1 Motivation and Aims

Recently, low-frequency coherent structures have been found in several technical flow processes, e.g. in stirred tank reactors [1]. These structures are of fundamental importance for process performance. Therefore, further investigations are highly desirable. Direct numerical simulations or large-eddy simulations allow detailed insight into the turbulent flows, but they demand high numerical efforts. In this situation numerical models, which are based on the unsteady Reynolds-averaged Navier-Stokes equations (URANS), seem to be a compromise between accuracy and efficiency.

In the present paper, low-frequency coherent oscillations of the flow in a continuous casting tundish are investigated experimentally as well as numerically. Water model experiments resolve the mean flow and identify low-frequency flow oscillations. URANS equations are employed in order to investigate the flow in the water model numerically. For that purpose two different turbulence models are implemented into the simulations: (i) a Reynolds stress model and (ii) a filter-based k-ϵ turbulence model.

2 Experiment

2.1 Water Model Facility

The main component of the water model facility is the 0.25-scale tundish model made of perspex. The dimensions of this model are displayed in Fig. 1. The model is fed by a closed water circuit. Here it is possible to tune the flow rate in the model by a valve between $\dot{V} = 1.6 \ldots 2.2\, l/s$.

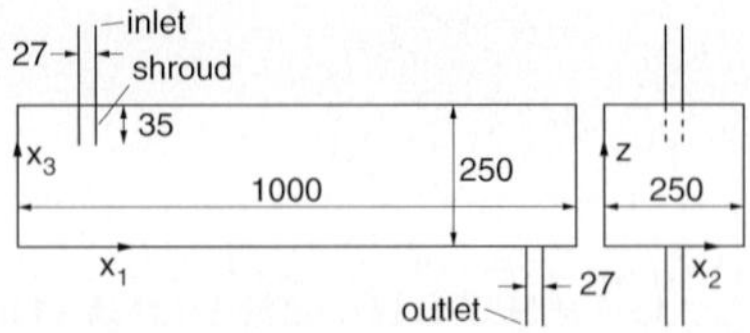

Fig. 1. Dimensions of the tundish model

Fig. 2. Water model facility

2.2 Instrumentation

Velocity data inside the tundish model are obtained by a LDA (2D Dantec Dynamics) system. These measurements are performed on a $10\,mm \times 10\,mm$ grid in different locations in the tundish. Figure 2 shows the set-up for the grid on a vertical plane near to the sidewall of the tundish model.

The pressure was measured by a pressure tube (orifice diameter 3 mm) connected to a pressure transducer (Kistler). Time series of the pressure were recorded at different locations within the flow.

3 Numerical Model

3.1 Theory

The numerical model is based on the unsteady, time-filtered (cut off) Navier-Stokes equations

$$\frac{\partial \langle u_i \rangle}{\partial x_i} = 0 \tag{1}$$

$$\frac{\partial \langle u_i \rangle}{\partial t_i} + \frac{\partial}{\partial x_j} \left(\langle u_j \rangle \langle u_i \rangle \right) = -\frac{1}{\rho} \frac{\partial p}{\partial x_i} + \frac{\partial}{\partial x_j} \left(\nu \frac{\partial \langle u_i \rangle}{\partial x_j} + \tau_{ij} \right) \tag{2}$$

The unknown Reynolds stresses τ_{ij} in (2) are modelled by (i) a Reynolds stress model and (ii) a filter-based k-ε model [3] with a filter width $\Delta = 10\,cm$. In the near-wall region, the turbulent flow quantities are approximated by non-equilibrium wall functions.

3.2 Numerics

The model equations are solved by the finite-volume method. The numerical schemes are QUICK upwind interpolation, CDS differencing and second-order BDF for the time integration. The solution domain contains 220.000 hexahedral grid cells.

The boundary conditions of the numerical simulation fit to the parameters of the experimental configuration. Here it has to be emphasized that the boundary conditions are steady-state conditions even in the unsteady simulation. Initial values for this flow field are taken from the corresponding steady-state solution (RANS) of the model equations.

4 Results

Figure 3 displays velocity profiles along a vertical line through the pouring jet downstream of the shroud. Mean and rms velocity profiles of the numerical simulations obtained by the Reynolds stress (RS) as well as by the filter-based k-ϵ (FKE) turbulence model fit well to the experimental values (EXP). Here, the mean and rms values are derived from a long-term time average of the numerically and experimentally determined time-series of the velocities. Note, that the Reynolds stress model slightly overestimates the rms values in the shear layers of the jet. Contrary, the filter-based k-ϵ model underestimates the rms values in these regions. Profiles along other lateral and longitudinal lines in the tundish show similar agreement between experimental and numerical data.

Long-term oscillations of the flow are detected in the spectra, which are deduced from the pressure measurements, Fig. 4 (left). Here, a marked peak occurs at a frequency $f^p_{peak} \simeq 0.2\,Hz$. It is also possible to identify significant peaks in the spectra of long-term LDA measurements [2]. However, pressure measurements performed at a fixed data sampling rate are easier to analyze than LDA data. LDA measurements yield data at a varying sampling rate.

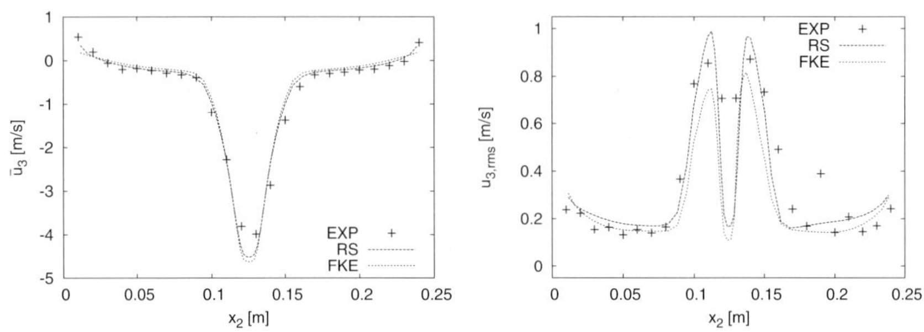

Fig. 3. velocity profiles ($\overline{u}_3$, $u_{3,rms}$) at $x_1 = 250\,mm$ and $x_3 = 100\,mm$

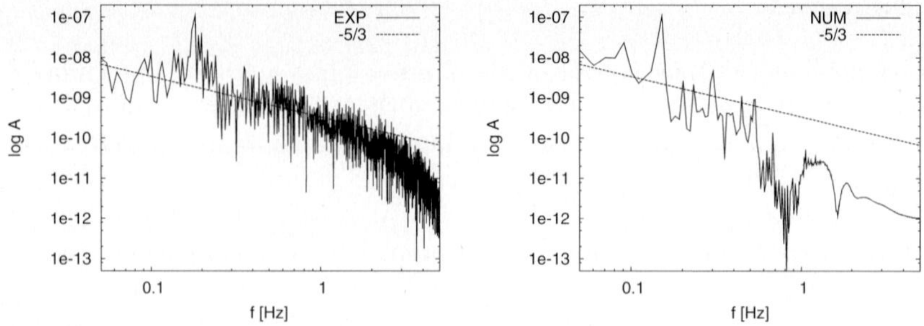

Fig. 4. FFT of p in experiment (left) and FFT of u_3 in our simulation (right)

The interpretation of these data sets is more complicated, because varying sampling rates may lead to additional but erroneous peaks in the spectra.

The numerical simulation yields similar oscillations, i.e. for both turbulence models. Here, a marked peak is identified at a frequency $f^u_{peak} \simeq 0.15\,Hz$ in the spectra of the vertical velocity u_3, Fig. 4 (right).

5 Conclusions

Long-term oscillating flows in continuous casting tundishes are investigated by a model experiment as well as by numerical simulations. The results of the simulation runs fit well to the experimental observations (mean flow data and frequencies). Especially, long-term oscillations of the tundish flow are resolved by the simulations.

References

1. Nikiforaki L, Montante G, Lee K C, Yianneskis M (2004) Chem Eng Sci 58: 2937–2949
2. Schwarze R, Obermeier F (2005) Numerical modeling of large scale oscillations in the continuous casting process. In: GAMM Jahrestagung 2005, Section S09, Computational Fluid Mechanics. Luxemburg
3. Johansen ST, Wu JY, Shyy W (2004) Int J Heat Fluid Flow 25:10–21

Part IV

Turbulent Boundary Layers

Large-eddy Simulation of Accelerating Boundary Layers

G. De Prisco, A. Keating, U. Piomelli and E. Balaras

Department of Mechanical Engineering, University of Maryland, College Park, MD, USA

1 Introduction

Turbulent boundary layers subjected to a favorable pressure gradient (FPG) (i.e. ones that result in freestream acceleration) are common in many engineering applications, such as airfoils and curved ducts. While the canonical zero-pressure gradient (ZPG) boundary layer is relatively well understood, less is known about FPG boundary-layers. Reviews of current knowledge can be found in several articles by Narasimha and Sreenivasan [1, 2, 3]; here we summarize the findings of these and other experimental [4, 5, 6, 7, 8, 9, 10] and numerical [11] studies. A decrease of the turbulence production is a significant cause of re-laminarization in accelerating flows. This phenomenon is related to a reorientation of the quasi-streamwise vortices in the near-wall region and a decrease in the wall-normal momentum transport. The fact that the re-laminarization is due to the structural features of the boundary-layer eddies has significant implications for Reynolds-stress models, which generally have difficulty dealing with problems of this type. The turbulence in the outer layer remains frozen through most of the acceleration, and does not equilibrate with either the local freestream or friction velocity. Close to the wall, the flow undergoes a process of laminarization, in which the skin friction coefficient drops sharply. Finally, after the acceleration is completed, the flow quickly re-transitions to an equilibrium boundary layer. Several questions remain open; among them are: (1) if, and why, the outer layer turbulence remains frozen, (2) how the inner and outer layers interact during and after the acceleration, and (3) how the retransition to turbulence takes place (and why it takes place so abruptly).

To answer some of these questions, large-eddy simulations (LES) of boundary layers in FPG were performed with strong acceleration parameters, equal to 3 and 4×10^{-6}. The latter calculation matches one of the cases of the experiment in [8, 9]. For comparison, on a 0.3m-long NACA0012 airfoil at $Re = 1.5 \times 10^6$, K has values between 3×10^{-6} and 0 in the region between 2% of the chord and the point of maximum thickness.

2 Problem Formulation

Simulations were carried out using a second-order finite-difference LES code. The filtered Navier-Stokes equations were solved on a Cartesian staggered grid. Conservative second-order finite differences were used for spatial discretization while a semi-implicit fractional-step method [12] was used for time integration. The subgrid-scale stresses were parameterized using the dynamic eddy-viscosity subgrid-scale model [13, 14] with the eddy-viscosity coefficient averaged over Lagrangian flow pathlines [15]. An unsteady inflow boundary condition was obtained from a separate simulation that used the recycling/rescaling method [16], while a convective outflow boundary condition was used at the downstream boundary. At the top boundary, a variable freestream velocity[1] $U_\infty(x)$ was used [16]. The simulations were performed on a domain of size $476\delta_o^* \times 20\delta_o^* \times 20\delta_o^*$, using $1024 \times 64 \times 128$ grid points. The results obtained using this resolution compare well with coarser calculations. The Reynolds number, based on freestream velocity at the inflow, U_o, and on the displacement thickness at the inflow, δ_o^*, is 1,260.

3 Results and Discussion

Figure 1*(a)* shows the acceleration parameter, K and the resulting freestream velocity, U_∞ for the two cases. In the first case the freestream velocity at the outflow is almost three times that at the inflow. The skin-friction coefficient $C_f = 2\tau_w/\rho U_\infty^2$ is shown in Fig. 1*(b)*. Although the mean freestream velocity increases in both cases, C_f begins to decrease near the location of maximum K. After the pressure gradient is relaxed, rapid retransition towards an equilibrium turbulent value occurs. The chain-dotted line shows the ZPG value of C_f obtained from the correlation [17] $C_f = 0.0576Re_x^{-0.2}$. At the inflow the FPG experiments [8, 9] have lower skin friction than the correlation value, suggesting that pressure gradient effects are already significant on the upstream boundary layer. The region of relaminarization, on the other hand, is predicted well by the LES. The retransitioning occurs more abruptly than in the experiment, and the downstream C_f is higher in the LES. This may be due to the coarser resolution in the recovery region. Note that the streamwise and spanwise grid sizes were constant throughout the domain, while u_τ increases by a factor of approximately three; this results in grid sizes in wall units that give excellent resolution in the upstream region ($\Delta x^+ \simeq 28$, $\Delta z^+ \simeq 9.4$), but are marginal in the recovery region ($\Delta x^+ \simeq 82$, $\Delta z^+ \simeq 28$).

The lower acceleration case shows similar behavior to the high-acceleration one; however the relaminarization is less severe, with C_f reduced by a much smaller amount in the acceleration region, and recovery taking place earlier than for the high-K case.

[1] In the following, U_∞ denotes the (variable) freestream velocity, whereas $U_o = U_\infty(0) = 1$ is the reference velocity in the flat-plate upstream of the FPG.

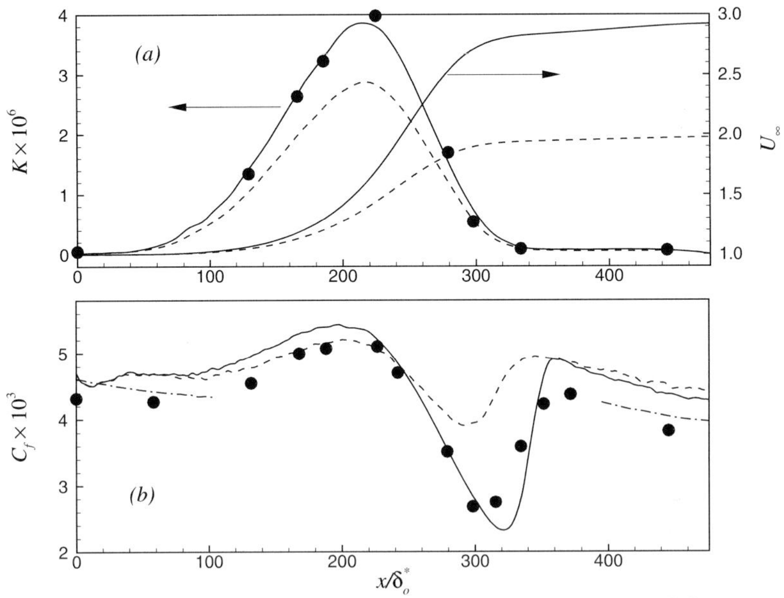

Fig. 1. Streamwise development of acceleration parameter K and freestream velocity *(a)*, and Skin friction coefficient *(b)*. • Experiments [8, 9]; ——— high-K case; - - - low-K case; —·— ZPG boundary-layer correlation

Figure 2 shows the mean velocity profiles in wall coordinates at several locations in the flow. One observes the existence of a logarithmic layer (following the standard law, $U^+ = 2.5 \log y^+ + 5$) at the inflow. As the FPG becomes significant, the slope of the logarithmic region decreases (a well-known effect of acceleration [18]). The two cases are in good agreement up to the point of maximum K. Thereafter, the high-K case departs significantly from the equilibrium boundary layer profile, becoming more laminar-like. The recovery of the inner layer to an equilibrium logarithmic law occurs quite rapidly, between $x/\delta_o^* \simeq 330$ and 370. The agreement with the experimental data is very good in the equilibrium and relaminarization regions. In the recovery region, it is affected by the errors in the prediction of u_τ discussed earlier. In outer coordinates (not shown), in fact, the agreement is excellent at all locations.

The effects of the acceleration on the flow can be illustrated by examining the development of turbulent quantities experienced by a particle moving along a mean streamline. To this end we have selected four streamlines, one that originates in the buffer layer at $x/\delta_o^* = 0$, one in the logarithmic layer, one roughly in the middle of the boundary layer, and another one in the outer region of the boundary layer ($y/\delta_{99} = 0.4$ and 0.85, respectively). The streamlines are shown in Fig. 3*(a)*. Figures 3*(b)* and *(c)* show the development of the Reynolds shear stresses normalized by the freestream velocity and the correlation coefficient $C_{uv} = - < u'v' > /(< u'^2 >< v'^2 >)^{1/2}$.

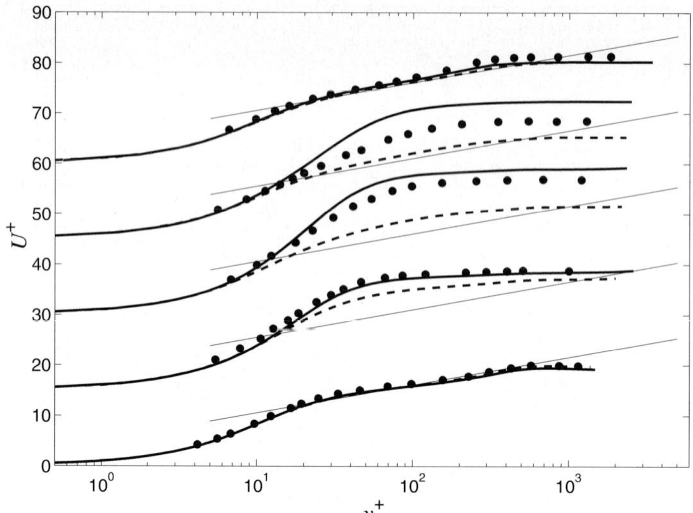

Fig. 2. Wall-normal profiles of the mean streamwise velocity at $x/\delta_o^* = 166, 276, 315, 333, 369$ (from bottom to top). • Experiments [8, 9]; ——— high-K case; - - - low-K case; ——— logarithmic law

In this normalization, $- < u'v' >$ decreases significantly in the acceleration region. The decrease, however, is not due to a decorrelation of the u' and v' fluctuations: the correlation coefficient, in fact, is nearly constant (or increases

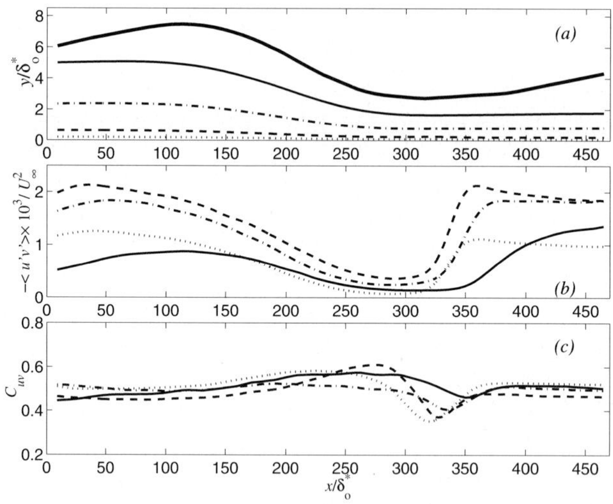

Fig. 3. Development of Reynolds shear stress and correlation coefficient along streamlines. *(a)* Streamline coordinates; *(b)* C_{uv}; *(c)* $- < u'v' > \times 10^3/U_\infty$. The thick line corresponds to the boundary-layer edge. Streamline originating in: ——— outer-layer; —·— middle of the boundary layer; ··· log region; - - - buffer layer

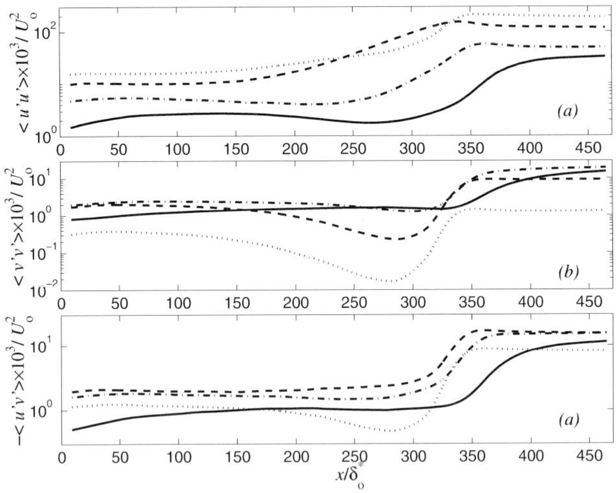

Fig. 4. Development of Reynolds stresses along selected streamlines, normalized by U_o. *(a)* $< u'u' >$; *(b)* $< v'v' >$; *(c)* $- < u'v' >$. Same legend as Fig. 3

slightly) in the region where the shear stress begins to decrease. If we examine the Reynolds stresses normalized by U_o, however, we observe that the streamwise stress remains constant, but the wall-normal one decreases significantly in the near-wall region, consistent with the observations of Blackwelder and Kovasznay [6]. This affects the shear stress in two ways: first, lower fluctuation levels are available to generate shear stress, even with constant correlation coefficient; secondly, the production of $- < u'v' >$ is decreased. Notice, however, that in the outer layer the turbulence levels remain constant. In the relaminarization region, therefore, we observe frozen turbulence in the outer layer, and inactive turbulence in the inner one. The turbulence intensity remains relatively high, so that once the stabilizing influence of the pressure gradient is removed, transition occurs very rapidly, following a process that resembles bypass transition due to high freestream turbulence.

The reasons for these changes in the Reynolds stresses can be explained by flow visualizations: In the outer region we observe that the hairpin eddies are turned into the streamwise direction by the strain. They induce strong incursions of more quiescent, outer-layer fluid into the wall region (Fig. 5), and strong ejections of inner-layer fluid into the outer flow. This results in the formation of a well-mixed layer that spans more than 50% of the boundary-layer thickness. In this region the turbulence production is decreased. These observations are consistent with those of Falco [7]. Below the outer-layer quasi-streamwise vortices, the inner layer becomes more orderly, and fewer bursts are observed.

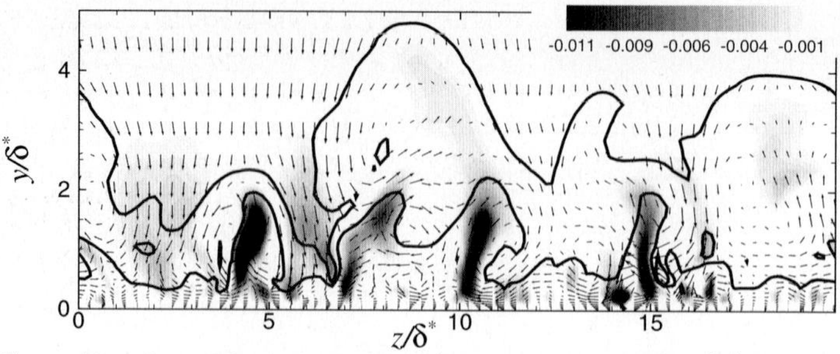

Fig. 5. Secondary-flow velocity vectors and countours of the $u'v'$ velocity correlation. The two lines correspond to the local δ_{95} and δ_{99} boundary-layer thicknesses

4 Conclusions

The accelerating boundary layer simulations carried out highlighted several phenomena that affect the flow. The acceleration affects the outer-layer eddies significantly, changing their structure and shape. This results in the formation of a well-mixed layer, in which the turbulence production is decreased, and the turbulence advected from upstream remains frozen. The inner layer is also affected: the flow becomes more orderly, with longer, more two-dimensional streaky structures and decreased frequency of bursts.

The authors acknowledge the financial support of the Air Force Office of Scientific Research under Grant No. F496200310112, Monitored by Dr. Thomas J. Beutner and Lt. Col. Rhett Jefferies.

References

1. Narasimha R, Sreenivasan KR (1973) J Fluid Mech 61:417–447
2. Narasimha R, Sreenivasan KR (1979) Adv Appl Mech 19:221–309
3. Sreenivasan KR (1982) Acta Mech 44:1–48
4. Launder BE, (1964) Gas Turbine Lab Report 77, MIT
5. Kline SJ, Reynolds, WC, Schraub FA, Runstadler PW (1967) J Fluid Mech 30:741–773
6. Blackwelder RF, Kovasznay LSG (1972) J Fluid Mech 53:61–83
7. Falco RE (1980) J Fluids Eng 102:174–182
8. Fernholz HH, Warnack D (1998) J Fluid Mech 359:329–356
9. Warnack D, Fernholz HH (1998) J Fluid Mech 359:357–381
10. Ichimiya M, Nakamura I, Yamashita S (1998) Exp Fluid Sci 17:37–48
11. Piomelli U, Balaras E, Pascarelli A, (2000) J Turbulence 1:1–16
12. Kim J, Moin P (1985) J Comput Phys 59:308–323
13. Germano M, Piomelli U, Moin P, Cabot WH (1991) Phys Fluids A 3:1760–1765

14. Lilly DK (1992) Phys Fluids A 4:633–635
15. Meneveau C, Lund TS, Cabot WH (1996) J Fluid Mech 319:353–385
16. Lund TS, Wu X, Squires KD (1998) J Comput Phys 140:233–258
17. Schlichting H (1979) Boundary layer theory. McGraw Hill, New York
18. Spalart PR (1986) J Fluid Mech 172:307–328

Generation of Turbulent Inlet Conditions for Velocity/Thermal Boundary Layer Simulations

Guillermo Araya, Elaine Bohr, Kenneth Jansen and Luciano Castillo

Rensselaer Polytechnic Institute, Troy NY-12180 US arayaj@rpi.edu

1 Introduction

Several techniques for turbulent inflow generation in an evolving boundary layer have been put in practice with different degrees of success and an extensive review of these methods is encountered in [1]. Lund et al. [1] proposed a modification to the concept of Spalart and Leonard [2] for accounting spatial growth in the inflow condition based on the similarity of the velocity profiles at different locations in a turbulent boundary layer. They considered an auxiliary simulation where a velocity field was extracted from a plane near the inlet of the principal domain, called recycle plane, rescaled it and reintroduced it as a boundary condition at the inlet of the auxiliary zone. In a posterior study, Kong et al. [3] extended the previous concept for thermal inflow generation predictions by performing Direct Numerical Simulations of turbulent thermal boundary layer on a flat plate with isothermal and isoflux boundary conditions at the wall. In this paper, an alternative method for generation of turbulent inlet data is presented based on the works carried out in [1] and [3]. The velocity scaling is based on the analysis performed in [4] and the temperature scaling is founded in investigations performed in [5] and [6]. The advantage of these novel scaling laws is the use of different parameterizations for the outer and inner zones based on a similarity analysis applied to the governing equations of the flow. The new models are tested in DNS of a flat plate at zero pressure gradient, considering an isothermal wall boundary condition.

2 Description of the Method for Predicting Turbulent Inflow Data

Table 1 summarizes the principal scaling parameters employed in the present research for the re-scaling step during the process of generating turbulent inflow conditions. The instantaneous temperature can be expressed as the summation of a mean value plus a fluctuation:

Table 1. Velocity and temperature scaling

Parameter	Type	Classical scaling		Proposed scaling	
		Inner	Outer	Inner	Outer
U	Mean velocity along x	u_τ	u_τ	u_τ	U_∞
V	Mean velocity along y	U_∞	U_∞	u_τ	$U_\infty \frac{d\delta}{dx}$
u'	Fluctuating velocity along x	u_τ	u_τ	u_τ	U_∞
v'	Fluctuating velocity along y	u_τ	u_τ	u_τ	$U_\infty \frac{d\delta}{dx}$
T	Mean temperature	T_τ	T_τ	$\mathrm{Pr}\sqrt{S_t}\,(T_w - T_\infty)$	$(T_w - T_\infty)\,\frac{\delta_T^*}{\delta_T}$
t'	Fluctuating temperature	T_τ	T_τ	$\mathrm{Pr}\sqrt{S_t}\,(T_w - T_\infty)$	$(T_w - T_\infty)\,\frac{\delta_T^*}{\delta_T}$

$$\tilde{T}(x, y, z, t) = T(x, y) + t'(x, y, z, t) \tag{1}$$

In the inner region, the mean temperature follows a thermal law of the wall:

$$T_w - T^{inner} = T_{si}(x) g_{si\infty}(y_T^+, \mathrm{Pr}, *) \tag{2}$$

In the outer region, a defect law is applied:

$$T^{outer} - T_\infty = T_{so}(x) g_{so\infty}(\bar{y}_T, \mathrm{Pr}, *) \tag{3}$$

Where Pr is the Prandtl number and * accounts any upstream condition effect.

Applying (5) and (3) to the inlet and recycle planes according to Fig. 1 (functions $g_{si\infty}(y_T^+, \mathrm{Pr}, *)$ and $g_{so\infty}(\bar{y}_T\, \mathrm{Pr}, *)$are universal and equal at the inlet and recycle planes, therefore they can be cancelled out);

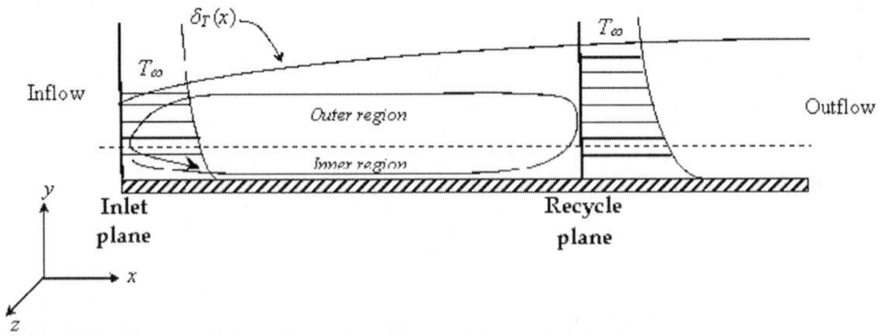

Fig. 1. Schematic of the domain

$$T_{inlet}^{inner} = \lambda_1 \, T_{recy} + T_{w,inlet} - \lambda_1 T_{w,recy} \tag{4}$$

$$T_{inlet}^{outer} = \lambda_2 \, T_{recy} + (1 - \lambda_2) T_\infty \tag{5}$$

Where the λ_n-parameters (n=1,2) are the relationships between the mean temperature scales at the inlet and at the recycle planes; for inner and outer zones, respectively: $\lambda_1 = \frac{T_{si,inlet}}{T_{si,recy}}$ and $\lambda_2 = \frac{T_{so,inlet}}{T_{so,recy}}$. Equations (4) and (5) give us expressions for calculating mean temperatures at inlet plane based on the solution at recycle plane. A similar procedure is used for the temperature fluctuations in the inner and outer regions.

3 Numerical Model

The dimensions of the computational domain are selected as follows:$0 < x < 10\delta_{inlet}$, $0 < y < 3\delta_{inlet}$ and $0 < z < \frac{\pi}{2}\delta_{inlet}$ where δ_{inlet} is the inlet boundary layer thickness. The incompressible version (1.6) of PHASTA code (Parallel Hierarchic Adaptive Stabilized Transient Analysis)developed at Rensselaer is used, which considers a Finite Element approach and a SUPG (Streamline Upwind Petrov-Galerkin) stabilization. Details about numerical procedures, mesh and boundary conditions can be found in [7].

4 Results and Discussion

In this section, preliminary results are showed related to the turbulent velocity boundary layer, as a previous step for solving the energy equation where temperature is regarded as a passive scalar. Fig. 2 shows the mean streamwise velocity in inner variables obtained at Re_θ= 1900. Comparison with other

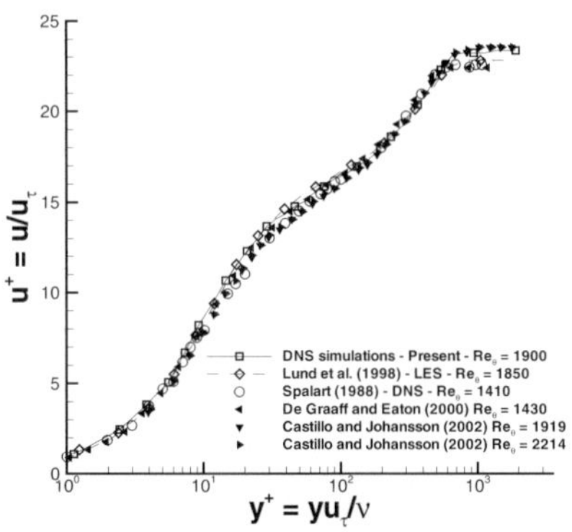

Fig. 2. Velocity profiles in inner variables

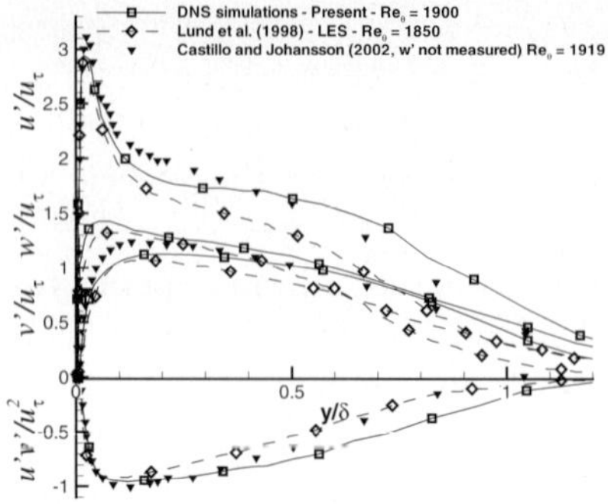

Fig. 3. Normal and Shear Reynolds stresses in inner variables

numerical predictions([1] and [8]) as well as experimental data ([9]) exhibited a fairly agreement. In Fig. 3, the root-mean square of velocity fluctuations (u_{rms}, v_{rms} and w_{rms}) normalized by the local friction velocity are shown, it is observed a good agreement with the experimental data of [9] and numerical simulations of [1] in the inner region, but some disagreement can be seen in the outer layer.

5 Conclusions

It was demonstrated that the rescaling recycling method gives promising results for turbulent boundary layers over flat plates. The new scaling used for the velocity field performed fairly well in the inner region, but some adjustments must be done in the outer region.

References

1. T. Lund, X. Wu, and K. Squires. Generation of turbulent inflow data for spatially-developing boundary layer simulations. *J. Comp. Phys.*, 140:233–258, 1998.
2. P. Spalart and A. Leonard. Direct numerical simulation of equilibrium turbulent boundary layers. *Proc. 5th Symp. on Turbulent Shear Flows, Ithaca, NY*, 1985.
3. H. Kong, H. Choi, and J. Lee. Direct numerical simulation of turbulent thermal boundary layers. *Phys. of Fluids*, 12(10):2555–2568, 2000.
4. W. George and L. Castillo. Zero-pressure-gradient turbulent boundary layer. *Appl. Mech.*, 50:689–729, 2000.
5. X. Wang. *Similarity analysis for turbulent boundary layers subject to pressure gradient and forced convection heat transfer.* PhD thesis, Rensselaer Polytechnic Institute, 2003.

6. X. Wang and L. Castillo. Asymptotic solutions in forced convection turbulent boundary layers. *J. of Turbulence*, 4:1–18, 2003.
7. E. Bohr. *Inflow generation technique for Large Eddy Simulation of turbulent boundary layers.* PhD thesis, Rensselaer Polytechnic Institute, 2005.
8. P. Spalart. Direct simulation of a turbulent boundary layer up to re_θ= 1410. *J. Fluid Mech.*, 187:61–98, 1988.
9. L. Castillo and G. Johansson. The effects of upstream conditions in a low reynolds number turbulent boundary layer with zero pressure gradient. *J. of Turbulence*, 3, 2002.

Experimental Analysis of a Turbulent Boundary Layer at High Reynolds Numbers

G. Aloisio[1], A. Dolcini[1], F. Di Felice[1] and G.P. Romano[2]

[1] INSEAN, The Italian Ship Model Basin, Rome, Italy difelice@insean.it
[2] University "La Sapienza", Roma, Italy romano@dma.ing.uniroma1.it

1 Introduction

In a turbulent boundary layer (TBL), the mean flow generates a variety of large and small scale turbulent structures. In the last years, researchers have tried to understand which kind of vortical structures makes this process possible and how their interaction takes place. A wide debate arises both on the characterization of the TBL statistics profiles and on the existence of the so called 'packets' of vortical structures. For the first argument, Barenblatt [1] conjectured the so called "incomplete self similarity", i.e. the mean velocity profile dependence on Reynolds number decreases with increasing Reynolds numbers. For sufficiently high Reynolds numbers, the profile always follows a power law. On the other side, Zagarola [2] proposed the "complete self-similarity law". The mean velocity profile was found to depend on the Reynolds number; while at low values the power law is a good approximation, at higher numbers the external part follows the universal logarithmic law, with a overlap region at about $y^+ = 500$ (hereafter wall distances are normalized using wall variables). The second main argument is related to what kind of vortical coherent structures characterizes a TBL, how these structures are generated, sustained (or self-sustained), and if and how they depend on Reynolds number. The recent Adrian model [3] is based on auto-generation and self-sustenance of coherent structures: hairpin vortices are created from quasi-streamwise vortices in the buffer layer and new 'legs' are generated when the 'head' of the previous vortex moves away from the wall. This continuous process allows the vortices to form the so called 'packets'. On the other hand, non-planar spanwise velocity component is important in the Hussain model [4]. These are not the only models proposed in the last decade; notwithstanding the model, many questions are still open, in particular the extension of models to high Reynolds numbers. This is the aim of the present work, in which time resolved PIV is applied to a TBL in a very large facility; near-wall statistics and space-time evolution of structures are investigated to give some views on extensions to large Reynolds numbers.

2 Experimental Set-up and Analysis of Data

PIV measurements were performed at INSEAN free-surface large circulating water tunnel. The test section is 10 m long, 3.6 m wide with a depth of 2.25 m. Measurements have been performed with mean velocities equal to 0.5 and 1.05 m/s at the centre of the test section, 0.6 m. below the free surface and 8 m downstream the inlet; the Reynolds numbers, based on momentum thickness, are 3940 and 7257. Time resolved PIV system consists of a high-speed camera (up to 2000 frame per second at full resolution of 1024x1024), in a waterproof perspex box partially submerged. A continuous Argon Laser was employed; hollow glass spheres with a mean particle diameter of 10 mm are used as tracers.

3 Results

As mentioned, all quantities are scaled by inner wall variables using friction velocity u^*. The friction velocity is estimated in two different empirical ways. The first involves (see, for example, [2]):

$$u^* = \frac{\nu}{x}\left[\frac{e^{\frac{3}{2\alpha}} 2^\alpha \alpha(1+\alpha)(2+\alpha)}{\sqrt{3+5\alpha}}\right]^{\frac{1}{1+\alpha}}, \qquad \alpha = \frac{3}{2\ln Re_x}, \tag{1}$$

(deduced for a pipe), while the second involves the Osterlund formula [5]:

$$u^* = U_\infty\left(\frac{c_f}{2}\right)^{1/2}, \qquad c_f = 2\left[\frac{1}{k}\ln Re_\theta + C\right]^{-2}. \tag{2}$$

Differences computed with the two formulas were negligible (2.03 cm/s *vs* 2.01 cm/s at Re=3940 and 4.03 cm/s *vs* 3.98 cm/s at Re=7257). Velocity profiles are compared with logarithmic and power laws:

$$U^+_{pow} = C_1 (y^+)^\gamma, \qquad C_1 = 8.7, \qquad \gamma = 0.137. \tag{3}$$

$$U^+_{log} = \frac{1}{k}\ln y^+ + B, \qquad k = 0.436, \qquad B = 6.15. \tag{4}$$

The previous laws are overlapped with experimental data in Fig. 1. In the first plot (Re=3940), the power law well fits data when $y^+ > 500$, while closer to the wall both laws overlap with experimentas. At Re=7257, the whole profile in the range $100 < y^+ < 1100$ is better described by the power law; this result is in agreement with Barenblatt and Zagarola conjectures (significant differences appear only at high Reynolds numbers).

After the previous averages, it is possible to analyze snapshots; the interesting zone is between y=3mm and y=9mm ($y^+ = 70$ and $y^+ = 250$), where vortical structures develop, interact and eventually re-generate new vortices. Two cases are presented, in order to study Reynolds numbers effects. The first

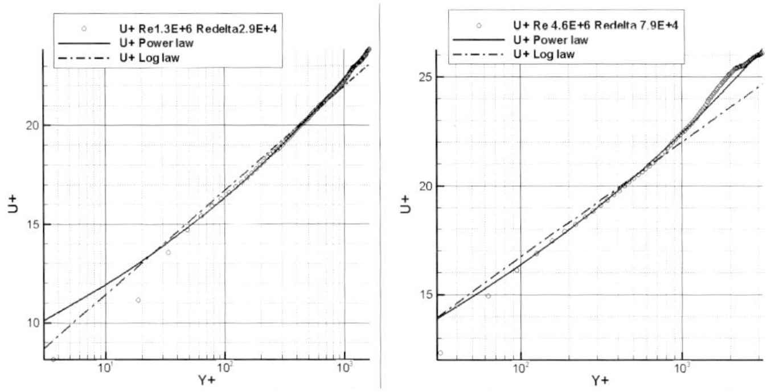

Fig. 1. Mean velocity profiles for the low Reynolds number (on the left) and for the high Reynolds number (on the right); comparison with logarithmic and power laws

one, Re=3940, is shown in the left of Fig. 2, where the contour plots of Jeong and Hussain [6] indicator and of vorticity are given. An hairpin-shaped vortex linked to a bursting event (signed as A) is observed. Moreover, a counter-rotating vortex appears close to the wall; this generates a new structure similar to the first one (B). A 'packet' of smaller vortices aligned onto the same direction is also observed (C). This process is similar to the self-sustaining model proposed by Adrian [3]. The second snapshot, obtained at Re=7257, is reported on the right of Fig. 2; even at such high Reynolds numbers, aligned large scale vortices are detectable. However the size and the number of such structures seems to be Reynolds number dependent (it is important to stress that the number of vortices is much smaller than at lower Reynolds number).

In order to achieve a more quantitative definition of the characteristics of these vortical structures, a wavelet-based analysis has been performed. The algorithm used in the present work is based on the complex continuous wavelet

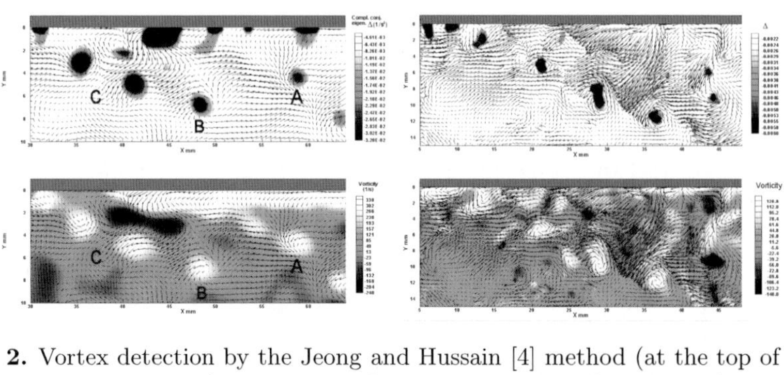

Fig. 2. Vortex detection by the Jeong and Hussain [4] method (at the top of each plot) and by vorticity (at the bottom); snapshot at low (on the left) and high Reynolds number (on the right). The wall is at the top of each image and the flow is from left to right

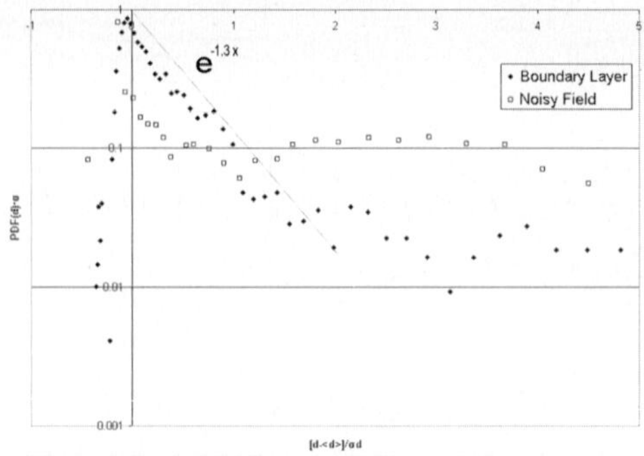

Fig. 3. Normalised probability density functions of vortex size obtained from LIM; comparison between data from velocity field measured by PIV and noisy PIV non-physical data

transform [7]; Mexican Hat has been used as mother wavelet. Structure identification has been performed by LIM (Local Intermittency Measure), which reaches a maximum in correspondence of a vortical structure. The correlation between LIM peaks and vortices in a 2D PIV field has already been showed [8]. LIM allows to evaluate both the position and the dimension of the vortical structure. Thus, it is possible to verify, through the probability density function (PDF) of vortex size, that the TBL is characterised by a scale distribution, which follows an exponential law (in agreement with Camussi [8]) i.e. a hierarchy of length scales (Fig. 3). Similarly, a noisy non-physical velocity field was analysed; the result given in Fig. 3 confirms that the vortex size PDF is an effective TBL peculiarity and not an artefact.

References

1. Barenblatt G. I. (1993) Scaling laws for fully developed turbulent shear flows. Part 1. Basic hypotheses and analysis J. Fluid Mech. 248: 513-520.
2. Zagarola M. V., Smits A. J. (1998) Mean-flow scaling of turbulent pipe flow. J. Fluid Mech. 373: 33-79.
3. Adrian R. J., Meinhart C.D., Tomkins C.D. (2000) Vortex organisation in the outer region of the turbulent boundary layer. J. Fluid Mech. 422:1-54.
4. Jeong J, Hussain AKMF, Schoppa W, Kim J (1997) Coherent structures near the wall in a turbulent channel flow. J. Fluid Mech. 332:185-214.
5. Osterlund J. M., Johansson A. V., Nagib H. M., Hites M. H. (2000) A note on the overlap region in turbulent boundary layers. Phys. Fluids 12:1-4.
6. Jeong J., Hussain F. (1995) On the identification of a vortex. J. Fluid Mech. 285:69-94.
7. Farge M. (1992) Wavelet transforms and their applications to turbulence. Ann. Rev. Fluid Mech. 24:395-457.
8. Camussi R. (2002) Coherent structure identification from wavelet analysis of particle image velocimetry data. Exp. Fluids 32:76-86.

Hybrid RANS/LES of Neutral Atmospheric Boundary Layer: Simple Terrain

A. Hansen, J. Mann, J. Johansen, and N. N. Sørensen

Risø National Lab., P.O.Box 49, DK4000-Roskilde, Denmark

1 Introduction

At present Large Eddy Simulation (LES) of most boundary layer flows are unfeasible due to limited computer resources. To reduce computational cost, hybrid methods have been proposed where the unsteady RANS equations (URANS) solved in the near-wall region are joined with LES performed in the well resolved outer region. Despite the advantages of the method, results show a characteristic mismatch between the RANS and LES-regions [2].

We propose a hybrid method, which is used to simulate the neutrally stratified atmospheric boundary layer (ABL) over flat and rough terrain. To improve results a backscatter model is introduced similar to the one proposed by Mason and Thomson [4] and Piomelli et al. [5]. By generating stochastic forcing, the backscatter of energy from subgrid scales (SGS) to resolvable scales at a rate corresponding to the local dissipation rate is modeled as argued by Leslie & Quarini [3], greatly improving results.

2 Hybrid Model

Filtering or time averaging of the incompressible momentum equations lead to stresses, which need modeling. Here, this modeling is simply denoted SGS-modeling. The method we propose for SGS-modeling is similar to Detached-Eddy-Simulation (DES) formulated by Spalart et al. [10]. They use the traditional eddy-viscosity hypothesis and transition from URANS to LES is controlled by a characteristic turbulent length scale - we use a similar scale:

$$\tilde{l} = \min\left(\frac{k^{3/2}}{\varepsilon}, l_f\right); \quad l_f = C_{DES}\triangle \tag{1}$$

C_μ is a model constant, k is the SGS turbulent kinetic energy (TKE), ε is the dissipation rate of TKE and $\triangle$ is the local grid spacing taken as the maximum distance over the three directions, $\triangle = \max(\triangle x, \triangle y, \triangle z)$. The SGS-viscosity is found by:

$$\nu_{sgs} = C_\mu \frac{k^2}{\varepsilon} \propto C_\mu k^{1/2} \tilde{l} \qquad (2)$$

The length scale automatically controls the switch from the standard high Reynolds number $k - \varepsilon$ model [9] close to the wall to LES away from the wall with a spatial filter that scales with $\triangle$. ε and k are obtained in the entire domain by solving two transport equations, where the dissipative term of the k-equation is modified to incorporate the new length scale [11]. By using the $k - \varepsilon$ model together with the logarithmic law-of-the-wall in the URANS region the method can be used for flow over rough surfaces. The SGS-model was calibrated by simulating decaying isotropic turbulence (valid for the LES-region) and $C_{DES} = 0.65$ ($C_\mu = 0.09$) was found when the convective terms where solved using a second-order central differencing scheme [1].

3 Stochastic Backscatter Model

It has been shown that even though the forward energy transfer (modeled by eddy-viscosity) is large there is a substantial backscatter of energy ($\partial K/\partial t$) from subgrid-scales to resolved scales. In LES this is observed as stochastic fluctuations in the SGS-stresses. As proposed by Leslie & Quarini [3] the stochastic backscatter should be modeled separately. The backscatter was modeled by adding stochastic forcing to the momentum equation at each timestep and relating the level to the local dissipation [3]:

$$\frac{\partial K}{\partial t} = C_b \varepsilon = |f_i|^2 \triangle t \qquad (3)$$

C_b is a parameter that depend on spatial filter, $\triangle t$ is the time step and f_i are the three components of forcing. For a Top-hat filter the value of C_b was found to be about 0.489 [3]. ε was obtained from the solved transport equation.

The forcing should be correlated and contain length and time scales characteristic for the SGS. The time scale was applied by adding gaussian noise at a prober level to the forcing used at previous timestep: $f_i^{n+1} = f_i^n \alpha_T + w$, where the correlation coefficient $\alpha_T = \exp(-\triangle t / T_{SGS})$. T_{SGS} was chosen as the local velocity divided by $\tilde{l}$ giving $\alpha_T = \exp(-u \triangle t / \tilde{l})$. Spatial correlation was added by applying a 1:2:1 smoothing operator in all three directions and the forcing was made anisotropic by scaling each forcing component with the local velocity: $f_i \propto |u_i| / (u_i u_i)^{1/2}$. No correction for continuity was made.

As the surface is approached the dominant eddies become small and the SGS-stresses approach a deterministic value. This effect both reduces the time scale and the level of backscatter. To model this a factor is defined that expresses the "RANS-to-LES-ratio", $\alpha_{LES} = (\tilde{l}/l_f)^2$. The factor is one in the LES region and zero at the wall. Both C_b and α_T is multiplied with this factor to remove backscatter and time correlation in the RANS region.

4 Numerical Approach

The CFD code EllipSys3D developed by Michelsen [6] and Sørensen [7] was used in all calculations. It is a multiblock finite volume discretization of the incompressible momentum equations. The code uses structured grids with cell variables stored in the cell centers and is run in parallel using MPI. The PISO algorithm was used to enforce the pressure/velocity coupling and convective terms were solved using a second-order central differencing scheme, based on deferred correction [8]. Simulations were performed with a second order accurate iterative time stepping algorithm and the global time-step for all simulations was chosen to give a maximum CFL (Courant-Friedrich-Levy-number) of no more than 0.25.

All lateral boundaries were specified as being periodic and the simulation was run until the boundary layer was fully developed. At the top boundary a symmetry boundary condition was used ($\partial u/\partial z = \partial v/\partial z = 0, w = 0$) to mimic an inversion layer, and at the lower boundary the logarithmic law-of-the-wall was applied with a roughness height of $z_0 = 0.1$m. The flow was driven by a constant pressure gradient and Coriolis forcing was applied ($f_c = 10^{-4}s^{-1}$).

For the neutral stratified ABL the boundary layer height, h, (= domain height, H) is often taken to be proportional to the Ekman layer depth, $h_e = u_*/f_c$. A proportionality factor of 0.3 is often chosen, $h = 0.3h_e$. Simulations were performed with $u_* \simeq 0.43ms^{-1}$ and $H = 1200m$. The domain width was about four times the domain height ($W = 4H$), and a domain length of about 6 times the height was used ($6H, 4H, 1H$). To capture near wall gradients the grid was refined in the wall-normal direction. The mesh was stretched from the wall (with a maximum stretch of $S = 1.2$) to a height from where constant meshing was applied. The height of the first computational cell was set equal to the roughness height z_0. The results shown are made on a relative coarse mesh with (144×96×48) cells, $\triangle t = 1$s and $\triangle = 44.2$m.

5 Results

Figure 1 (left) shows the mean velocity profile of the surface layer. The commonly found mismatch from the logarithmic law-of-the-wall is reproduced when simulating without the backscatter model. The mismatch is due to the unphysical turbulent structures from the RANS-region, which do not carry any significant Reynolds stresses. When stochastic backscatter is applied the velocity closely follows the logarithmic profile. The backscatter model breaks up the unphysical velocity streaks otherwise observed in the mismatch region and generates resolvable Reynolds-stress carrying eddies.

Figure 1 (right) shows the mean distribution of the three force components. The streamwise component is the dominant while the wall-normal component is very weak. The amplitude of the forcing is lower than the model proposed by Piomelli et al. [5].

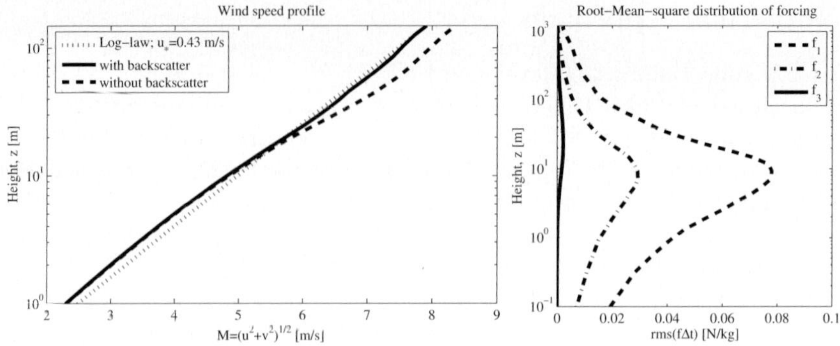

Fig. 1. (left) Surface layer velocity profile with and without backscatter. (right) Root-mean-square distribution of the three force components

6 Conclusion

The standard high Reynolds number k-ε model [9] used in many atmospheric applications was in a simple manner modified to a hybrid RANS/LES model. The SGS-model is similar to DES [10] but adapted to handle rough walls. Furthermore a simple and cheap (CPU increase of less than 5%) backscatter model was proposed that backscatter energy at a level corresponding to the local dissipation as proposed by Leslie & Quarini [3]. The hybrid- and backscatter-model were used to simulate the ABL and showed promising results, more tests are however necessary.

References

1. Hansen, A., Sørensen, N.N., Johansen, J. and Michelsen, J.A. (2005) AIAA paper 2005-0885.
2. Nikitin, N.V., Nicoud, F., Wasistho, B., Squires, K.D., and Spalart, P.R. (2000) Phys. Fluid, 12, 1629-1632.
3. Leslie, D.C., and Quarini, G.L. (1979) J. Fluid Mech., 91, 65-91.
4. Mason, P.J., and Thomson, D.J. (1992) J. Fluid Mech. 242, 51-78.
5. Piomelli, P., Balaras, E., Pasinato, H., Squires, K.D., Spalart, P.R. (2003) Int. J. Heat and Fluid Flow, 24, 538-550.
6. Michelsen, J.A. (1994) Technical report AFM, 94-06, Technical University of Denmark, Lyngby.
7. Sørensen, N.N. (1995) Risø-R-827(EN), Risø National Lab., Roskilde, Denmark.
8. Khosla, P.K. and Rubin, S.G. (1974) Computers Fluids, 2, 207-209.
9. Launder, B.E., and Spalding, D.B. (1974) Comput. Meths. Appl. Mech. Eng., 3, 269-289.
10. Spalart, P.R., Jou, W-H., Strelets, M., and Allmaras, S.R. (1997) 1st AFOSR Int. Symp. Eng. Turb. Modelling and Measurements, May 24-26, Corsica.
11. Travin, A., Shur, M., Strelets, M. (2000) 412 EUROMECH colloqvium on LES of complex transitional and turbulent flows, Munich, Oct. 2000

Evidence on Non-Universality of Kármán Constant

Kapil A. Chauhan[1], Hassan M. Nagib[2] and Peter A. Monkewitz[3]

[1] Illinois Institute of Technology, 10 West 32nd Street, Chicago, IL 60616
chaukap@iit.edu
[2] Illinois Institute of Technology, 10 West 32nd Street, Chicago, IL 60616
nagib@iit.edu
[3] Swiss Federal Institute of Technology, CH-1015, Laussane, Switzerland
peter.monkewitz@epfl.ch

The logarithmic form of mean velocity profile for the overlap region in classical turbulent boundary layer theory is given as, $U^+ = \kappa^{-1}\ln(y^+) + B$. The coefficient of the logarithmic term κ, known as the von Kármán constant, is considered *universal* in the classical theory. This has often been interpreted as the same value of κ can be used for all wall bounded flows, such as pipes, channels and boundary layers, irrespective of the pressure gradient. The different kinds of wall bounded flows are only differentiated by a different outer layer or "wake" part, where they have a different parameter Π. In the present study we focus on experimental evidence from results for flat plate boundary layer experiments at high Reynolds numbers, $Re_\theta > 10,000$, in presence of various adverse or favorable pressure gradients[1]. The experiments were performed in the National Diagnostic Facility (NDF) at IIT under five different streamwise pressure gradients; namely : Zero Pressure Gradient (ZPG), Adverse Pressure Gradient (APG), Favorable Pressure Gradient (FPG), Strongly Favorable Pressure Gradient (SFPG) and Complex Pressure Gradient (CPG), which consists of flow under SFPG for about half the fetch followed by return to a ZPG downstream. Detailed measurements of the velocity profiles using single hot-wire anemometry and skin-friction using oil-film interferometry were carried out at six downstream stations with freestream velocities of 40, 50 and 60m/s measured at a reference position near the leading edge of the plate.

1 Skin-Friction Coefficient

The wall shear stress in the NDF was measured independently and directly by oil-film interferometry with an accuracy of $\pm 1.5\%$. Such measurements are the first of their kind for boundary layer flows with pressure gradient.

Figure 1(left) shows oil-film data for all pressure gradients with Re_θ ranging from 6,000 to 70,000. The variation of C_f with Re_θ for ZPG is represented by the Coles-Fernholz logarithmic skin-friction relation[4] with the values of κ and C found to be 0.384 and 4.127, respectively from the best fit to both data sets. The increase in Reynolds number for KTH and NDF data was obtained by increasing both the freestream velocity for a fixed x and the streamwise distance on the flat plate for a fixed U_∞, and all the data follow the same C_f behavior. Hence, the ZPG flow exhibits self-similarity in the C_f variation and can be represented by a single curve for any Reynolds number. Also shown in the figure is a similar curve plotted utilizing the conventionally used values of $\kappa = 0.41$ and $B = 5.0$. Therefore, any flow with C_f deviating from the trend of ZPG data will have a different κ and or B if the logarithmic form still applies. For the pressure gradient cases we can see that the FPG and SFPG data are above the ZPG curve, and the APG results are below them, as expected. However, in general there is no self similar behavior for these cases and we see in each case a family of curves corresponding to different x or U_∞. This is especially clear for the SFPG case where three distinct curves (solid lines) are seen corresponding to U_∞=40, 50 & 60m/s at the upstream reference position, as we move through different x locations along the plate. On the other hand, if we fix the x location, the effect of increasing velocity is shown by the dashed lines. This SFPG condition definitely exhibits a non-equilibrium behavior as we see different trends for varying x and varying U_∞. Such a trend is also weakly present in the FPG and APG cases, but diminishes as Re increases indicating the flow is approaching a self-similar state.

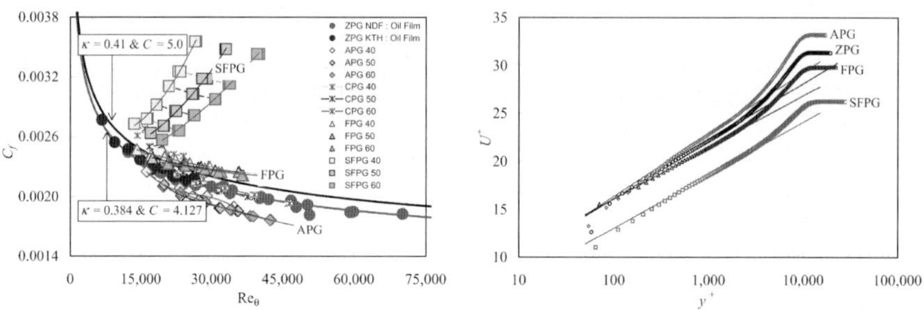

Fig. 1. (left) The variation skin-friction coefficient, C_f with Re_θ for various pressure gradients. Data for constant U_∞ near the leading edge of the flat plate. (right) Mean velocity profiles for different pressure gradients at x=6.4m and U_∞=60m/s freestream velocity at a reference upstream position plotted in inner scaling. u_τ obtained from oil-film used to normalize data

2 Mean Velocity Profiles

The profiles for mean velocity exhibit the logarithmic behavior in the overlap region in a similar way to the well documented case of zero pressure gradient (ZPG)[2]. Figure 1(right) shows the mean velocity profiles obtained at x=6.4m with U_∞=60m/s, normalized with u_τ obtained from oil-film interferometry. There is obvious deviation from the ZPG profiles with adverse moving up and favorable shifting down. Hence, when plotted in inner scaling with *true* u_τ there is a change in the additive constant B. More interesting is the change in slope signifying a changing κ which is clearly observed when APG, ZPG and FPG profiles are carefully examined. Hence, the presence of pressure gradient not only affects the additive constant but also the von Kármán *coefficient*. If the skin-friction velocity is found using the Clauser fit in the overlap region with pre-determined values of κ and B we would get a different u_τ resulting in a collapse of all the profiles on one log law in the overlap region and with a different wake parameter. Obviously C_f determined from Clauser fit is not appropriate. The resulting error from such estimates are significant and as large as 50% for the highest Reynolds number of SFPG.

3 Variation of κ

The behavior of the C_f seen in Fig. 1(left) will not be represented by any correlation without accounting for changes in parameters of mean velocity. Almost all the skin-friction correlations found in the literature are derived assuming a constant κ and those applicable for boundary layer with pressure gradients account for a changing wake parameter Π as the measure of pressure gradient effects. One such correlation is given by Prof. Coles in the 1968 Stanford Conference database [3], derived for two-dimensional self similar flows assuming constant values of κ and C(the additive constant) with varying Π to represent the degree of pressure gradient changes and given as,

$$U_\infty^+ = \kappa^{-1}[\ln \kappa + \kappa C + 2\Pi - \ln|1+\Pi| + \ln Re_{\delta^*}] \qquad (1)$$

The values of C_f found using this equation but with varying κ, C and Π, utilizing the values extracted from the experimental velocity profiles, are compared with results from oil-film measurements in Fig. 2(right). There is a remarkable agreement between the measured values of C_f and those obtained using (1) with varying parameters. We note that, not only the equation duplicates the behavior of C_f for the mild favorable and adverse pressure gradients, which tend to attain self-similar state, but it also follows the very distinct non-equilibrium behavior of strongly favorable pressure gradient when the measured values of the coefficients are used. Also notable, is the collapse for the complex pressure gradient where we see the equation following the trend of experiments in the regimes of strong acceleration through the downstream

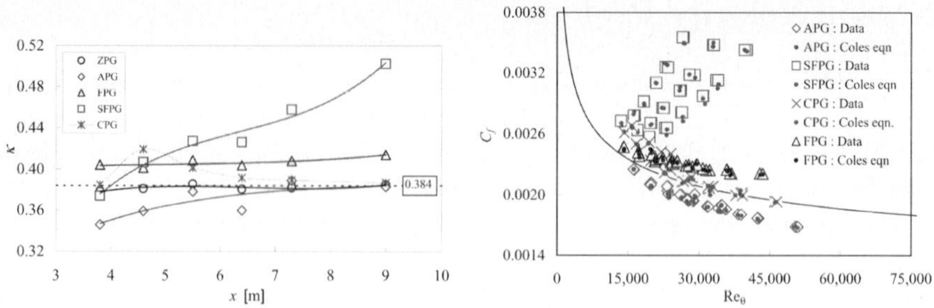

Fig. 2. (left) The variation von Kármán coefficient κ with x for various pressure gradients. Solid lines are representative of trends only. (right) Comparison of C_f obtained from Coles equation with oil-film data

relaxation to a zero pressure gradient. The good agreement of experiments with (1), with varying κ and C, confirms the observations in the velocity profiles seen in Figs. 1(right) & 2(left). To further evaluate changes in log-law parameters with pressure gradient, we directly look into the variation of κ found using a diagnostic function, $\kappa = 1/\Xi = [y^+ dU^+/dy^+]^{-1}$. In the overlap region, the behavior of the diagnostic function is constant implying a logarithmic behavior [2], and using appropriate inner and outer limits, an averaged κ is found for each velocity profile. The variation of κ with x for the different pressure gradients is plotted in Fig. 2(left). The values plotted are averaged for the three velocity cases and the solid lines shown are representative of global behavior only. The κ for ZPG remains fairly constant with x near the value of 0.384 which is also the same obtained from fitting the oil-film data. The κ values for FPG remain consistently higher than ZPG throughout x and constant near a value of 0.41. For FPG, the pressure gradient parameter β decreases with x while κ appears to be relatively constant. On other hand, there is an increasing trend of κ for SFPG with x although β is almost a constant (see Fig.6 in [1]). Opposite to FPGs, the APG case shows a lower value of κ. There is also an increase in κ with x, finally attaining the ZPG value near the last station. These trends are consistent with the trends in the C_f plots at high Reynolds number. A correspondingly similar behavior is also found for the additive constant B.

References

1. Nagib, H. M., Chauhan, K. A., and Monkewitz, P. A., *Scaling of High Reynolds Number Turbulent Boundary Layers Revisited*, AIAA paper 2005-4810, 4th AIAA Theoretical Fluid Mechanics Meeting, Jun 6-9, Toronto, Canada, 2005.
2. Nagib, H. M., Christophorou, C., and Monkewitz, P. A., *High Reynolds Number Turbulent Boundary Layers Subjected to Various Pressure-Gradient Conditions*, IUTAM Symposium on One Hundered Years of Boundary Layer Research, DLR Göttingen, Germany, Aug 12-14, 2004.

3. Coles, D. E., *The Young Person's Guide to the Data*, Computation of Turbulent Boundary Layers - 1968 AFOSR-IFP-Stanford Conference, vol. 2, Stanford University, 1968.
4. Österlund, J. M., Johansson, A. V., Nagib H. M., and Hites, M. H., *A Note on the Overlap Region in Turbulent Boundary Layers*, Phys. Fluids, 12:1-4, 2000.

Turbulent Marginal Separation: A Novel Triple-Deck Problem for Turbulent Flows

B Scheichl and A Kluwick

Institute of Fluid Mechanics and Heat Transfer, Vienna University of Technology, Resselgasse 3/E322, A-1040 Vienna, Austria, bernhard.scheichl@tuwien.ac.at

A new rational theory of an incompressible nominally steady and two-dimensional turbulent boundary layer (TBL) along a smooth impermeable surface, which is exposed to an adverse pressure gradient (APG) impressed by the prescribed external inviscid and irrotational free-stream flow, has been elaborated. It represents an asymptotic description of the flow by taking the limit $Re \to \infty$ where Re denotes a suitably defined global Reynolds number, which is characteristic for the external bulk flow.

1 General Asymptotic Structure of an APG-TBL

In the following all quantities are nondimensionalised on a global length and velocity scale, respectively, and on the uniform fluid density. The newly developed general flow description is essentially based on the Hypothesis. *In the limit $Re \to \infty$ in the Reynolds-averaged Navier–Stokes equations the turbulent motion is represented locally by a single velocity scale.* Let u_* denote that local reference velocity. The (here and in the following capitalised) Hypothesis above then states that at each distance normal to the surface under consideration the Reynolds stresses are all of $O(u_*^2)$ such that the rotational motion induced by the velocity fluctuations is of $O(u_*)$. Guided by some subtle evidence of this characteristic which is provided by first investigations, the authors will further attempt to substantiate the Hypothesis by a rigorous analysis of the unsteady equations of motions in a separate study.

A comprehensive asymptotic treatment of turbulent boundary layers which relies on the Hypothesis and elucidates how the classical flow description (see the pioneering paper [1], for instance) is included and even extended is presented in [2]. There it is demonstrated that, in general, the streamwise velocity defect with respect to the external flow of a TBL does not necessarily depend on Re but must be regarded as asymptotically small in order to formally allow for a thin-shear-layer approximation of the Reynolds equations. That is, the Hypothesis at a first glance excludes the existence of boundary layers having a velocity defect of $O(1)$, which, however, is indispensable to separate an initially firmly attached TBL from the surface under the action of a smooth streamwise pressure distribution, since it would indicate an unacceptable boundary layer thickness of $O(1)$. As a highly remarkable remedy, an asymptotic boundary-layer-type description of such flows then is inevitably associated with the existence of a positive small number, here denoted by α, which measures the (non-dimensional) slenderness of the turbulent shear layer and, most important, is essentially independent of Re. Hence, that number is treated as the principal perturbation parameter, and the formal limit to be investigated primarily is written as

$$\alpha \to 0 \quad \text{as} \quad Re^{-1} = 0. \tag{1}$$

In turn, in the limit (1) the TBL closely resembles a turbulent free shear layer as it is characterised by a wake-type streamwise velocity profile which, at least in the case of attached flows, does not satisfy the no-slip condition at its base. Of course, the wake limit is perturbed as sublayers close to the wall emerge due to high but finite values of Re. The following aspects of the new approach which has been established by adopting the limit (1) are crucial for its applicability:

(a) It in principle includes the classical small-defect formulation and its extension (see [2]) as a special case;
(b) It is fully compatible with the asymptotic structure of any commonly applied Reynolds shear stress closure which is seen to contain a small number that can be identified with α (for example, $\sqrt{\alpha} \approx 0.085$ in case of the well-known algebraic mixing-length closures);
(c) It yields a self-consistent description of boundary layers which may even undergo marginal separation.

The four-tiered splitting of a TBL which is prone to separate is sketched in Fig. 1: Let x, y, and u denote natural coordinates, respectively, along and

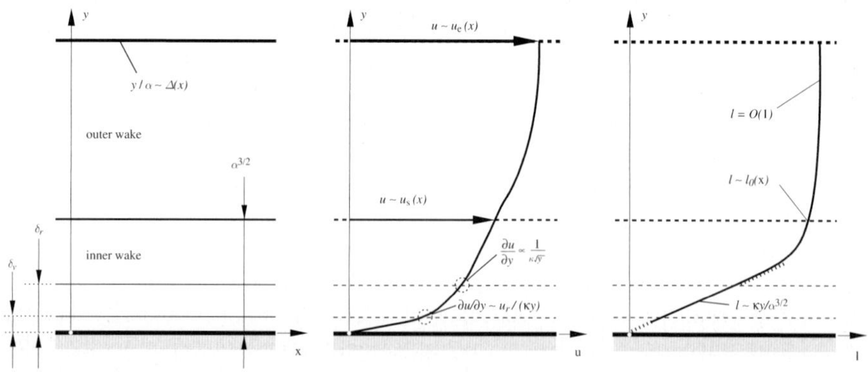

Fig. 1. Asymptotic structure of a TBL having a velocity defect of $O(1)$

perpendicular to the surface, and the streamwise velocity. We at first focus on the primary limit (1). By introducing an inner coordinate $Y = y/\alpha$, to leading order the outer wake then is characterised by the value $u = u_e(x)$ at the boundary layer edge, here given by $Y = \Delta(x)$, and the slip velocity $u = u_s(x)$ at its base, that is at $Y = 0$. The mixing length for the shear stress is written as $\alpha^{3/2}\,\ell(x,Y)$ where ℓ is a quantity of $O(1)$. Since the outer wake is unaffected by the wall, ℓ is assumed to take a finite value $\ell_0(x) = \ell(x,0)$ of $O(1)$, which is the only closure information needed for the further investigation.

2 Locally Strongly Interacting Boundary Layers

Here the notion marginal separation refers to boundary layer solutions where $u_s(x)$ vanishes at a single location but immediately recovers. Then an analogous marginal singularity is observed in the distribution of $\Delta(x)$. In that case the flow state and the associated given pressure distribution are termed critical. Taking into account the interaction process of the boundary layer with the external potential flow due to the locally strong displacement effect of the boundary layer thickness then allows for an uniformly valid asymptotic description of marginally separating flows on basis of the Reynolds-equations in the limit (1). The resulting so-called triple-deck formulation is well-known to govern strongly interacting laminar boundary layers in the limit $Re \to \infty$. Here we only present some key results of the first turbulent case known in literature, which is here of interest; for an extensive discussion we refer to [3].

The flow is seen to be described locally by an one-parameter family of solutions of the underlying interaction problem. This coupling parameter entering the problem accounts for the deviation of the imposed pressure gradient from its critical distribution in relation to the value of α. The local triple-deck structure of the TBL together with numerical solutions of the interaction problem for a specific value of that parameter are presented in Fig. 2: Let X denote a local streamwise variable appropriate for the interaction region the streamwise extent of which is of $O(\alpha^{3/5})$. The upper deck (UD) accounts for the pressure feedback $P(X)$ in the external flow induced by the boundary layer displacement. The latter is described by the function $A(X)$ where $-\alpha^{1/5}A(X)$ measures the local deviation of the boundary layer thickness $\Delta(x)$ from its critical value. It is exerted by the lower deck (LD) and transferred to the upper deck in the middle deck (MD). Most interestingly, both the induced and the imposed pressure gradient come into play in leading order in the streamwise momentum equation governing the lower deck. In the right-hand figure the top abscissa refers to $P(X)$, the one at the bottom to, respectively, $A(X)$ and the local representation $U_s(X) = u_s/\alpha^{3/10}$ of the slip velocity. The graph for U_s shows a closed recirculating flow regime at the base of the outer wake. Furthermore, S and R denote the streamwise positions of separation and reattachment, respectively. In addition, the asymptotes $U_s = O(X^{1/2})$, $A = O(X^{1/3})$ as $X \to \infty$, and $P = O(X^{-2/3})$ as $|X| \to \infty$ are plotted as dashed curves.

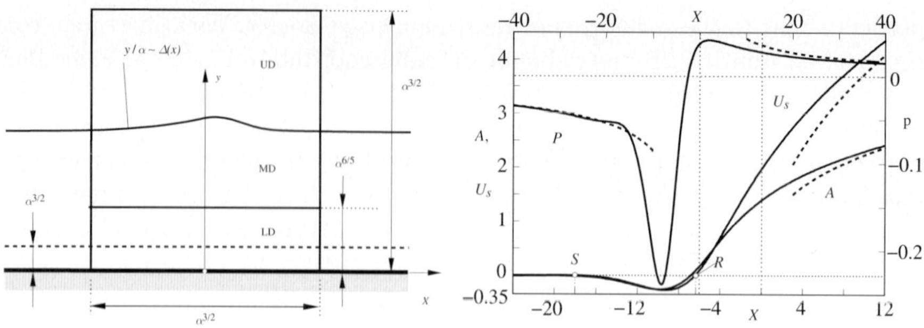

Fig. 2. Triple-deck structure in the limit (1) and solutions of the interaction problem

3 Effect of Finite Values of the Reynolds Number

The singular perturbation of the wake limit (1) due to high but finite values of Re gives rise to the emergence of several sublayers as sketched in Fig. 1. Their scalings are found to be determined uniquely by the Hypothesis (see also [2]): The classical viscous wall layer accounting for the usual no-slip condition holding at the surface has a thickness $\delta_v \sim 1/(\sqrt{|\tau_w|}Re)$ where τ_w is the local skin friction. Furthermore, the so-called intermediate layer of thickness $\delta_\tau \sim |\tau_w|$ exhibits a streamwise velocity defect with respect to u_s of $O(\sqrt{|\tau_w|})$. Matching reveals the logarithmic law of the wall and the skin friction formula

$$u_\tau/u_s \sim \kappa/\ln|u_\tau^3 Re|, \quad u_\tau = \mathrm{sgn}(\tau_w)\sqrt{|\tau_w|}, \quad Re \to \infty. \tag{2}$$

Here κ denotes the v. Kármán constant. The velocity defect admits the square-root law known from a separating TBL on top of the intermediate layer. In turn, the Re-independent inner wake (also indicated by the dashed line in Fig. 2) has to be introduced where convective terms are again linearised about $u = u_s$. Note that both the logarithmic and the square-root law are included in the familiar near-wall closure $\ell \sim \kappa y/\alpha^{3/2}$ applied to the intermediate layer.

The friction law (2) ceases to be valid, and the Re-dependent layers collapse near the locations of detachment as well as reattachment, when both u_s and u_τ are of $O(Re^{-1/3})$. Then the logarithmic portion of u is superseded by the square-root behaviour. The associated asymptotically correct continuation of the friction law (2) is presently under investigation.

This work was granted by the Austrian Science Fund (FWF), project # P16555-N12.

References

1. Mellor, GL (1972) Int J Engn Sci 10:851–873
2. Scheichl B, Kluwick A (2005) Non-unique turbulent boundary layer flows having a moderately large velocity defect: A rational extension of the classical asymptotic theory. Theor Comp Fluid Dyn (submitted in revised form)
3. Scheichl B, Kluwick A (2005) Turbulent Marginal Separation and the Turbulent Goldstein Problem. AIAA paper 2005-4936

Symmetry-preserving Regularization Modeling of Turbulent Channel Flow

Roel Verstappen

Institute of Mathematics and Computing Science, University of Groningen, P.O.Box 800, 9700AV Groningen, The Netherlands; R.W.C.P.Verstappen@rug.nl

1 Introduction

Most turbulent flows can not be computed directly from the (incompressible) Navier-Stokes equations,

$$\partial_t u + \mathcal{C}(u, u) + \mathcal{D}(u) + \nabla p = 0, \tag{1}$$

because they possess far too many scales of motion. The computationally almost numberless small scales result from the nonlinear convective term $\mathcal{C}(u, v) = (u \cdot \nabla)v$ which allows for the transfer of energy from scales as large as the flow domain to the smallest scales that can survive viscous dissipation. As the full energy cascade can not be computed, a dynamically less complex mathematical formulation is sought. In the quest for such a formulation, we consider smooth approximations (regularizations) of the convective term:

$$\partial_t u_\epsilon + \widetilde{\mathcal{C}}(u_\epsilon, u_\epsilon) + \mathcal{D}(u_\epsilon) + \nabla p_\epsilon = 0, \tag{2}$$

where the variable name is changed from u to u_ϵ to stress that the solution of (2) differs from that of (1). The idea is to smooth the convective term directly to set bounds to the creation of smaller and smaller scales of motion and thus to confine the cascade of energy. Equation (2) provides a basis for a simulation shortcut if the low modes of u_ϵ approximate the corresponding low modes of the Navier-Stokes-solution u, whereas the high modes of u_ϵ vanish faster than those of u [1].

The first outstanding approach in this direction goes back to Leray [2], who took $\widetilde{\mathcal{C}}(u, u) = \mathcal{C}(\bar{u}, u)$ and proved that a moderate filtering of the convective velocity is sufficient to regularize a turbulent flow. The Navier-Stokes-α-model forms another example of regularization modeling, see for instance [3],[4]. In this model, the convective term becomes $\widetilde{\mathcal{C}}_r(u, u) = \mathcal{C}_r(u, \bar{u})$, where $\mathcal{C}_r$ denotes the convective operator in rotational form: $\mathcal{C}_r(u, v) = (\nabla \times u) \times v$.

The regularization method basically alters the nonlinearity to control the convective energetic exchanges. In doing so, one can preserve certain fundamental properties of (the convective operator in) the Navier-Stokes equations exactly, e.g. symmetries, conservation properties, transformation properties, Kelvin's theorem, Bernouilli's theorem, etc.

2 Symmetry-preserving Approximations

We purpose to consider approximations that conserve the energy, the enstrophy (in 2D) and the helicity in the absence of viscous dissipation, among others because they are intrinsically stable (in the energy-norm; in 2D: enstrophy-norm). The Leray model conserves the energy, but not the enstrophy or helicity, whereas the Navier-Stokes-α-model conserves the enstrophy and helicity, yet not the energy. Since the conservation of energy, enstrophy and helicity results from the symmetries of the convective operator $\mathcal{C}$ (see e.g. [5]), we propose to approximate $\mathcal{C}$ in such manner that the underlying symmetries are preserved. This criterion yields the following class of (energy-, enstrophy- (2D) and helicity-conserving) approximations

$$\partial_t u_\epsilon + \mathcal{C}_n(u_\epsilon, u_\epsilon) + \mathcal{D}(u_\epsilon) + \nabla p_\epsilon = 0, \tag{3}$$

($n = 2, 4, 6$) where the convective term is smoothened according to

$$\mathcal{C}_2(u, v) = \overline{\mathcal{C}(\overline{u}, \overline{v})} \tag{4}$$
$$\mathcal{C}_4(u, v) = \mathcal{C}(\overline{u}, \overline{v}) + \overline{\mathcal{C}(\overline{u}, v')} + \overline{\mathcal{C}(u', \overline{v})} \tag{5}$$
$$\mathcal{C}_6(u, v) = \mathcal{C}(\overline{u}, \overline{v}) + \mathcal{C}(\overline{u}, v') + \mathcal{C}(u', \overline{v}) + \overline{\mathcal{C}(u', v')} \tag{6}$$

Here, a bar denotes a filtered quantity and a prime indicates the residual. The three approximations $\mathcal{C}_n(u, u)$ are consistent with $\mathcal{C}(u, u)$, where the error is of the order of ϵ^n ($n = 2, 4, 6$) for symmetric filters with filter length ϵ. Both the Leray model and the alpha model are second-order accurate in terms of ϵ.

3 Nonlinear Transport Mechanism

The evolution of the vorticity

$$\partial_t \omega_\epsilon + \mathcal{C}_n(u_\epsilon, \omega_\epsilon) + \mathcal{D}(\omega_\epsilon) = \mathcal{C}_n(\omega_\epsilon, u_\epsilon), \tag{7}$$

resembles that of the Navier-Stokes equations: the only difference is that $\mathcal{C}$ is replaced by $\mathcal{C}_n$. The approximations $\mathcal{C}_n$ reduce the high frequencies in the vortex stretching term. The Navier-Stokes equations yield $\mathcal{C}(\omega, u) = \overline{S}\overline{\omega} + \overline{S}\omega' + S'\overline{\omega} + S'\omega'$ (where $S = \frac{1}{2}(\nabla u + \nabla u^{\mathrm{T}})$ is the deformation tensor), while here the vortex stretching term becomes $\mathcal{C}_2(\omega, u) = \overline{\overline{S}\overline{\omega}}$, $\mathcal{C}_4(\omega, u) = \overline{S}\overline{\omega} + \overline{\overline{S}\omega'} + \overline{S'\overline{\omega}}$,

$\mathcal{C}_6(\omega, u) = \overline{S\overline{\omega}} + \overline{\overline{S}\omega'} + \overline{S'\overline{\omega}} + \overline{S'\omega'}$. Qualitatively, vortex stretching leads to the production of smaller and smaller scales, i.e. to a continuous, local increase of both S' and ω'. Consequently, at the positions where vortex stretching occurs, the terms with S' and ω' will eventually amount considerably to $\mathcal{C}(\omega, u)$. Since the approximations $\mathcal{C}_n(\omega, u)$ diminish these terms, the symmetry-preserving regularization counteracts the production of smaller and smaller scales by means of vortex stretching and may eventually stop the continuation of the vortex stretching process. In conclusion, the approximations $\mathcal{C}_n(u, u)$ restrain the convective production of smaller and smaller scales of motion by means of vortex stretching, while ensuring that the solution cannot blow up (in the energy-norm; 2D: enstrophy-norm).

A detailed study of the triadic interactions shows that $\mathcal{C}_4(u, u)$ (for example) approximates the local interactions between large scales of motion (meaning that $\epsilon|k| < 1$) up to fourth-order, i.e. the triadic interactions between large scales are only slightly altered. All interactions involving longer wavevectors (smaller scales of motion) are reduced. The amount by which the interactions between the wavevector-triple (k, p, q) are lessened depends on the length of the legs of the triangle $k = p + q$. Triadic interactions for which at least two legs are (much) longer than $1/\epsilon$ are (strongly) attenuated, whereas interactions for which at least two legs are (much) shorter than $1/\epsilon$ are reduced to a small degree only.

4 Results for Turbulent Channel Flow

As a first step in the application of symmetry-preserving regularization, the approximation $\mathcal{C}_4$ is tested for a turbulent channel flow by means of a comparison with the direct numerical simulations performed by Kim et al. [6]. Based on the channel half-width and the friction velocity the Reynolds number is 180. The numerical algorithm used to solve (3) preserves the considered symmetries, too, [7] and is therefore well-suited to test the proposed simulation shortcut. We consider two, coarse, computational grids consisting of $16\times16\times8$ and $32\times32\times16$ grid points, respectively. The filtering is performed by means of a Helmholtz operator. The least to be expected from a simulation shortcut is a good prediction of the mean flow. As can be seen in Fig. 1, the conservative approximation $\mathcal{C}_4$ satisfies that minimal requirement already at the very coarse $16\times16\times8$ grid (when the filter length ϵ is taken equal to about three times the grid width h). Overall good agreement between the $\mathcal{C}_4$-calculation at the $32\times32\times16$ grid and the DNS is observed for both the first- and second-order statistics, see Fig. 1. Heuristic arguments as well as computational results (Fig. 2) show that the energy spectrum of the solution of (3)+(5) follows the DNS for large scales of motion, whereas a much steeper (numerically speaking: more gentle) power law is found for small scales, which is precisely what a simulation shortcut is ought to do.

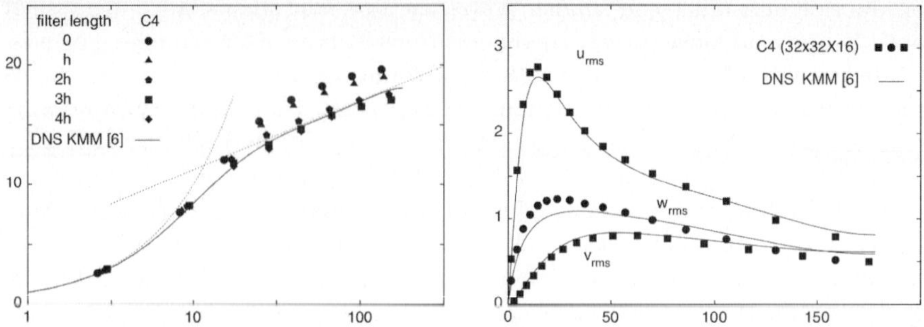

Fig. 1. The least to be expected from a simulation shortcut is a good prediction of the mean flow. The left-hand figure shows that the symmetry-preserving regularization model satisfies that minimal requirement already at the very coarse $16\times16\times8$ grid (when the filter length ϵ is about three times the grid width h). Overall good agreement for the second-order statistics is obtained with $32\times32\times16$ gridpoints (see right-hand figure)

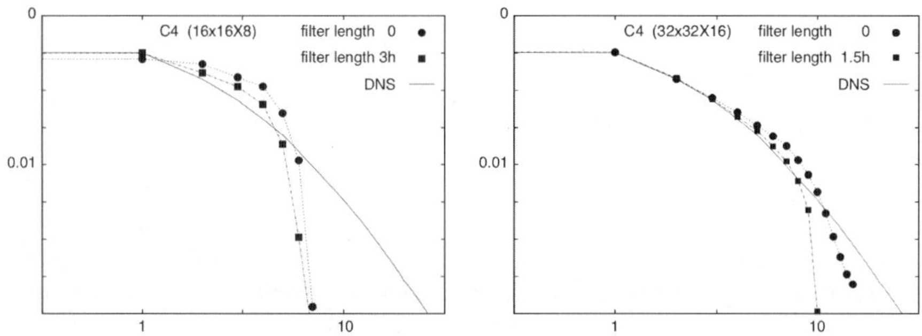

Fig. 2. One-dimensional (streamwise) energy spectra at $y^+ \approx 5$. The spectra illustrate that the regularization improves the agreement with the DNS for the low-modes, whereas the tail vanishes faster

References

1. Guermond JL, Oden JT, Prudhomme S (2004) J Math Fluid Mech 6:194–248
2. Leray J (1934) Acta Math 63:193–248
3. Holm DD, Marsden JE, Ratiu TS (1998) Adv Math 37:1-81

4. Foias C, Holm DD, Titi ES (2001) Physica D 152:505–519
5. Foias C, Manley O, Rosa R, Temam R (2001) Navier-Stokes equations and turbulence. Cambridge University Press
6. Kim J, Moin P, Moser R (1987) J Fluid Mech 177: 133–166
7. Verstappen RWCP, Veldman AEP (2003) J Comp Phys 187:343–368

Part V

Convection

lateral wall and on the upper/lower plates), various degrees of deviation from the Boussinesq approximation, Prandtl number variation and control of the temperature boundary conditions on the heated and cooled plates [1, 2, 3]. For these reasons numerical simulations could be seen as ideal experiments, where all flow parameters are under strict control.

2 Governing Equations and Set–up

Under the Boussinseq approximation, thermal convection is governed by the three dimensional unsteady Navier-Stokes equations as follows:

$$\frac{Du}{Dt} = -\nabla p + \theta \hat{x} + \left(\frac{Pr}{Ra}\right)^{\frac{1}{2}} \nabla^2 u, \qquad \nabla \cdot u = 0, \tag{2}$$

$$\frac{D\theta}{Dt} = \left(\frac{1}{RaPr}\right)^{\frac{1}{2}} \nabla^2 \theta. \tag{3}$$

Here $\hat{x}$ is the vector pointing in the opposite direction with respect the gravity and θ is the non-dimensional temperature ($0 < \theta < 1$). A second-order accurate finite-difference scheme in cylindrical coordinates discretized on a staggered mesh as in [4] has been used for the integration of the equations. In a direct numerical simulation it is mandatory to adequately resolve both velocity and thermal boundary layers near the walls and the smallest between the Kolmogorov and Corrsin scales (η) in the bulk, hence ask for a *very fine* spatial resolution. For this reason an over-resolved simulation at $Ra = 2 \cdot 10^{11}$ was performed to check the effect of grid resolution. With the mesh in the bulk equal to 1.2η and 10 points in the thermal boundary layer, a Nusselt number of $Nu = 440.3 \pm 10$ was obtained, to be compared with the previous result of $Nu = 447.2 \pm 11.7$ obtained with a grid in the bulk 4 times larger than η and only 5 points in the thermal boundary layer [4]. *A posteriori* checks of the thickness of both boundary layers and additional integral quantities confirmed the adequacy of the discretization used. Three simulations at $Pr = 0.7$ and $Ra = 2 \cdot 10^{12}, 2 \cdot 10^{13}, 2 \cdot 10^{14}$ have been performed. In Table 1 the discretization for the three simulations performed are presented. They were performed using a IBM p690 Power4 32-CPU node (for $Ra = 2 \times 10^{12}$ and $Ra = 2 \times 10^{13}$)

Table 1. Variation of the Kolmogorov scale (η) and thermal boundary layer thickness (δ_θ) with Ra; these values are computed from present simulations

Ra	η/h	δ_θ/h	$N_\theta \times N_r \times N_x$
2×10^{12}	8.2×10^{-4}	3.76×10^{-4}	$257 \times 193 \times 769$
2×10^{13}	4.3×10^{-4}	1.73×10^{-4}	$385 \times 301 \times 1381$
2×10^{14}	2.4×10^{-4}	7.00×10^{-5}	$513 \times 401 \times 1801$

and using a NEC SX-6+ 8-CPU node (for $Ra = 2 \times 10^{14}$) asking for about 150'000 CPU hours and producing about 1 TB of raw data. To our knowledge these are the biggest simulations of fully confined turbulence ever performed.

3 Results, Comments and Conclusion

In Fig. 1 we report Nu as function of Ra for the present results and those by [4] in comparison with some experiments performed in the same geometry and under apparently identical conditions [1, 2, 3]. Although the agreement might seem satisfactory for a first look, a different representation of the same data (Fig. 2a), clearly shows that differences up to 20% are present. Some of the differences between the results can be explained in terms of different Pr numbers. In the experiments, in fact, Pr is constant only for $Ra < 10^{12}$ while for higher Ra it attains higher values. In Fig. 2b the Prandtl numbers

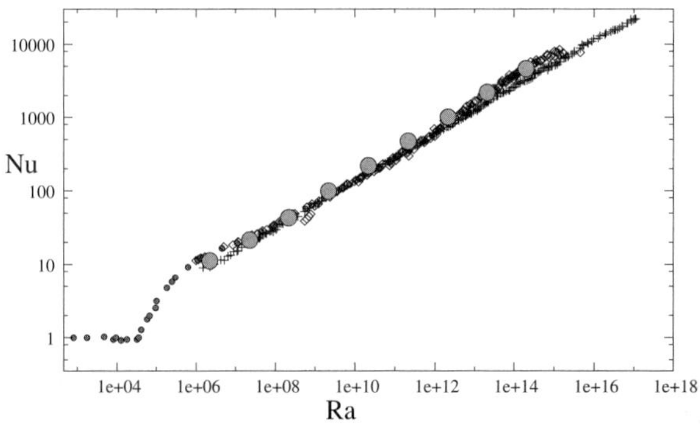

Fig. 1. Nu vs. Ra for numerical simulation (large bullets) and experiments by [1] (+), [2] (◇) and [3] (circles)

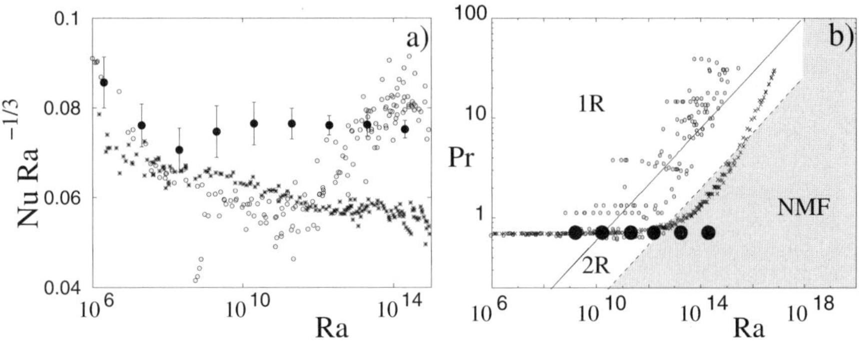

Fig. 2. (a): Compensated $Nu \cdot Ra^{-1/3}$ for numerical simulation (circles) and experiments by [1] (x) and [2] (◇). (b): Pr vs. Ra for numerical simulation (bullets) and experiments by [1] (x) and [2] (◇)

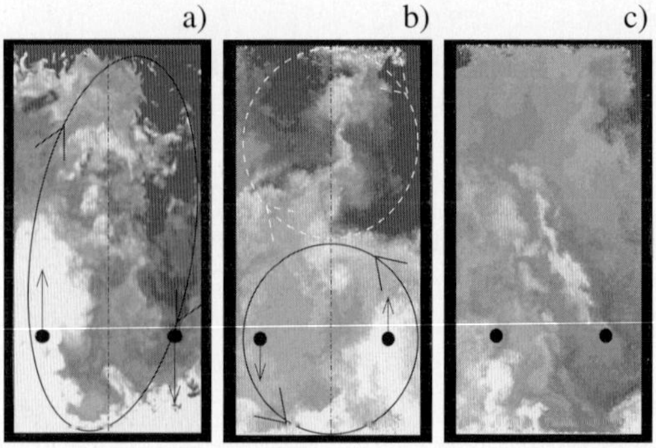

Fig. 3. Snapshots of temperature showing the possible mean flow configurations: *a)* $Ra = 2 \times 10^{10}$, *b)* $Ra = 2 \times 10^{12}$, *c)* $Ra = 2 \times 10^{14}$, Only the temperature range $0.475 \leq \theta \leq 0.525$ is represented with $\Delta\theta = 0.005$

for experiments and simulation are reported in a $Ra - Pr$ plane divided into different regions, according to the model by Stringano & Verzicco [5], which indicate the most likely mean flow structure. The $Pr - Ra$ plane is divided in a region where only a single roll can be found (1R region), where the mean flow is characterized by two rolls (2R region) and region where no mean flow is found (NMF region); in the latter the upper and lower thermal boundary layers do not 'communicate', and, following Malkus [6], we have $Nu \sim Ra^{1/3}$. These findings are confirmed by the flow visualizations of Fig. 3 where snapshots of the temperature field for points of the $Pr - Ra$ plane belonging to regions 1R, 2R and NMF confirm the predictions of Fig. 2b. Further details on the flow dynamics can be found in [7]. The main result of this paper is to show how numerical simulations can help the understanding of laboratory experiments thus allowing for a deeper comprehension of the problem. In this particular case the possibility of keeping Pr strictly constant for every value of Ra allowed us to find a no-mean-flow (NMF) region where, according to [6], we might expect $Nu \sim Ra^{1/3}$. In the experiments, probably owing to difficulties in keeping Pr constant, the above power-law was not observed.

References

1. Niemela J, Sreenivasan KR (2003) J. Fluid Mech. 481:355–384
2. Roche P, Castaing B, Chabaud B, Hebral B (2002) Europhys. Lett. 58:693–698
3. Chavanne X, Chillà F, Chabaud B, Castaing B, Hebral B (2001) Phys. Fluids 13:1300–1320
4. Verzicco R, Camussi R, (2003) J. Fluid Mech. 477:19–49
5. Stringano G, Verzicco R, (2005) J. Fluid Mech (in press).
6. Malkus W.V.R. (1954) Proc. Roy. Soc. Lond. A225:196–212
7. Amati G, Koal K, Massaioli F, Sreenivasan KR, Verzicco R (2005) Phys. of Fluids (in press).

Homogeneous Rayleigh-Bénard Convection

E. Calzavarini[1], D. Lohse[1] and F. Toschi[2]

[1] Dept. of Applied Physics, University of Twente, 7500 AE Enschede, The Netherlands. e.calzavarini@tnw.utwente.nl, lohse@tnw.utwente.nl

[2] CNR-IAC, Viale del Policlinico 137, I-00161 Roma, Italy and INFN, Sezione di Ferrara, Via Paradiso 12, I-43100 Ferrara, Italy. toschi@iac.cnr.it

1 The Problem

Much effort has been expended in recent decades in addressing the problem of heat transfer in Rayleigh-Bénard (RB) thermal convection cells. There is increasing agreement that in general there are no clean scaling laws for $Nu(Ra, Pr)$ and $Re(Ra, Pr)$, apart from asymptotic cases. One of these asymptotic cases has been doped the *ultimate state of thermal convection*, i.e. $Ra \to \infty$, where the heat flux becomes independent of the kinematic viscosity ν and the thermal diffusivity κ. The physics of this regime is that the thermal and kinetic boundary layers are broken down or do not play a role any more for the heat flux and the flow is bulk dominated. Scaling laws for this regime were first suggested by Kraichnan [1] and later by Spiegel [2]. The recent Grossmann-Lohse (GL) theory [3] also gives such an asymptotic regime which is bulk dominated and where the plumes do not play a role, namely

$$Nu \sim Ra^{1/2} Pr^{1/2}, \qquad (1)$$

$$Re \sim Ra^{1/2} Pr^{-1/2}. \qquad (2)$$

While current experimental data for high Rayleigh numbers are controversial, see [3] for an overview, numerical simulations have not been very effective in studying this regime because of difficulties in dealing with the huge number of degrees of freedom and scale separation engendered when Rayleigh numbers reach the order of at least 10^{12}.

In order to bridge such difficulties, we study a tri-periodic convective cell, or homogeneous Rayleigh-Bénard (HRB) system, to investigate the properties of the convective cell once the effect of boundary layers has been eliminated [5]. A model system such as this was first introduced by Borue and Orszag [4]. While these authors focused especially on turbulent spectra and correlation functions behavior, in the present context we address mainly the scaling of integral quantities, such as heat flux and mean velocity fluctuations, respect to the dimensionless parameters Ra and Pr.

The physical relevance of the HRB model, in particular the expected similarities and differences with respect to real experiments of fully developed turbulent convection, will be shortly discussed at the end.

2 The Model

The system to be studied is described in terms of the following partial differential equations

$$\boldsymbol{u}_t + \boldsymbol{u} \cdot \nabla \boldsymbol{u} + \nabla p = \nu \Delta \boldsymbol{u} + \hat{z}\, \alpha g \theta\,, \tag{3}$$

$$\theta_t + \boldsymbol{u} \cdot \nabla \theta = \kappa \Delta \theta + \frac{\Delta T}{L} u_z\,, \tag{4}$$

where $\boldsymbol{u} = (u_x, u_y, u_z)$ is an incompressible velocity field, $\nabla \cdot \boldsymbol{u} = 0$, and αg is the thermal expansion coefficient times the acceleration due to gravity. These equations are used to describe the evolution of the velocity field in a triply-periodic cubic volume $[0, L]^3$ in the presence of a temperature field $T(\boldsymbol{x}, t) = \overline{T}(\boldsymbol{x}) + \theta(\boldsymbol{x}, t)$. The temperature is expressed as a fluctuation θ with respect to a mean profile $\overline{T}(\boldsymbol{x})$ that is *imposed* to be equal to the mean conductive temperature profile in such a Rayleigh-Bénard cell; i.e. linear and of the form $\overline{T}(\boldsymbol{x}) = -z\Delta T/L + \Delta T/2$.
We performed a high-resolution DNS of the above set of equations, see [6] for details on the implementation, at changing both the Ra and Pr number. Statistical analysis were then performed on the database of flow configurations that were collected, in statistically stationary conditions, over a time interval of order 10^2 large eddy turnover times.

3 Results

In the HRB system the Nusselt number is defined as the dimensionless heat flux

$$Nu = \frac{\langle u_z \theta \rangle}{\kappa \Delta T\ L^{-1}} + 1 \tag{5}$$

where the average $\langle ... \rangle$ is over volume and over time. From (3)–(4) one can derive two exact relations for the volume averaged thermal dissipation rate $\epsilon_\theta = \kappa \left\langle (\partial_i \theta)^2 \right\rangle$ and the volume averaged kinetic dissipation rate $\epsilon_u = \nu \left\langle (\partial_i u_j)^2 \right\rangle$, namely

$$\epsilon_u = \nu^3\ L^{-4}\ Nu Ra Pr^{-2} \tag{6}$$

$$\epsilon_\theta = \kappa\ \Delta T^2\ L^{-2}\ Nu. \tag{7}$$

One can therefore numerically compute Nu in three different ways: (i) from its direct definition (5), (ii) from the volume averaged kinetic dissipation rate (6), (iii) from the volume averaged thermal dissipation rate (7).
The results are shown in Fig. 1(a) as a function of Ra for $Pr = 1$. There is very good agreement of Nu obtained from the three different methods for all Ra. Fitting all data points beyond $Ra = 10^5$ with an effective power law, we

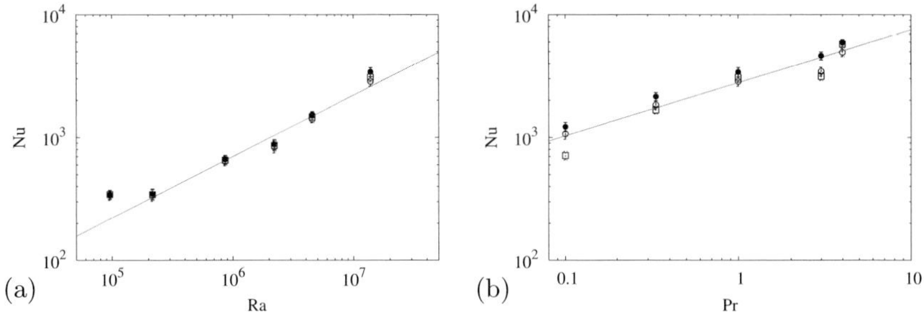

Fig. 1. (a) $Nu(Ra)$ for $Pr = 1$, computed in three different ways: (•) using (5), (□) using (6), and (○) from (7). The power law fits, performed on the mean value of the three different estimates and for $Ra > 10^5$, gives a slope 0.50 ± 0.05. (b) $Nu(Pr)$ for $Ra = 1.4 \cdot 10^7$, fit performed as before, with a resulting slope of 0.43 ± 0.07

obtain $Nu \sim Ra^{0.50 \pm 0.05}$, consistent with the asymptotically expected law (1). In Fig. 1(b) we display Nu as function of Pr for fixed $Ra = 1.4 \cdot 10^7$. For the cases with $Pr \neq 1$ the convergence of the three different methods to calculate Nu is not perfect. This may be due to numerical errors in the resolution of the small scale differences, especially when ν and κ are considerably different. However, one can clearly notice a strong increase of Nu with Pr. A fit with an effective power law gives $Nu \sim Pr^{0.43 \pm 0.07}$, which is again consistent with the asymptotic power law suggested by the GL theory.
Similar conclusions follow from Fig. 2, in which we show the Reynolds number, $Re = \frac{u'L}{\nu}$, scaling versus Ra (a) and as function of Pr for fixed Ra (b). We want to stress here that this is the *fluctuation* Reynolds number, defined by the rms velocity fluctuation $u' = \langle \boldsymbol{u}^2 \rangle^{1/2}$: in HRB no large scale wind exists. Therefore, if one identifies the wind Reynolds number in GL with the fluctuation Reynolds number here, $Re(Ra, Pr)$ display scaling law consistent with the prediction for the ultimate regime. In conclusion, we confirm that both the Ra- and the Pr-scaling of Nu and Re in HRB convection are consistent with the suggested scaling laws for the bulk-dominated regime.
Although apparently physically unrealizable, because of the boundary conditions, the measures from the highly turbulent HRB model may be tested

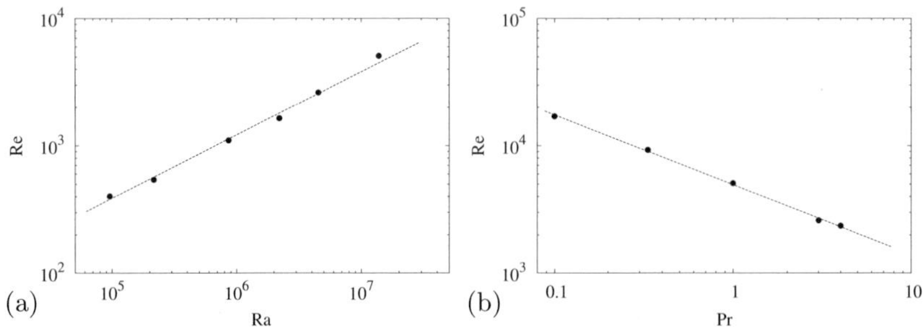

Fig. 2. (a) $Re(Ra)$ for $Pr = 1$, with a fitted slope 0.50 ± 0.02. (b) $Re(Pr)$ for $Ra = 1.4 \cdot 10^7$, with a fitted slope -0.55 ± 0.01

against a recent series of experimental results obtained from an RB setup especially designed to reduce the influence of top and bottom plates on the physical core of thermal convection [7]. In these experiments the temperature gradient in the bulk of the cell is not imposed but rather, as in fixed flux convection, measured as a dependent parameter. Interestingly these experiments show, that the Nu and Re versus Ra scalings observed are consistent with our bulk simulations.

On the other hand, a peculiar properties of the HRB model, that becomes particularly relevant to its dynamics in the *low Rayleigh number regime*, is the formation of accelerating antiparallel vertical jets in the system. These flow patterns, that where already observed in the former study [4], can be associated to the existence of a particular class of unstable solution for the full non-linear sets of differential equations that we want to sketch out in the following.

Due to the periodic boundary conditions, the coupled system of (3)–(4) admits the particular solution $\theta = \theta_0 e^{\kappa\lambda t} \sin(\boldsymbol{k} \cdot \boldsymbol{x})$, $u_z = u_0 e^{\kappa\lambda t} \sin(\boldsymbol{k} \cdot \boldsymbol{x})$ and $u_x = u_y = 0$, which is independent from the vertical coordinate z (here $\boldsymbol{k} = (k_x, k_y)$) and with:

$$\lambda = -\tfrac{1}{2}\,(Pr+1)k^2 + \tfrac{1}{2}\,\sqrt{(Pr+1)^2 k^4 + 4Pr\,(Ra\;L^{-4} - k^4)} \tag{8}$$

From (8) one finds that the first unstable mode appears for $Ra \geq Ra_c = (2\pi)^4 \sim 1558$, corresponding to the instability of the smallest possible wavenumber in the system, i.e. $\boldsymbol{k} = \frac{2\pi}{L}(1,0)$ [8].

These solutions are clearly manifest in direct numerical simulations at Rayleigh numbers slightly above the critical value ($Ra \gtrsim Ra_c$) [8] and in general for $Ra \lesssim 10^5$ were only few unstable modes are active [6]. Despite the presence of these exact exploding modes, the system clearly shows that due to some secondary instability mechanism, these solutions do not survive indefinitely, resulting always in a statistically stationary behavior. Is the interplay between the active exploding modes and the destabilization that sets the value of the Nusselt number, i.e. the heat transfer through the cell in the low-Ra regime.

References

1. R. H. Kraichnan, (1962) *Phys. Fluids* **5**, 1374.
2. E. A. Spiegel, (1971) *Ann. Rev. Astron. Astrophys.* **9**, 323–352.
3. S. Grossmann and D. Lohse, (2000) *J. Fluid. Mech.* **407**, 27; (2001) *Phys. Rev. Lett.* **86**, 3316; (2004) *Phys. Fluids* **16**, 4462-4472.
4. V. Borue and S. Orszag, (1997) *J. Sci. Comp.* **12**, 305.
5. D. Lohse and F. Toschi, (2003) *Phys. Rev. Lett.* **90,** 034502.
6. E. Calzavarini, D. Lohse, F. Toschi and R. Tripiccione, (2005) *Phys. Fluid* **17**, 055107.
7. M. Gibert, H. Pabiou, F. Chilla and B. Castaing (2005) *present proceedings.*
8. E. Calzavarini, C. R. Doering, J. D. Gibbon, D. Lohse, A. Tanabe and F. Toschi, (2005) *unpublished, under consideration for publication on Phys. Rev. Lett.*

A RANS Model for the Pressure-velocity Fluctuation Correlation

L. Chandra and G. Grötzbach

Institut für Kern- und Energietechnik, Forschungszentrum Karlsruhe GmbH, 76021 Karlsruhe, Germany
groetzbach@iket.fzk.de

1 Introduction

Unstable and stable thermal stratifications play an important role in technical, atmospherical and astrophysical fluid flows. Comparisons between Reynolds Averaged Navier-Stokes models (RANS) with Direct Numerical Simulation (DNS) data show that current models in CFD codes lead to serious uncertainties in buoyant flow [5]. Investigations in meteorology [9] demonstrate that buoyancy significantly influences the turbulent diffusion of the turbulent kinetic energy $\overline{E'}$ and that the corresponding gradient diffusion model needs improvement for buoyant flows. The turbulent diffusion of $\overline{E'}$ consists of the velocity-fluctuation triple correlation $\overline{u'_j E'}$ and the velocity-pressure fluctuation correlation $\overline{u'_j p'}$. They are usually modeled together in the gradient-diffusion approximation [7]. DNS data of an internally heated fluid layer (IHL) and Rayleigh-Bénard convection (RBC) show different behavior and importance of these terms in both flow types due to differences in the underlying mechanisms [14]. These reveal that both terms should be modeled separately. In this paper it will be shown that the standard model fails completely to reproduce the turbulent diffusion in RBC and IHL. Therefore, as a part of an improved model, starting with the approximation as in [4], a modified RANS model for $\overline{u'_j p'}$ will be obtained. It will be validated using DNS data.

2 Modeling Requirements

The following investigations are based on DNS data for RBC and IHL. In case of RBC we consider a fluid layer between two infinite horizontal plates which is heated from below. In case of IHL, the plates are at a lower temperature than the fluid confined in-between which is having a uniform volumetric energy source. In RBC the fluid layers are unstably thermally stratified throughout the height of the channel [11], whereas in IHL the fluid layers are mostly stably

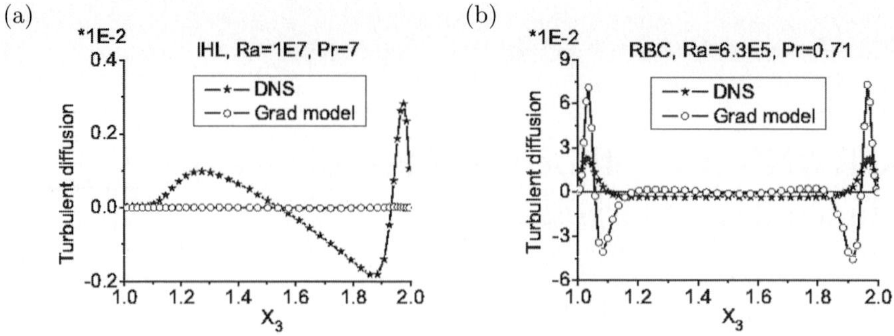

Fig. 1. Vertical profile of the turbulent diffusion of $\overline{E'}$, DNS ($\star$) and gradient model (o) for IHL (a) and RBC (b)

stratified with an unstably stratified thin upper thermal boundary layer [13]. IHL with an internal Rayleigh number $Ra_i = 1 \times 10^7$ and Prandtl number $Pr = 7.0$ and RBC with an external Rayleigh Number $Ra_e = 6.3 \times 10^5$ and $Pr = 0.71$ are used for analysis and validation. Hereafter, Ra_i and Ra_e (see [2] and [13]) are referred to as Ra.
The analysis of the DNS data in Fig. 1 shows that the existing classical gradient diffusion model (1) for the turbulent diffusion of $\overline{E'}$ is inadequate for both flow types. Moreover, it even fails to reproduce roughly the required distribution. This indicates that the model needs improvement for a more accurate description of both flows.

3 RANS Modeling

Both the standard as well as the advanced RANS models use the transport equation for $\overline{E'}$. The gradient diffusion approximation is usually employed for calculating the turbulent diffusion of $\overline{E'}$ (see e.g. [7]):

$$\overline{u'_j E'} + \overline{u'_j p'} \approx -\frac{\nu_t}{\sigma_E} \frac{\partial \overline{E'}}{\partial x'_j} \tag{1}$$

Here ν_t and σ_E are the isotropic eddy viscosity and turbulent Prandtl number for $\overline{E'}$, respectively. According to [8], $-\overline{u'_j p'} = \frac{1}{5}\overline{u'_j E'}$ holds in homogeneous turbulence. On the other hand, strong deviations from this approximation have been observed in both flows, [2] and [14]. This reveals that the two terms should be modeled separately. Following [6] a RANS model for $\overline{u'_j E'}$ has been derived which includes the effect of buoyancy and anisotropy [2].
In this paper the derivation of a modified RANS model for $\overline{u'_j p'}$ will be discussed. For this deduction the approximation as in [4] has been considered as the starting point which is as follows:

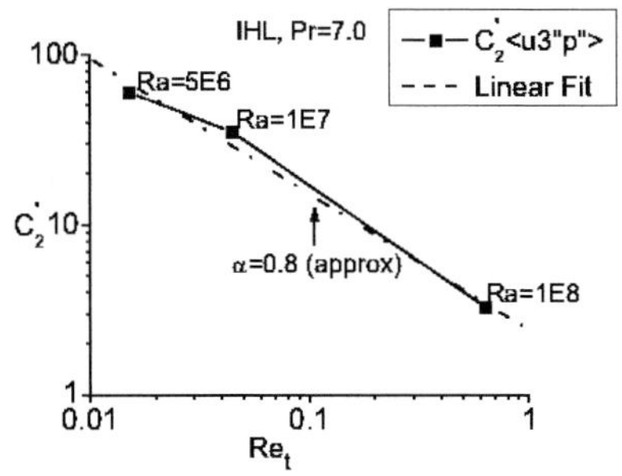

Fig. 2. The coefficient C_2' in IHL at $x_3 = 1.18$

$$\overline{u_j'p'} \approx -\, C_2' \, \overline{E'}^{1/2} \, l \, \frac{\partial \overline{u_l'u_j'}}{\partial x_l}. \tag{2}$$

Here C_2' is a coefficient and l is a length scale, with $l = \frac{\overline{E'^{3/2}}}{\overline{\varepsilon'}}$ from [12]. In order to simplify this model an approximation has been introduced that $\overline{u_l'u_j'} \ll \overline{u_l'^2}$. According to [3] the model coefficient may depend on the turbulent Reynolds number $Re_t = \frac{\overline{E'^2}}{\nu\overline{\varepsilon'}}$. Thus, some model coefficients for closure terms in the transport equations for $\overline{E'}$ and the turbulent heat fluxes have to be increased by a large factor in IHL [13]. Fig. 2 indeed shows that C_2' depends on Re_t e.g. in IHL, with $C_2' = \frac{C_2}{Re_t^{\alpha}}$. Here the coefficient is $1.5 \le C_2 \le 7$ in both IHL and RBC over most of the height. The exponent $\alpha = 0.8$ has been calculated in IHL using the DNS data. This value will be applied over the full height also in RBC. The above results in the modified model:

$$\overline{u_j'p'} \approx -\frac{C_2}{Re_t^{\alpha}} \left(\frac{\overline{E'^2}}{\overline{\varepsilon'}} \frac{\partial \overline{u_j'^2}}{\partial x_l} \right) \delta_{jl} \tag{3}$$

4 Model Validation

For validation of the modified RANS model for $\overline{u_j'p'}$ as given by (3), C_2 is set to 1.5 for IHL and 3.0 for RBC and $\alpha = 0.8$. The comparisons between the vertical profiles of $\overline{u_j'p'}$ and its model analyzed from the DNS data are shown in Fig. 3. These results indicate an acceptable qualitative and quantitative agreement between $\overline{u_j'p'}$ and its modeled values in IHL and in RBC.

5 Conclusions

In order to improve the turbulent diffusion model of $\overline{E'}$, its closure terms $\overline{u_j'E'}$ and $\overline{u_j'p'}$ are modeled separately. Here, a modified model for $\overline{u_j'p'}$ has

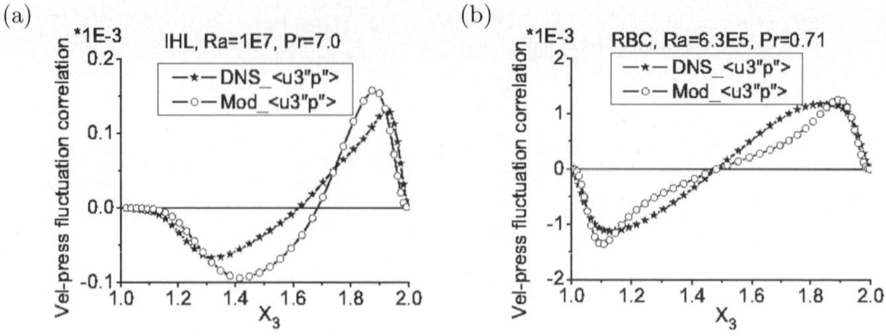

Fig. 3. Vertical profiles of $\overline{u_j' p'}$, DNS($\star$) and model(o) for IHL (a) and RBC (b)

been proposed. It shows acceptable qualitative and quantitative agreement with DNS data. Coupling of this model with models for $\overline{u_j' E'}$ from [2] results in extended models for the turbulent diffusion of $\overline{E'}$. A simplified model is available in [1]. All show better predictive capability compared to the standard gradient diffusion model. Incorporating the models into current CFD codes allows us expecting a more accurate description of buoyant flows at the price of using an additional transport equation for the energy of the vertical velocity fluctuations $\overline{u_3'^2}$. Really sufficient results may be expected from combinations with algebraic heat flux approximations including improved modelling of the temperature variance [11].

References

1. Chandra L, Grötzbach G (2005) In JKT2005, Nürnberg, Kerntechnische Gesellschaft e. V.(eds.), INFORUM Bonn
2. Chandra L (2005) "A model for the turbulent diffusion of turbulent kinetic energy in natural convection," PhD thesis, Universität Karlsruhe, FZKA 7158
3. Daly B J, Harlow F H (1970) Phys. Fluids, 18 : 2634-2649
4. Donaldson C duP (1969) AIAA Journal, 7 : 272-278
5. Grötzbach G, Batta A, Lefhalm C-H, Otić I (2004) Nuthos-6, Nara, Japan
6. Hanjalić K, Launder B E (1972) J. Fluid Mech., 52 : 609-638
7. Launder B E, Spalding D B (1972) "Lectures in Mathematical Models of Turbulence," Academic Press, London
8. Lumley J L (1978) Adv. Appl. Mech., 18 : 123-176
9. Moeng C-H, Wyngaard J C (1989) J. Atmospheric Sc., 46 : 2311-2330
10. Monin A S, Yaglom A M (1971) "Statistical Fluid Mechanics: Mechanics of Turbulence," The MIT Press, Cambridge, Massachusetts
11. Otić I, Grötzbach G, Wörner M (2005) J. Fluid Mech., 525 : 237-261
12. Rotta J (1951) 1. Mitteilung, Zeitschrift für Physik, 129 : 547-572
13. Wörner M, Schmidt M, Grötzbach G (1997) J. Hydraulic Research, 35 : 773-797
14. Wörner M, Grötzbach G (1998) Int. J. of Heat and Fluid Flow, 19 : 150-158

Temperature Measurements in a Rectangular Rayleigh-Bénard Cell

A. Maystrenko, C. Resagk, A. Thess

Department of Mechanical Engineering, Ilmenau University of Technology, P.O. Box 100565, 98684 Ilmenau, Germany

1 Introduction

The object of our investigation is the temperature distribution in a rectangular air-filled Rayleigh–Bénard(RB)cell. We describe the dependence of the thickness and the structure of the thermal boundary layer on the Rayleigh number (Ra) at constant aspect ratio.

2 Experiment

We report about the turbulent Rayleigh–Bénard convection inside a rectangular box with 2.5 m length, 0.5 m width and 0.5 m height. The box is filled with air (Pr = 0.7). In order to initiate the convective flow it is heated from below and cooled from above by means of two water-heated/cooled aluminum plates. The temperature distribution in both plates is controlled by 40 PT-100 temperature probes. There are three windows for access of measurement probes in the longitudinal axes of both plates. The cell has adiabatic walls of Perspex with 8 mm thickness and a 180 mm isolation layer made out of Styrofoam. To measure local temperature profiles a sensor system based on a glass encapsulated NTC micro–thermistor was developed. The sensor is fixed on a 4 mm thick support and is moved by a traverse system in normal direction to 34 positions up to a distance of 150 mm below the center of the cooling plate. The sensor signals are recorded over a low-current resistance bridge in combination with a high-precision multimeter HP3458. By means of moving the micro–thermistor from point to point the temperature profile in the boundary layer was measured. The measurement was done over 1.5 hours for each position with a sampling rate of 200 Hz. The total number of temperature data was 1,080,000 samples for each measurement point. We performed a series of measurements in a range from $Ra = 5.2 \times 10^7$ to $Ra = 6 \times 10^8$ with constant aspect ratio. The complete set of experimental parameters of the temperature measurements is listed in Table 1.

Table 1. Experimental parameters of temperature measurements

Ra	5.7×10^7	8.4×10^7	1.5×10^8	2.3×10^8	3.7×10^8	4.7×10^8	6×10^8
ΔT	5.2 K	7.4 K	12.9 K	21.4 K	35.8 K	50.2 K	71.9 K

3 Results and Discussion

3.1 Temperature Time Series

Three examples of temperature time series at three different z-positions for $Ra = 10^9$ are shown in Fig. 1. In the diagrams the normalized temperature is plotted over a time interval of 50 seconds. The value $\Theta(\mathrm{t}) = 0$ corresponds to the temperature of the cooling plate and the value $\Theta(\mathrm{t}) = 1$ represents the bulk temperature. Very close to the cooling plate (left figure) at z/h = 0.003 (h is the height of the cell) the fluctuations are very large and distributed symmetrically around the mean value. At a distance z/h = 0.03 the normalized mean temperature is nearly equal the bulk temperature but shows strong negative spikes. These spikes indicate cold plumes in this area, the so called mixing zone [1]. If we go further to the bulk region, the temperature fluctuations at the position z/h = 0.3 are smaller and represent a relatively constant bulk temperature. As expected, at lower Ra–number, the fluctuations at the near wall position (z/h = 0.003) became smaller. The temperature curve at $Ra < 10^8$ is very smooth and shows the typical behavior of the temperature field in the boundary layer very close to the wall. At the second position z/h = 0.03mm it looks similar to high Ra–number, but now less cold plumes occurs and at z/h = 0.3 we observe a practically unchanged behavior of temperature data.

3.2 Temperature Profiles

From the time series profiles of the normalized mean temperature $\Theta(z)$, standard deviation $std(z)$, skewness $s(z)$ and kurtosis $k(z)$ were computed and

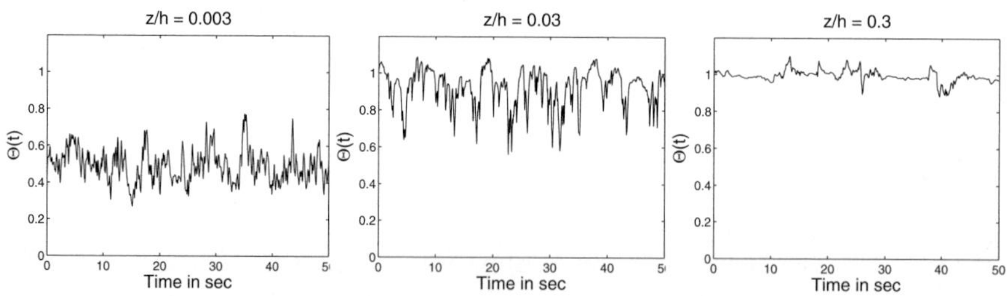

Fig. 1. Temperature time series at three different z–positions for $Ra = 6 \times 10^8$

displayed in Fig. 2 for three different Ra–numbers. In all cases we get typical boundary layer profiles: the temperature rises very quickly in a thin layer below the plate. The close up of the region near the wall shows different slopes, but all profiles approach the bulk temperature more far away from the wall. The high temporal resolution of the measured data allows a precise statistical analysis. The skewness describes the symmetry of the sampled data around the mean value. Inside the boundary layer (skewness approach the value 0) one can see a symmetrical fluctuation of the temperature values, but skewness below zero indicates cold plums, if we are approaching the bulk region. Kurtosis is a degree of the coincidence of the temperature distribution. In the boundary layer area kurtosis approach value 3 and shows a normal Gaussian distribution. In direction of the bulk region kurtosis rise and becomes higher than 3. The data distribution is here less coincident than in the boundary layer area. One interesting point related to the heat transport in the cell is the thickness of the boundary layer. Following Tilgner et al. [2] we analyze the boundary layer thickness δ_σ defined as the value of z at which the rms

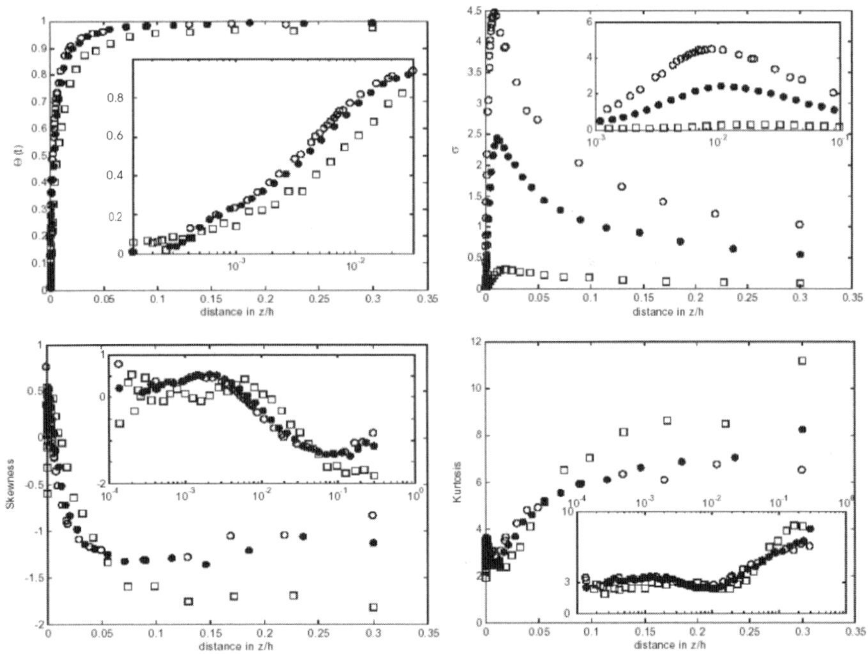

Fig. 2. Temperature profiles for $Ra = 5.7 \times 10^7$ (empty squares), $Ra = 3.7 \times 10^8$ (full circles) and $Ra = 6 \times 10^8$ (empty circles): mean temperature $\Theta(z)$, standard deviation σ, skewness and kurtosis

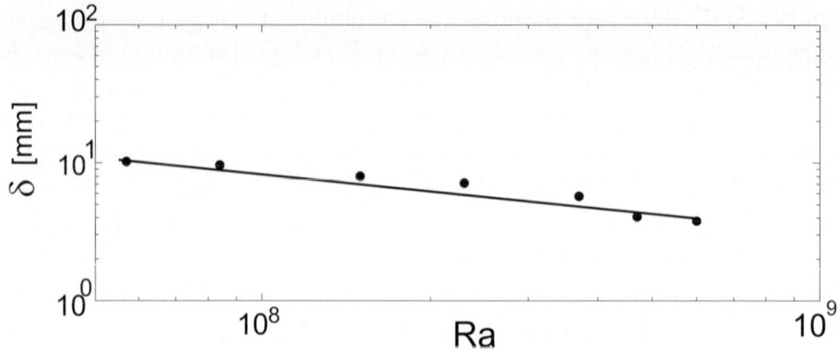

Fig. 3. Boundary layer thickness for different Ra-numbers

temperature fluctuations attain their maximum. In fact we can describe δ_σ as a function of Ra by a power law:

$$\delta_\sigma = 893 \times Ra^{-0.26}. \tag{1}$$

The comparison our boundary layer thickness results with experiments in RB cells with cylindric geometry shows a different behavior. For example for a 1m high cell we get $\delta_\sigma \sim Ra^{-0.29}$ [2] and for a 7m high cell $\delta_\sigma \sim 450700 \times Ra^{-0.40}$[3]. In Fig. 3 the variation of the boundary layer thickness for the whole measured Ra-number range is plotted.

Acknowledgments

We gratefully acknowledge the support of the Helmholtz association of national research centers.

References

1. Castaing B, Gunaratne G, Heslot F, Kadanoff L, Libchaber A, Thomae S, Wu X Z, Zaleski S, Zanetti G (1989) J. Fluid Mech. 204, 1-30
2. Tilgner A, Belmonte A, Libchaber A (1993) Phys. Rev. E 47, R2253–R2256
3. R. du Puits, C. Resagk, A. Tilgner, F.H. Busse, and A.Thess (2005) J. Fluid Mech. to be published

Inertial Convertion in a Vertical Channel

M. Gibert, H. Pabiou, F. Chillà, and B. Castaing

École Normale Supérieure de Lyon, 46 allée d'Italie, 69364 Lyon Cedex 07, France
Mathieu.Gibert@ens-lyon.fr

1 Introduction

We present here the first experimental study of the bulk of turbulent convection in a vertical channel which mass flow averaged over the cross section is equal to zero. This well corresponds to convection in astrophysical objects, like stars or giant planets. Contrary to the flows in Rayleigh-Bénard (RB) cells for which boundary layers over the top and bottom plates have a great influence on the global heat transfert, our experimental setup is designed to create a flow that is poorly influenced by the boundary layers. Theoretically, it can be modelled by turbulent convection in an infinitely long channel. Two recent works have triggered our curiosity about this situation: the study of cold avalanches in a quarrel pit, by Perrier et al. [1], and recent numerical studies [2, 3]. Related works [4, 5] study the buoyant mixing of miscible fluids in a vertical tube, with unstable density gradient. The ideal case, with its assumed translational invariance along the vertical direction, also asks for sligthly different definitions of the Nusselt (Nu) and the Rayleigh (Ra) numbers, when compared to the RB case. β being the average vertical temperature gradient:

$$Nu = \frac{Q}{\lambda\beta}, \qquad Ra = \frac{g\alpha\beta d^4}{\nu\kappa}, \tag{1}$$

where λ is the thermal conductivity of the fluid, and Q is the heat flux, α is the isobaric thermal expansion coefficient, ν is the kinematic viscosity, κ the heat diffusivity, and g the gravity acceleration. d is a characteristic length, which could be the transverse size of the channel.

2 Experimental Setup

In a traditionnal rectangular RB cell of height 40 cm and section 40×10 cm^2, we narrow the flow in a channel of height 20 cm and section 10×10 cm^2

by means of a PVC structure, as schematically represented on figure 1(a). The fluid is water and the walls of the RB cell are 25mm thick PMMA. Measurements are made by four thermistors located at the corners of a rectangle (10 cm high, 5 cm wide) situated at the center of the channel. The resistors are connected as a Wheaston bridge, in such a way that the voltage output is proportional to the temperature difference between the lower thermistors and the upper ones. (sensitivity 2 mVK^{-1}). The cell is thermally insulated and surrounded by a cooper screen. The screen is temperature regulated at the average temperature between top and bottom plates, which is maintained constant on a whole set of measurements. The top plate is cooled through a regulated water bath. The bottom plate is heated by Joule effect in 5 resistors, regularly spaced. Heat losses have been roughly estimated before the runs with the empty cell. Three different global informations can be obtained from recording the output of the bridge (acquisition frequency: 1 Hz). The average on 24 hours recording gives the average gradient in the channel. The root mean squared amplitude of fluctuations gives a characteristic temperature amplitude θ. Finally, the power spectrum of the fluctuations shows a characteristic cut-off frequency f_c, or time $\tau_c = 1/2\pi f_c$. This time is interpreted as a large scale correlation time, giving information on the velocity of thermal structures. Using the width of the channel ($d = 10$ cm) as a characteristic length and the measured temperature gradient β, we calculate the Nusselt and the Rayleigh numbers (Eq. 1), where Q is the difference between the power supplied and the heat losses, divided by the cross section area

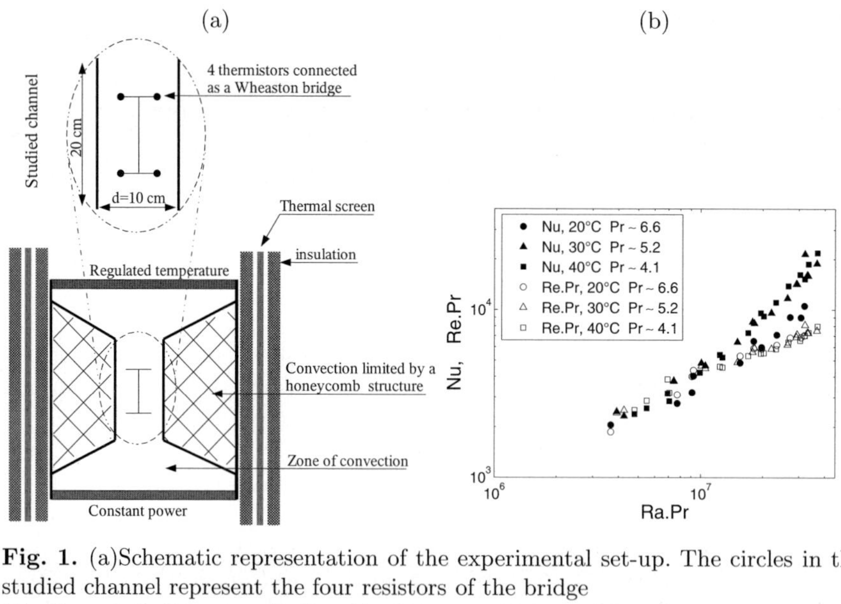

Fig. 1. (a)Schematic representation of the experimental set-up. The circles in the studied channel represent the four resistors of the bridge
(b) Nu and $RePr$ versus $RaPr$. d is chosen as the typical length

of the channel (10^{-2}m^2). We also define the Reynolds number : $Re = \frac{d^2}{\nu\tau_c}$. Three sets of measurements were made, at different average temperatures, 20°C, 30°C and 40°C. They correspond to three different Prandtl numbers $Pr(20^o\text{C}) = 6.6$, $Pr(30^o\text{C}) = 5.2$ and $Pr(40^o\text{C}) = 4.1$. The Prandtl number is the ratio of the kinematic viscosity ν to the heat diffusivity κ.

3 Results

Figure 1(b) shows both Nu and $RePr$ versus $RaPr$. While the three sets of values correctly merge for $RePr$, they do not for Nu. This is particularly clear for the highest Ra values, for which experimental uncertainties are negligible. Thus, while the $RePr$ versus $RaPr$ convergence seems to indicate that the kinematic viscosity has no influence (pure inertial behaviour), the Nu results seems to push the opposite way. However, there exists no fundamental reason imposing d as the correlation length. Indeed, the root mean square of the bridge output (θ) has the meaning of a temperature only if the correlation length of the temperature fluctuations is smaller than the distance between the upper and lower thermistors (10cm). Otherwise, θ would represent a gradient rms. Assuming a short correlation length, we define a natural length $L = \theta/2\beta$. Moreover, identifying this length L with the correlation length gives a value for the typical velocity: $v = L/\tau_c$. Using PIV measurements facilities, we made a direct observation of the flow without the thermal screen. For this measure, we worked with the mean temperature of the fluid being equal to the room temperature (30°C), and a relatively high heat power (200W) much bigger than losses. The observed rms velocity (0.5cm/s) is in close agreement with the value of v for the same heat flux and mean temperature, which fully justifies our interpretation. Finally, the measured values of L for our whole sets of experiments are less than 7 cm which is less than the height between thermistors (10 cm). Using L as characteristic length, we define natural Reynolds and Rayleigh numbers Re_n and Ra_n:

$$Re_n = \frac{L^2}{\nu\tau_c} = Re\left(\frac{L}{d}\right)^2, \qquad Ra_n = \frac{g\alpha\beta L^4}{\nu\kappa} = Ra\left(\frac{L}{d}\right)^4. \tag{2}$$

Using these natural values further improves the merging of the three sets of data in a Re_nPr versus Ra_nPr plot, as shown in figure 2(a). Moreover, as shown in figure 2(b), these three sets reasonably merge for Nu versus Ra_nPr, confirming the inertial character of the convection. The slope of the loglog plot of Re_nPr versus Ra_nPr is very close to 0.5, in agreement both with expectations and with numerical results [3]. The corresponding slope for Nu is closer to 0.6. However, the difference with 0.5 cannot be certified. We present on the graph an estimation of error bars due to a 1 W uncertainty on the heat losses. This merging of data suggests that all the influence of the viscosity (and thus of Re_n) is hidden in the behaviour of L. If we look at the

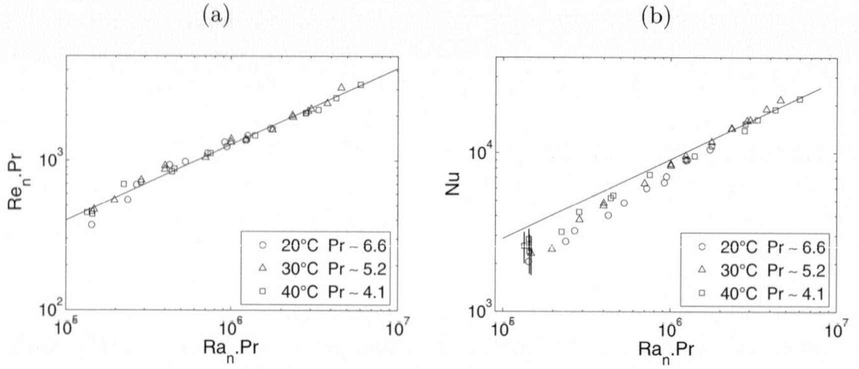

Fig. 2. (a)$Re_n Pr$ versus $Ra_n Pr$ with Re_n and Ra_n based on the coherence length L, the experiments at different Prandtl merge. The line corresponds to $1.1\sqrt{Ra_n Pr}$ (b) Nu versus $Ra_n Pr$. The line corresponds to $9.1\sqrt{Ra_n Pr}$

dependence L with Re_n (not shown here), all data seem to merge on a single curve. It shows a kind of cross-over for a Reynolds number of order 250. For $Re_n < 250$, the length L is constant. For $Re_n > 250$, the length L seems to take a linear behaviour with $\ln Re_n$. This linear dependance together with the relation $Re_n Pr \propto \sqrt{Ra_n Pr}$ lead to $v \propto \sqrt{g\alpha\beta d^2} \ln Re_n$. This logarithmic correction could be related to the logarithmic boundary layer which develops along the walls and which are not present in the numerical experiment [3].

4 Conclusion

To conclude, the agreement between the present experimental results and simple ideas on inertial convection can only be obtained if we consider that $L = \theta/2\beta$ is the natural coherence length controling the convection. In our experiment, it indeed appears as a coherent time. The existence of this self-constructed vertical length L, different from the imposed horizontal one d, is an important result of this experiment, with eventual consequences for other examples of free homogeneous inertial convection.

References

1. Perrier F, Morat P, Le Mouel JL, Phys. Rev. Letters **89** 134501 (2002)
2. Lohse D, Toschi F, Phys. Rev. Letters **90** 034502 (2003)
3. Calzavarini E, Lohse D, Toschi F, Tripiccione R, Phys. Fluids **17** 055107 (2005)
4. Debacq M, Fanguet V, Hulin JP, Salin D, Perrin B, Phys. Fluids **13** 3097 (2001)
5. Arakeri J.H., Avila F.E., Dada J.M., Tovar R.O., http://eprints.iisc.ernet.in/archive/00001685/01/convection.pdf

Applying the Leray-α Model to Rayleigh-Bénard Convection

M. van Reeuwijk[1], H.J.J. Jonker[1], and K. Hanjalić[1]

Department of Multi-Scale Physics and J.M. Burgers Center for Fluid Dynamics
Delft University of Technology, Lorentzweg 1, 2628 CJ Delft, The Netherlands
m.van.reeuwijk@ws.tn.tudelft.nl

Summary. The Leray-α model is applied to Rayleigh-Bénard convection. Profiles of velocity variance, spectra and budgets of turbulent kinetic energy are shown for a reference DNS at $Ra = 10^5$, $Pr = 1$ and three filter sizes α. The results show that the model seriously overpredicts the velocity variances, especially for large filter sizes. This overprediction may be caused by the isotropic filtering of the boundary layers, the free-slip boundary conditions or unanticipated side-effects of the modification of the energy-cascade.

The Leray-α model [1] is a promising method for simulating three-dimensional turbulent flows on a relatively coarse grid. The model is inspired by the Lagrangian averaged Navier-Stokes-α (LANS-α) model [2], which can be derived using variational principles from a Lagrangian that has been averaged along fluid particle trajectories. The governing equations are

$$\partial_t u_i + \widetilde{u}_j \partial_j u_i = \nu \partial_j^2 u_i - \partial_i p + f_i, \tag{1}$$

$$\partial_i u_i = 0, \tag{2}$$

$$\widetilde{u}_i - \alpha^2 \partial_j^2 \widetilde{u}_i = u_i, \tag{3}$$

with u_i the velocity, p the pressure, ν the kinematic viscosity and f_i a body force. The rationale of the Leray-α model is to introduce a second, smoother velocity field $\widetilde{u}_i$ (resembling the Lagrangian average velocity) that advects the fluid, thereby reducing the nonlinearity of the Navier-Stokes equations. The smoothed velocity $\widetilde{u}_i$ is obtained by applying a smoothing filter (3) with filter size α to the unfiltered velocity u_i. This principle has recently been proposed as a regularization model for large-eddy simulation [3] that allows a systematic derivation of the implied subgrid model. Note that the Leray-α model is purely dispersive in character i.e it is non-dissipative.

The action of the filter (3) is roughly to damp wavenumbers $k > \alpha^{-1}$. For $\alpha k \ll 1$, $\widetilde{u} \approx u_i$ and nonlinear interactions are not affected. However, for $\alpha k \gg 1$, the two velocity scales slowly decouple and the unfiltered velocity u_i is advected as a passive scalar by $\widetilde{u}_i$. The energy cascade is modified and downscale transfer is the only mechanism to generate variance at scales $k > \alpha^{-1}$. It can be shown that the energy-density function in terms of the unfiltered velocity is not modified for $\alpha k \ll 1$, scaling as $E_{uu} \propto \epsilon^{2/3} k^{-5/3}$. For $\alpha k \gg 1$, the cascade slows down, resulting in a scaling as $E_{uu} \propto \epsilon^{2/3} k^{-1/3}$. Due to the change of slope of the spectrum, more variance will be present at the high wavenumbers, resulting in a move of the dissipation range towards lower wavenumbers, and thus enhanced computability.

So far the Leray-α model has been applied to isotropic homogeneous turbulence only [1]. Therefore, the aim of this paper is to asses the Leray-α model for a wall-bounded flow, namely Rayleigh-Bénard convection. Rayleigh-Bénard convection (R-B) comprises the fluid flow between two flat plates that is generated by heating the bottom plate and cooling the top plate. The system can be characterized by the Prandtl number $Pr = \nu\kappa^{-1}$ and the Rayleigh number $Ra = \beta g \Delta\Theta H^3 (\nu\kappa)^{-1}$, and reacts by convective motion which is characterized by the Reynolds number $Re = UH\nu^{-1}$ and by an enhanced heat transfer through the Nusselt number $Nu = \phi H(\kappa\Delta\Theta)^{-1}$. Here U is a characteristic velocity and ϕ the heat-flux. Both Re and Nu are non-trivial functions of Ra and Pr and are still the subject of ongoing research [4]. Fixed temperature and no-slip velocity boundary conditions are enforced on the top and bottom plates. Free-slip boundary conditions are applied for $\widetilde{u}$, which may seem counter-intuitive but is essential for maintaining a divergence-free $\widetilde{u}$ field [5].

Numerical simulations are performed at $Ra = 10^5$ and $Pr = 1$ for a $\Gamma = L/H = 4$ aspect ratio domain with $H = 0.15$ m, $\Delta\Theta = 2$ K, $\beta = 1.74 \times 10^{-4}$ K^{-1} and $g = 9.81$ ms^{-2}. For these parameters, the typical convective turnover-time $t^* = 44$ s and statistics are collected for 20 turnovers after the statistically steady state situation has been reached. Periodic boundary conditions are applied for the sidewalls. A mesh of $64 \times 64 \times 64$ cells is used, which is of sufficient resolution for direct numerical simulation (DNS). Introducing the non-dimensional filter size $\hat{\alpha} = \alpha/H$, the DNS results will be compared to Leray-α results with three filter sizes: $\hat{\alpha} = 0.05, 0.10$ and 0.15.

Comparing the Leray-α results for to the reference DNS, a significant growth of the mean squared vertical velocity fluctuations $\overline{w'^2}$ (Fig. 1a) occurs as $\hat{\alpha}$ increases. For $\hat{\alpha} = 0.15$ and in the bulk of the flow, $\overline{w'^2}$ increases by a factor 2, which is quite surprising, as one could expect that decreasing the non-linearities of the Navier-Stokes equations would decrease the fluctuations. When making a spectral decomposition of $\overline{w'^2}$ in the bulk of the flow (Fig. 1b), it can be seen that the increase is even more dramatic as it is mainly the variance at the low wavenumbers which is overpredicted by more than 100%.

The equation of turbulent energy $e = \frac{1}{2}\overline{u_i' u_i'}$ of the unfiltered velocity can be obtained by multiplying (1) by u_i, averaging, and using that $\overline{u}_i = \overline{\widetilde{u}}_i = 0$ This results in

$$\underbrace{\beta g \overline{w'\Theta'}}_{\mathcal{P}} - \underbrace{\nu\overline{(\partial_j u_i')(\partial_j u_i')}}_{\varepsilon} = \underbrace{\partial_z(\overline{\widetilde{w}'e'} - \nu\partial_z e + \overline{w'p'})}_{\mathcal{T}}, \tag{4}$$

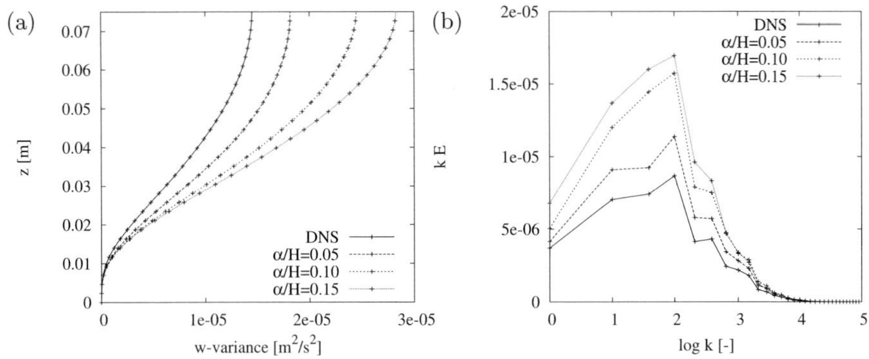

Fig. 1. Results for $Ra = 10^5$ and $Pr = 1$ for three values of $\hat{\alpha}$ and as a function of z. a) variance of vertical velocity component $\overline{w'^2}$; b) spectrum of $\overline{w'^2}$ at $z/H = 1/2$

where $e' = \frac{1}{2}u_i'u_i'$; $\mathcal{P}$, ε and $\mathcal{T}$ represent production, destruction and transport of turbulent kinetic energy respectively. Note that there is no production of turbulent kinetic energy by shear and that the only effect of the filtering is in a modified transport of the velocity fluctuations $\overline{\widetilde{w}'e'}$. The budget of e has been analyzed in [6, 7] and only the effect of $\hat{\alpha}$ will be discussed here. Shown in Fig. 2 are the production and dissipation terms for the DNS and for the two filter sizes $\hat{\alpha}$. It can be seen that the effect of $\hat{\alpha}$ is to enhance both $\mathcal{P}$ and ε. For $\hat{\alpha} = 0.15$, the production is overestimated by about 50% in the bulk of the flow and the dissipation by nearly 80% at the wall.

At this point it is not clear what is the mechanism for the increase of the variances. Focusing on the boundary layer, it may be that the filtering in this

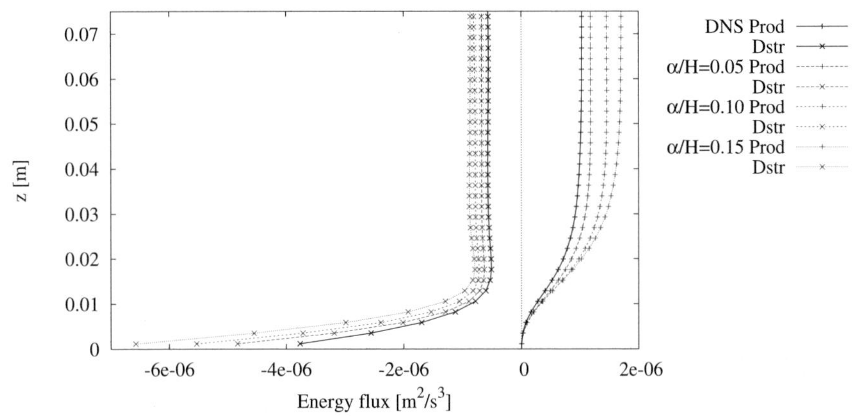

Fig. 2. production $\mathcal{P}$ (Prod) and dissipation ε (Dstr) of turbulent kinetic energy as a function of z

region causes the problem, as it is well known that properly resolving the boundary layers is crucial to proper simulation of R-B. Filtering -which can roughly be compared to using a coarser mesh - will therefore have adverse effects on the results, specifically as the filter (3) is isotropic, and of comparable size as the boundary layers. Secondly, it may be that the use of free-slip conditions for $\widetilde{u}_i$ changes the near-wall dynamics, although it seems to us that the isotropic filter is more likely to cause trouble. Focusing on a bulk-mechanism, it may that the modification of the energy-cascade has unanticipated side-effects, perhaps effectively trapping the flow on the large scales in some way. The mechanism behind the increased variances needs more research and will be addressed in future work.

This work is part of the research programme of the Stichting voor Fundamenteel Onderzoek der Materie (FOM), which is financially supported by the Nederlandse Organisatie voor Wetenschappelijk Onderzoek (NWO). The computations were sponsored by the Stichting Nationale Computerfaciliteiten (NCF).

References

1. A. Cheskidov, D.D. Holm, E. Olson, and E.S. Titi. On a Leray-alpha model of turbulence. *Proc. R. Soc. Lond. Ser. A Math. Phys. Eng. Sci.*, 461(2055): 629–649, 2005.
2. C. Foias, D.D. Holm, and E.S. Titi. The Navier-Stokes-alpha model of fluid turbulence. *Physica D*, 152-153:505–519, 2001.
3. B.J. Geurts and D.D. Holm. Regularization modeling for large-eddy simulation. *Phys. Fluids*, 15(1):L13–L16, 2003.
4. R.M. Kerr. Rayleigh number scaling in numerical convection. *J. Fluid Mech.*, 310:139–179, 1996.
5. M. van Reeuwijk, H.J.J. Jonker, and K. Hanjalić. Incompressibility of the Leray-α model for wall-bounded flows. *Phys. Fluids*, 18:018103, 2006.
6. J. A. Domaradzki, W. Liu, C. Hartel, and L. Kleiser. Energy-transfer in numerically simulated wall-bounded turbulent flows. *Phys. Fluids*, 6(4):1583–1599, 1994.
7. R.M. Kerr. Energy budget in Rayleigh-Bénard convection. *Phys. Rev. Lett.*, 87(24):244502, 2001.

Analysis of Thermal Dissipation Rates Based on Direct Numerical and Large–Eddy Simulations of Turbulent Rayleigh–Bénard Convection

Olga Shishkina and Claus Wagner

DLR - Institute for Aerodynamics and Flow Technology,
Bunsenstrasse 10, 37073 Göttingen, Germany
Olga.Shishkina@dlr.de, Claus.Wagner@dlr.de

Two functions σ and τ are defined to determine quantitatively the role of the turbulent background in thermal convection. Evaluating these functions from the data generated in the Direct Numerical Simulations (DNS) and Large-Eddy Simulations (LES) of turbulent Rayleigh–Bénard convection in wide cylindrical containers it is shown that the turbulent background contribution to the volume averaged thermal dissipation rate increases with the Rayleigh number.

1 Introduction

Turbulent Rayleigh–Bénard convection, i.e. the thermally driven fluid motion between a lower heated horizontal plate and an upper cooled plate, has been the subject of many fundamental investigations [1]. Most of the flow characteristics in Rayleigh–Bénard convection strongly depend on the Rayleigh (Ra) and Prandtl (Pr) numbers and the aspect ratio (Γ). Studying these dependencies, Grossmann & Lohse [2–3] analysed the thermal dissipation rate and suggested to investigate separately the turbulent background and the thermal boundary layers together with the plumes. Considering the thermal plumes as detached boundary layers, the authors split the volume averaged thermal dissipation rate into two contributions. The first is the thermal dissipation rate due to the plumes together with the smooth parts of the boundary layers and the second is the thermal dissipation rate of the turbulent background. According to the Grossmann–Lohse ansatz the turbulent background part of the thermal dissipation rate must dominate for large Ra. In order to check this hypothethis we developed a new method [4] to investigate the turbulent background contribution.

2 Numerical Experiments

The governing dimensionless equations for the Rayleigh–Bénard problem in Boussinesq approximation can be written as follows

$$\mathbf{u}_t + \mathbf{u} \cdot \nabla \mathbf{u} + \nabla p = \Gamma^{-3/2} Ra^{-1/2} Pr^{1/2} \nabla^2 \mathbf{u} + T\mathbf{z}, \quad \nabla \cdot \mathbf{u} = 0, \tag{1}$$

$$T_t + \mathbf{u} \cdot \nabla T = \Gamma^{-3/2} Ra^{-1/2} Pr^{-1/2} \nabla^2 T \tag{2}$$

with $\mathbf{u}$ the velocity vector, T the temperature, $\mathbf{u}_t$ and T_t their time derivatives and p the pressure, $Ra = \alpha g H^3 \Delta T / (\kappa \nu)$, $Pr = \nu / \kappa$, $\Gamma = D/H$ with H the height and D the diameter of the cylindrical container. Further, α is the thermal expansion coefficient, g the gravitational acceleration, ΔT the temperature difference between the bottom and the top plates, ν the kinematic viscosity and κ the thermal diffusivity. The dimensionless temperature varies between +0.5 at the bottom plate and −0.5 at the top plate. An adiabatic lateral wall is prescribed by $\partial T / \partial r = 0$ and on the solid walls the velocity field vanishes according to impermeability and no-slip conditions.

The data base used in our analysis was generated in DNS and LES. For the well resolved LES we used the tensor-diffusivity subgrid scale model [5]. The simulations of turbulent Rayleigh–Bénard convection in cylindrical containers filled with air ($Pr = 0.7$) were conducted for $\Gamma = 10$ and the Rayleigh numbers 10^5, 10^6, 10^7 (DNS) and for $\Gamma = 5$ and the Rayleigh numbers 10^5, 10^6, 10^7 (DNS) and 10^8 (LES) and were performed with the fourth order accurate finite volume method developed for solving (1), (2) in cylindrical coordinates (z, φ, r) on staggered non-equidistant grids. For details on the method, namely the fourth order discretization, the calculation of the velocity field at the cylinder axis, the hybrid explicit/semi-implicit time stepping and the numerical stability of the method we refer to Shishkina & Wagner [6,7]. The computational mesh consists of (110, 512, 192) nodes clustered in the vicinity of the rigid walls and is fine enough to resolve all relevant turbulent scales of the flows in the DNS for Ra up to 10^7 and in the LES for $Ra = 10^8$ [4].

3 Thermal Dissipation Rate Analysis

Following Grossmann & Lohse [3], we associate the thermal boundary layers and plumes with large values of the thermal dissipation rate

$$\epsilon_\theta = \Gamma^{-3/2} Ra^{-1/2} Pr^{-1/2} (\nabla T)^2. \tag{3}$$

In their theory ϵ_θ is split into a plumes-plus-boundary-layers and a background components. The former can dominate for lower and the latter must dominate for higher Rayleigh numbers.

In order to check this we define the functions $\tau(\xi)$ and $\sigma(\xi)$ according to (4) and evaluate them from the DNS and LES data. The function $\tau(\xi)$ describes the percentage of the fluid volume, for which ϵ_θ does not exceed $\xi \times 100\%$

of its maximum value $\epsilon_{\theta,\max} = \max_V \epsilon_\theta$. The function $\sigma(\xi)$ describes the contribution to the volume averaged ϵ_θ from those parts of the domain, where ϵ_θ does not exceed $\xi \times 100\%$ of its maximum. Namely,

$$\tau(\xi) = \langle \delta(\xi) \rangle_V \,, \qquad \sigma(\xi) = \frac{\langle \delta(\xi)\epsilon_\theta \rangle_V}{\langle \epsilon_\theta \rangle_V}, \tag{4}$$

where $\delta(\xi)$ is the threshold-function,

$$\delta(\xi) = 1, \text{ if } \frac{\epsilon_\theta}{\epsilon_{\theta,\max}} \leq \xi, \quad \text{and} \quad \delta(\xi) = 0 \quad \text{otherwise.}$$

In Figs. 1 and 2 the functions $\tau(\xi)$ and $\sigma(\xi)$ are plotted as they were calculated from the DNS/LES data. From Fig. 1 (left) it is concluded, for example, that those parts of the domain, where the thermal dissipation rate is less than 0.1% of its maximum value (i.e. $\xi = 10^{-3}$), equals 25.3% ($Ra = 10^5$), 57.4% ($Ra = 10^6$) and 85.3% ($Ra = 10^7$) of the whole domain, while Fig. 1 (right) reveals that these parts contribute 0.4% ($Ra = 10^5$), 2.1% ($Ra = 10^6$) and 8.9% ($Ra = 10^7$) to the volume averaged thermal dissipation rate.

Since the turbulent background is indicated by lower values and the plumes together with the boundary layers by higher values of the thermal dissipation rate, there exists a certain value of ξ, which separates these two regions. Whatever the exact value of ξ is, the values of $\tau(\xi)$ and $\sigma(\xi)$ obtained for a certain Ra are always higher than the corresponding values for a lower Ra, as it is seen in Figs. 1 and 2. Thus, the results presented above show that both the portion of the whole domain, which corresponds to the turbulent background, and the background contribution to the volume averaged thermal dissipation rate increase with Ra. This is true at least for the wide containers

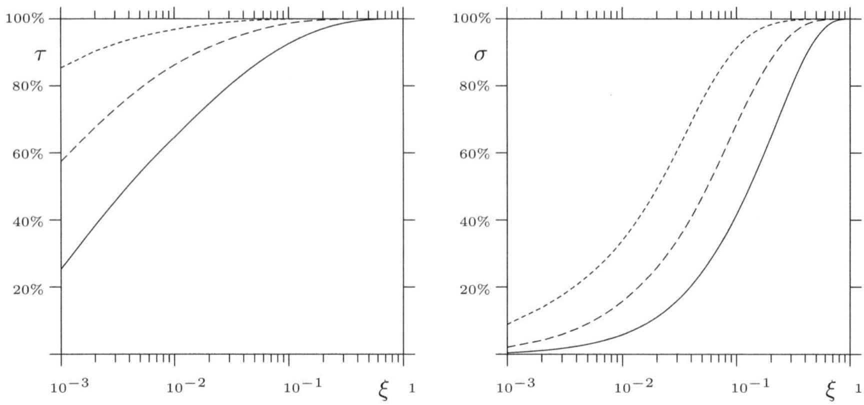

Fig. 1. Portion of the domain, where $\epsilon_\theta(\mathbf{x}) \leq \xi\epsilon_{\theta,\max}$ (left) and contribution to the volume averaged thermal dissipation rate from the parts of the domain, where $\epsilon_\theta(\mathbf{x}) \leq \xi\epsilon_{\theta,\max}$ (right), evaluated in the DNS for $Ra = 10^5$ (—), $Ra = 10^6$ (– –), $Ra = 10^7$ (- - -) and $\Gamma = 10$

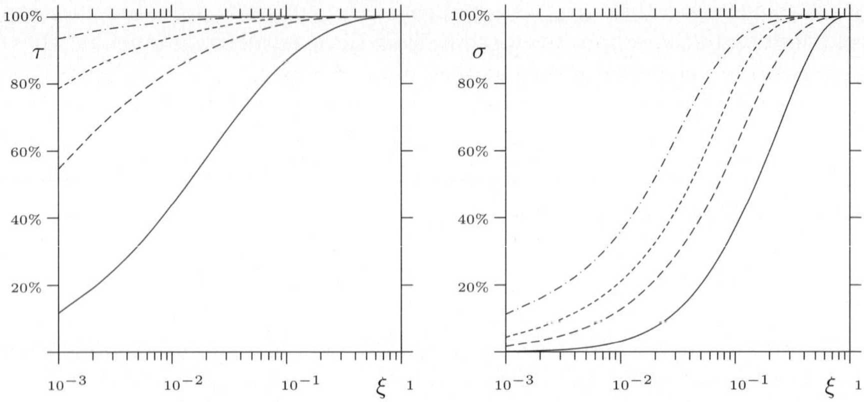

Fig. 2. Portion of the domain, where $\epsilon_\theta(\mathbf{x}) \le \xi\epsilon_{\theta,\max}$ (left) and contribution to the volume averaged thermal dissipation rate from the parts of the domain, where $\epsilon_\theta(\mathbf{x}) \le \xi\epsilon_{\theta,\max}$ (right), evaluated in the DNS for $Ra = 10^5$ (—), $Ra = 10^6$ (– –), $Ra = 10^7$ (- - -) and in the LES for $Ra = 10^8$ (– · –) and $\Gamma = 5$

and moderate Rayleigh numbers considered and supports the conjecture by Grossmann & Lohse [3] that the background part of the thermal dissipation rate dominates for large *Ra*.

Comparing the results obtained for different aspect ratios Γ (Figs. 1 and 2), we conclude that the turbulent background contributes more in the higher aspect ratio case.

Further, simulating Rayleigh–Bénard convection in a slender cylindrical domain, Verzicco & Camussi [8] observed that the contribution of the horizontal boundary layers (without plumes) to the volume averaged thermal dissipation rate increases with *Ra*. As shown above, the contribution of the turbulent background also increases. Therefore we conclude that the role of the plumes in thermal convection decreases with increasing Rayleigh number.

The authors gratefully acknowledge the financial support from the German Research Foundation under the contract WA 1510-1 and the interesting discussions with Prof. S. Grossmann during the Conference iTi2005 in Bad-Zwischenahn.

References

1. Ahlers G, Grossmann S, Lohse D (2001) Physik Journal 1:31–37
2. Grossmann S, Lohse D (2000) J Fluid Mech 407:27–56
3. Grossmann S, Lohse D (2004) Phys Fluids 16:4462–4472
4. Shishkina O, Wagner C (2006) J Fluid Mech: 546:51–60
5. Leonard A, Winckelmans G S (1999) Caltech ASCI tech. report cit-asci-tr043
6. Shishkina O, Wagner C (2004) Appl Numer Anal Comput Math 1:315–326
7. Shishkina O, Wagner C (2005) C R Mecanique 333:17–28
8. Verzicco R, Camussi R (2003) J Fluid Mech 477:19–49

Part VI

Scalar Mixing and Particles

Inertial Particles in Turbulence

L. Biferale[1], J. Bec[2], G. Boffetta[3], A. Celani[4], M. Cencini[5], A. Lanotte[6], S. Musacchio[7], and F. Toschi[8]

[1] Dept. Physics and INFN Univ. Tor Vergata, Via della Ricerca Scientifica 1, 00133, Rome, Italy biferale@roma2.infn.it

[2] CNRS Observatoire de la Côte d'Azur, B.P. 4229, 06304 Nice Cedex 4, France Jeremie.Bec@obs-nice.fr

[3] Dept. of Physics and INFN, University of Torino, Via Pietro Giuria 1, 10125, Torino, Italy boffetta@to.infn.it

[4] CNRS, INLN, 1361 Route des Lucioles, F-06560 Valbonne, France. antonio.celani@inln.cnrs.fr

[5] SMC-INFM c/o Dept. of Physics University of Rome "La Sapienza", Piazz.le A. Moro, 2, I-00185 Roma, Italy, and CNR-ISC via dei Taurini, 19 I-00185 Roma, Italy. massimo.cencini@roma1.infn.it

[6] CNR-ISAC, Sezione di Lecce, Str. Prov. Lecce-Monteroni km.1200, 73100 Lecce, Italy lanotte@roma2.infn.it

[7] Dept. of Physics, University of Rome "La Sapienza", Piazz.le A. Moro, 2, I-00185 Roma, Italy musacchio@roma1.infn.it

[8] CNR-IAC, Viale del Policlinico 137, 00161 Roma, Italy toschi@iac.cnr.it

In this contribution we review some results obtained from a direct numerical simulations of heavy particle transport in homogeneous, isotropic, fully developed turbulence, up to resolution 512^3 ($R_\lambda \approx 185$) [1]. We have followed the trajectories of up to 120 million particles with Stokes numbers, St, in the range from 0.16 to 3.5. We discuss the statistics of particle accelerations as a function of Stokes and as a function of the Reynolds numbers. We show that for small Stokes numbers the acceleration statistics is strongly influenced by particle clustering outside strong vorticity regions.

1 Heavy Particles Dynamics

Small impurities like dust, droplets or bubbles suspended in an incompressible flow are finite-size particles whose density may differ from that of the underlying fluid, and cannot thus be modelled as point-like tracers. The description of their motion must account for inertia whence the name *inertial particles.* At long times particles concentrate on singular sets evolving with the fluid motion, leading to the apparition of a strong spatial inhomogeneity dubbed *preferential concentration* (see [2, 3, 4, 5, 6, 7] for recent investigations).

Here we present a Direct Numerical Simulations (DNS) study of particles much heavier than the carrier fluid in high-resolution turbulent flows. In this limit, the equation of motion of particles take the particularly simple form [2, 12]:

$$\frac{d\mathbf{X}}{dt} = \mathbf{V}(t)\,, \qquad \frac{d\mathbf{V}}{dt} = -\frac{\mathbf{V}(t) - \mathbf{u}(\mathbf{X}(t), t)}{\tau_s}\,. \tag{1}$$

Here, $\mathbf{X}(t)$ denotes the particle trajectory, $\mathbf{V}(t)$ its velocity, $\mathbf{u}(\mathbf{x}, t)$ is the fluid velocity. The Stokes response time is $\tau_s = 2\rho_p a^2/(9\rho_f \nu)$ where a is the particle radius ρ_p and ρ_f are the particle and fluid density, respectively, and ν is the fluid kinematical viscosity. The Stokes number is defined as $St = \tau_s/\tau_\eta$ where $\tau_\eta = (\nu/\epsilon)^{1/2}$ is the Kolmogorov timescale and ϵ the average rate of energy injection. Equation (1) is valid for very dilute suspensions, where particle-particle interactions (collisions) and hydrodynamic coupling are not taken into account. The underlying fluid evolves according to the incompressible Navier-Stokes equations.

In [1] we have reported some results obtained by performing a DNS by solving Navier-Stokes equations on a cubic grid of size N^3 for $N = 128, 256, 512$ with periodic boundary conditions. Energy is injected by keeping constant the spectral content of the two smallest wavenumber shells. The viscosity is chosen so to have a Kolmogorov lengthscale $\eta \approx \Delta x$ where Δx is the grid spacing. Particles are seeded in the flow with homogeneously distributed initial positions and velocities equal to the local fluid velocity. We followed 15 sets of inertial particles with Stokes numbers in the range $[0.16 : 3.5]$. For each set, we dumped the position and the velocity of N_t particles every $dt = 1/10\tau_\eta$ with a maximum number of recorded trajectories of $N_t = 5{\cdot}10^5$ for the highest resolution. Along these trajectories we also stored the velocity of the carrier fluid. We have also followed fluid tracers ($St = 0$), that evolve according to the dynamics

$$\frac{d\mathbf{x}(\mathbf{t})}{dt} = \mathbf{u}(\mathbf{x}(\mathbf{t}), \mathbf{t})\,, \tag{2}$$

in order to systematically assess the importance of the phenomenon of preferential concentration at varying both St and R_λ.

2 Results

We mainly focus here on the statistics of particle acceleration $\mathbf{a}(t) = \frac{d\mathbf{V}}{dt}$. From previous studies on fluid tracers we know that acceleration statistics is very intermittent and strong fluctuations are associated to trapping events within vortical filaments ([9, 11, 13]). How does inertia impacts acceleration statistics? One may easily realize that in the limit $St \to 0$ the particle acceleration becomes indistinguishable from the acceleration of the underlying flow, the only difference between the two statistics may arise from a different sampling of the phase space induced by inertia. As soon as inertia is active, we know

Fig. 1. Snapshot of particle positions in a plane for small Stokes number, $St = 0.16$. The greyscale codes the modulus of the vorticity field. Notice that particles escape from high vorticity regions

that heavy particle tends to be expelled out of intense vorticity regions. We expect therefore that in the above limit the particle acceleration is identically to the fluid acceleration except for this "preferential concentration effect". On the other hand, in the limit of large Stokes, $St >> 1$, the particle velocity become more and more decorrelated with the underlying fluid velocity, particles tends to move ballistic inside the flow, the main effect on particle acceleration comes from the smoothing of high frequency induced by the Stokes drag force in (1). These issues are discussed in Figs. 2a and 2b. In these figures we show the root mean squared acceleration as a function of Stokes and for the three different Reynolds numbers (panel a). In the inset of the same figure we also show the clusterization properties as measured by the fractal correlation dimension measured on the set of particles for different Stokes. The correlation dimension is defined as $p(r) \sim r^{D_2}$ (for $r \ll \eta$) where $p(r)$ is the probability to find two particles at distance smaller than r [10]. In Fig. 2b we compare the root mean squared acceleration measured on the set of particles at the highest Reynolds numbers with the fluid acceleration measured on the same position of the particles, $\langle (\frac{D\mathbf{u}}{Dt})^2 \rangle^{1/2}$, and with the acceleration measured by filtering the fluid tracers trajectories over a time lags of the order of τ_s for each different Stokes. The first comparison is meant to support the statement that for small Stokes the main effect which drives particle motion is the preferential concentrantion, i.e. no mismatch with the statistics of the underlying flow is

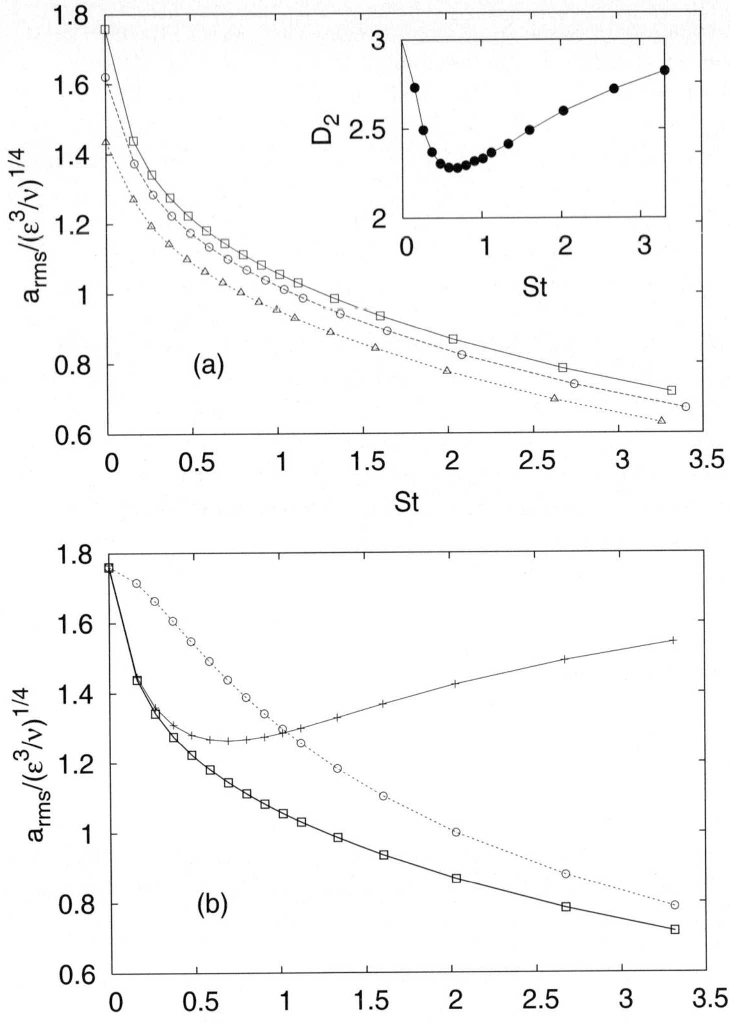

Fig. 2. (a) The normalised acceleration variance $a_{\mathrm{rms}}/(\epsilon^3/\nu)^{1/4}$ as a function of the Stokes number for $R_\lambda = 185$ (Squares); $R_\lambda = 105$ ($\circ$); $R_\lambda = 65$ ($\triangle$). The inhomogeneous distribution of particle is quantified for the highest Reynolds in the inset, where we plot the correlation dimension, D_2, as a function of St. (b) Comparison between the acceleration variance, a_{rms} (Squares), as a function of Stokes, with the acceleration of the fluid tracer measured on the particle position, $\langle (\frac{D\mathbf{u}}{Dt})^2 \rangle^{1/2}$ (+). The last curve ($\circ$), approaching the a_{rms} for large St, is the one obtained from the filtered tracers acceleration over a time lag of the order of τ_s –see [1] for more details. All data refer to $Re_\lambda = 185$

measured once the particle clustering is taken into account. Indeed the two curves coincide for small St. The second comparison is meant to show that in the other limit, $St \gg 1$, the main effect is just the Stokes filtering over a time lags of the order of τ_s. Indeed, in this second case the filtered acceleration tends to the true one only for $St \gg 1$.

An interesting aspect shown in Fig. 2a is the residual dependence of the normalized particle acceleration on Reynolds number. For the case of fluid tracers it is known that intermittent corrections to the dimensional estimate $a_{\rm rms} = a_0(\epsilon^3/\nu)^{1/2}$ may explain the Reynolds dependence ([8]). Data suggest that the fluid intermittency may be responsible of such deviations at $St > 0$ as well.

Besides the effect of inertia on typical particle accelerations it is also interesting to investigate the effects on the form of the probability distribution function $\mathbf{a}(t)$. As shown in Fig. 3a, the pdf's get less and less intermittent as St increases. In the inset of the same figure we show the behaviour of the flatness, $\langle \mathbf{a}^4 \rangle / \langle \mathbf{a}^2 \rangle^2$, as a function of St. The abrupt decreasing for $St > 0$ is even more evident here (notice that the y scale is logarithmic). Notice that the change in the shape of the normalized pdf means that intermittency affects acceleration statistics differently as a function of Stokes.

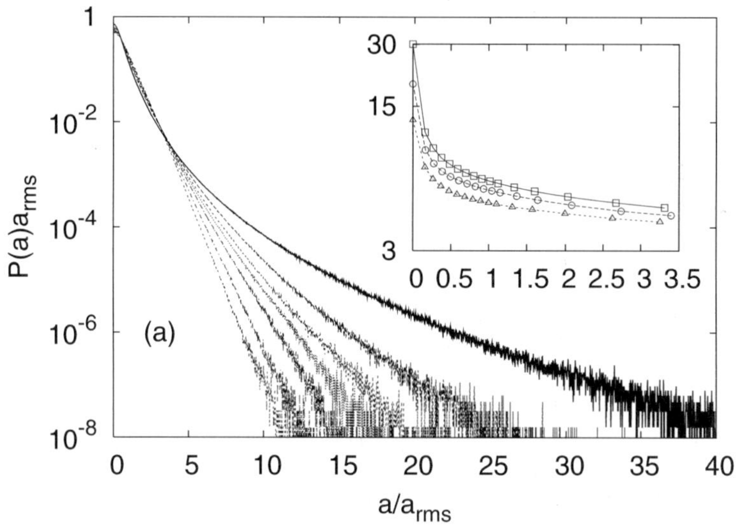

Fig. 3. (a) Acceleration pdf's for a subset of Stokes values ($St = 0, 0.16, 0.37, 0.58, 1.01, 2.03, 3.31$ from top to bottom) at $R_\lambda = 185$. The inset displays the acceleration flatness, $\langle \mathbf{a}^4 \rangle / \langle \mathbf{a}^2 \rangle^2$, at increasing R_λ from bottom to top. All curves are normalised to have unit variance

3 Conclusions

A systematic study of the acceleration statistics of heavy particles in turbulent flows, at changing both Stokes and Reynolds numbers has been reviewed [1]. We have shown that preferential concentration plays a *almost singular* role at small Stokes. Indeed, even a quite small inertia may suffice to expel particles from those turbulent regions (vortex cores) where the most intermittent and wild acceleration fluctuations would have been experienced; for small Stokes, a good quantitative agreement between the inertial particle acceleration and the conditioned fluid tracer acceleration is obtained; at large Stokes, the main effects is filtering of the velocity induced by the response Stokes times. For $St > 1$, the statistical properties of fluid tracers averaged over a time window of the order of τ_s are in a quite good agreement with the inertial particle properties.

One of the main open questions concerns with the possibility to build up a phenomenological approach that is able to describe the inertial particles acceleration as a function of both Stokes and Reynolds numbers. A naive generalization of the multifractal description, successfully used for fluid tracers ([8]), may be insufficient. In fact, it is not straightforward to include in such models the correlation between preferential concentration and the local topological properties of the carrier flow.

References

1. Bec J, Biferale L, Boffetta G, Celani A, Cencini M, Lanotte A, Musacchio S ans Toschi F. *Journ. Fluid Mech* submitted (2005). nlin.CD/0508012
2. Balkovsky E, Falkovich G, & Fouxon A, (2001) *Phys. Rev. Lett.* **86**, 2790–2793.
3. Bec J, (2003) *Phys. Fluids* **15**, L81-L85.
4. Bec J, Celani A, Cencini M, & Musacchio S. (2005) *Phys. Fluids* **17**, 073301.
5. Bec J, Gawedzki K, Horvai P, (2004) *Phys. Rev. Lett.* **92**, 224501.
6. Collins LR, Keswani A, (2004) *New J. Phys.* **6**, 119.
7. Eaton J.K, & Fessler (1994) *Int. J. Multiphase Flow* **20**, 169–209.
8. Biferale L, Boffetta G, Celani A, Devenish B.J, Lanotte A, & Toschi F. (2004) *Phys. Rev. Lett.* **93**, 064502.
9. Biferale L, Boffetta G, Celani A, Lanotte A, & Toschi F. (2005) *Phys. Fluids.* **17**, 021701.
10. Hentschel HGE, and Procaccia I, (1983) *Physica D* **8**, 435.
11. La Porta A, Voth, G.A, Crawford A.M, Alexander J, & Bodenschatz E. (2001) *Nature* **409**, 1017–1019.
12. Maxey M.R, & Riley J, (1983) *Phys. Fluids* **26**, 883–889.
13. Mordant N, Metz P, Michel O, & Pinton J.-P, (2001) *Phys. Rev. Lett.* **87**, 214501.

Turbulent Pair Dispersion: A PTV Experiment

Jacob Berg, Beat Lüthi, Jakob Mann and Søren Ott

Risø National Laboratory, 4000 Roskilde, Denmark

Particle Tracking Velocimetry (PTV) is an experimental technique used to obtain Lagrangian statistics in a turbulent flow[1, 2, 3]. Compared to the more widely used Particle Image Velocimetry (PIV) we track fluid elements in time and space and hence dispersion between more than one fluid element (particle hereinafter) can be observed. Dispersion of such particles is related to many applications and is governing the spreading of pollutants, combustion and prey-predator encounters in a turbulent sea, etc. In this paper we will focus on dispersion of two particles in a homogeneous and isotropic turbulent flow and present some preliminary results.

Dispersion of two particles in the inertial subrange where viscous effects can be neglected is described by the Richardson-Obukhov law [4],

$$\langle r(t)^2 \rangle = g\varepsilon t^3, \tag{1}$$

where $r(t)$ is the distance between particles at time t after the particles were initially close, $r(0) \equiv r_0$. ε is the kinetic energy dissipation. The law has been studied extensively [1, 5, 6, 8, 10] but still continues to puzzle. The law in its most definitive form has hardly been observed. We ask the question whether g is a function of r_0 in finite Reynolds number turbulence.

Passive tracer particles ($\sim 500\mu m$) are tracked in a turbulent water flow generated by eight rotating propellers placed in the corners of a tank with dimensions, $32 \times 32 \times 50 cm^3$. The data acquisition system consists of four commercial CCD cameras with a maximum recording rate of $50Hz$ at 1000×1000 pixels covering a measuring volume of $(15cm)^3$. With the current setup approximately 1000 particles are tracked simultaneously with a position accuracy of 0.05 pixels corresponding to less than $10\mu m$.

The propellers change direction in fixed intervals. The resulting large scale flow is slightly strained and axisymmetric while the smaller scales are isotropic. The large scale mean flow is, however, much smaller than the turbulent fluctuations, σ_u.

The flow characteristics are obtained by fitting the second order Eulerian longitudinal structure function to the corresponding structure function derived from the von Karman model [1]: $\varepsilon = 168mm^2/s^3$, and $\sigma_u = 23.3mm/s$, the integral scale, $L_{int} = 48mm$, Kolmogorov length and time scale, $\eta = 0.25mm$, $\tau = 0.07s$, and Taylor scale Reynolds number, $Re_\lambda = 172$ (Fig. 1 (a)).

In Fig. 1 (b) the second order Lagrangian structure functions are shown compensated for $C_0 = \langle(v_\alpha(t) - v_\alpha(t+\tau))^2\rangle/(\varepsilon\tau)$. The vertical component, which coincides with the symmetry axis, is the lowest curve in Fig. 1 (b). For the horizontal components the maximum value of C_0 plateaus at 5τ and $C_0 \sim 5$ in accordance with results from the simplest two time scale Lagrangian stochastic model taking finite Reynolds numbers into account [5].

Equation 1 is suppose to hold for all pairs with $\eta \ll r_0 \ll r(t) \ll L_{int}$. Figure 2 displays the mean square separation as a function of time for different millimeter-sized bins of initial separation, r_0. No clear scaling is observed for any of the bins as expected by (1). One reason could be that we have not taken the finite r_0 into account or simply that Re_λ is too low meaning that viscous effects at scales $\sim \eta$ and large scale effects at $\sim L_{int}$ are not well separated. Yeung [7] argued by extrapolating from DNS that true Lagrangian scaling can not be observed for $Re_\lambda < 700$.

To take the initial separation, r_0, into account we look at $\langle r(t)^2\rangle^{1/3}$ which according to (1) should scale linear with time. As shown in Fig. 2 (b) straight

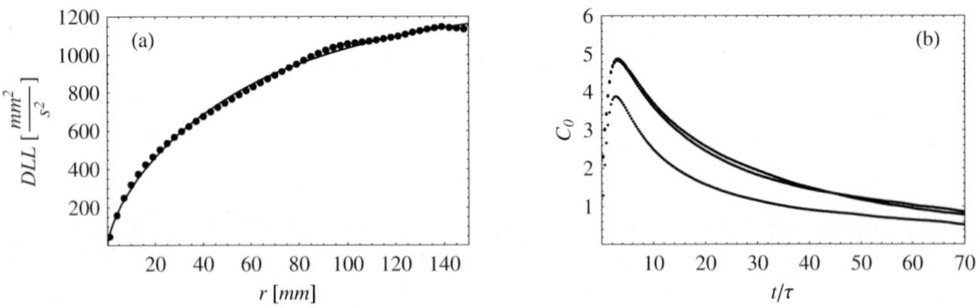

Fig. 1. (a) Eulerian second order structure function: Data (dots),von Karman model (solid). (b) Compensated Lagrangian second order structure functions

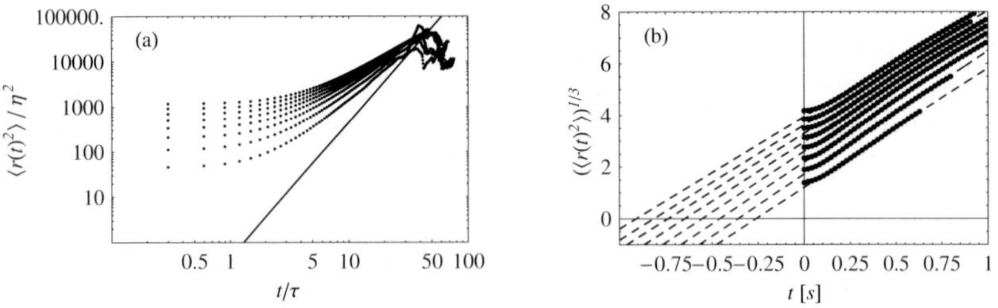

Fig. 2. (a) $\langle r(t)^2\rangle/\eta^2 = g(t/\tau)^3$. The slope of the straight line is three. The different curves correspond to different r_0. (b) $\langle r(t)^2\rangle^{1/3}$ with corresponding linear fits

lines fit the data and we can hence extrapolate the data back to the zero crossings , T_0, which we can understand as a time shift after which mechanisms of the initial separation has ceased out. The short time expansion, the ballistic regime, of the pair separation is also present and is reflected as the small kinks of the curves we observe at around $t = 0$.

By subtracting the zero crossings , T_0, from the time we observe in Fig. 3 (a) a collapse of the different curves corresponding to the different r_0.. The slope of the straight line is 0.55 ± 0.15. In Fig. 3 (b) a plot of $\langle r(t = 0)^2 \rangle$ versus $(-T_0)^3$ reveals a linear relationship with slope 0.5 ± 0.1. This value is thus a little smaller than the observed g of Fig. 3. This is expected because r_0 is higher than the intersect of $\langle r(t)^2 \rangle^{1/3}$ with the y-axis in Fig. 2 (b).

The distance neighbor function is the pdf of separations, r. Richardson proposed an exponential form based on the assumption of short time correlated velocities and self similar diffusion coefficient proportional to $r^{4/3}$. Figure 4 shows the functions for two different initial separations with the theoretical prediction by Richardson as the straight lines. While pairs with smallest initial separation certainly agree well with the theory the pairs with larger separation do not. This could be explained by the fact that we are not able to track pairs with large initial separation for long enough so that the final state can not be reached. On the other hand, it could also be indications of the failure of he Richardson diffusion model due to Lagrangian memory effects meaning that it is only well suited for pairs with small initial separations. Stochastic models as the one proposed by Borgas and Yeung [8] contain a physically richer description of the two pair separation problem.

By universal we mean that g should be independent of the forcing: Type and strength as well as r_0. Equation 1 is derived from inviscous K41 scaling, and g is only rigourously defined in this limit. g obtained from finite Reynolds number experiments can therefore be thought of only as asymptotic approaching values. Franzese and Cassiani [9] derived a theoretical Reynolds number dependent $g \equiv g(Re_\lambda)$ with asymptotic limit $g(\infty) = 0.64$.

Ott and Mann [1] found in an oscillating grid turbulent flow for $Re_\lambda \sim 100$, $g = 0.5$. This result has later been reproduced in DNS at $Re_\lambda = 280$

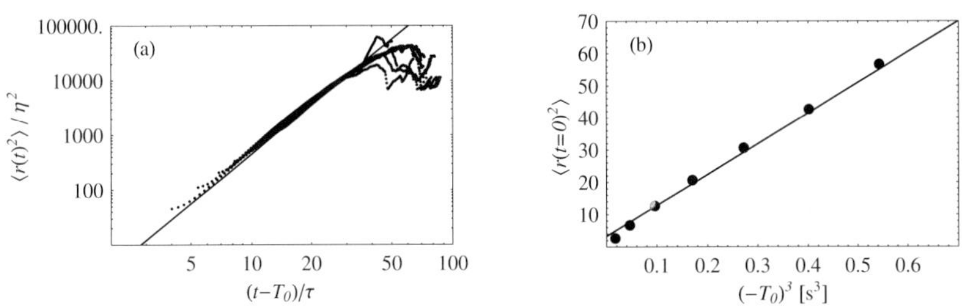

Fig. 3. (a) $\langle r(t)^2 \rangle$ as a function of $(t - T_0)/\tau$. The straight line is $0.47t^3$. Axes are in Kolmogorov units. (b) $\langle r(t = 0)^2 \rangle$ versus $(-T_0)^3$. The straight line is $0.45t^3$

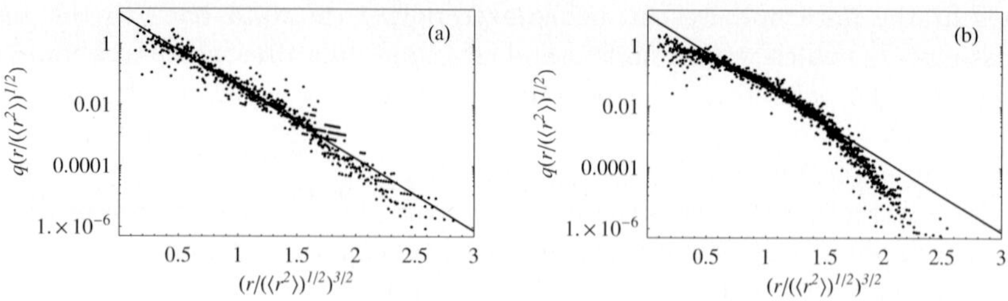

Fig. 4. (a) Distance neighbor function for initial separation, $2-3mm$. The straight line is the theoretical Richardson prediction. (b) Similar but for $8-9mm$

by Biferale et al.[6]. Our findings in this paper supports the findings in the earlier studies. In addition we have shown that g is independent of the initial separation which was not stressed in the earlier studies.

Bourgoin et al. [10] has at $Re_\lambda = 819$ reported a t^2 scaling depending on initial separations. This is the so called Batchelor scaling (the ballistic regime), for which the particle separation can fully be described by the Eulerian structure functions. The limiting time is the Batchelor time $t_0 = \varepsilon^{-1/3} r_0^{2/3}$. With increasing Reynolds number the inertial range grows meaning that r_0 can increase and therefore also t_0. The current study at a somewhat lower Reynolds number has, however, only focused on times larger than t_0.

References

1. Ott S and Mann J (2000) J. Fluid Mech. 422:207-223.
2. Lüthi B, Tsinober A and Kinzelbach W (2005) J. Fluid Mech. 528:87-118.
3. La Porta A, Voth G, Crawford A, Alexander J and Bodenschatz E (2001) Nature 409:1017-1019.
4. Monin AS and Yaglom AM (1975) Statistical Fluid Mechanics II, MIT Press.
5. Sawford BL (1991) Phys. Fluids A 3:1577-1588.
6. Biferale L, Boffetta G, Celani A, Devenish BJ, Lanotte A and Toschi F (2005) Phys. Fluids 17:021701.
7. Yeung PK (2002) Annu. Rev. Fluid Mech. 34:115-142.
8. Borgas MS and Yeung PK (2004) J. Fluid Mech. 503:125-160.
9. Franzese P and Cassiani M (2005) *submitted to J. Fluid Mech.*
10. Bourgoin M, Ouellette NT, Xu H, Berg J and Bodenschatz E (2005) *accepted for publication in Science*

Application of a Two-point Closure to the Dispersion of Particles in Isotropic and Sheared Turbulence

Wouter Bos and Jean-Pierre Bertoglio

Laboratoire de Mécanique des Fluides et d'Acoustique
Ecole Centrale de Lyon, Ecully, France

1 Introduction

Batchelor [1] was the first to formally describe the link between particle dispersion and the mixing of a passive scalar. Since then, several authors have used this link to address the problem of dispersion. For example, Larcheveque and Lesieur [2] used the Eddy Damped Quasi-Normal Markovian closure (EDQNM) for the scalar spectrum to study the relative dispersion of particle pairs. A recent review of dispersion and mixing can be found in Falkovich et al. [3]. In the present work it is shown that a recent extension of EDQNM to the case of homogeneous turbulence with a uniform mean scalar gradient can be applied to single-particle dispersion in both isotropic turbulence and homogenous shear flow.

2 Single Particle Dispersion in Isotropic Turbulence

We first consider an isotropic turbulent velocity field on which we impose a constant uniform mean temperature gradient Γ in the z-direction. The scalar is assumed non-diffusive so that the equation for the scalar fluctuation reads:

$$\frac{d\theta}{dt} = -\Gamma w', \tag{1}$$

with w' the z-component of the fluctuating velocity field. It can be easily seen by integrating the scalar equation over a particle trajectory that in this case the scalar fluctuation θ is directly related to the vertical separation distance z' of the particle from its initial position:

$$\theta = -\Gamma z' \tag{2}$$

The mean squared displacement then obeys the equation:

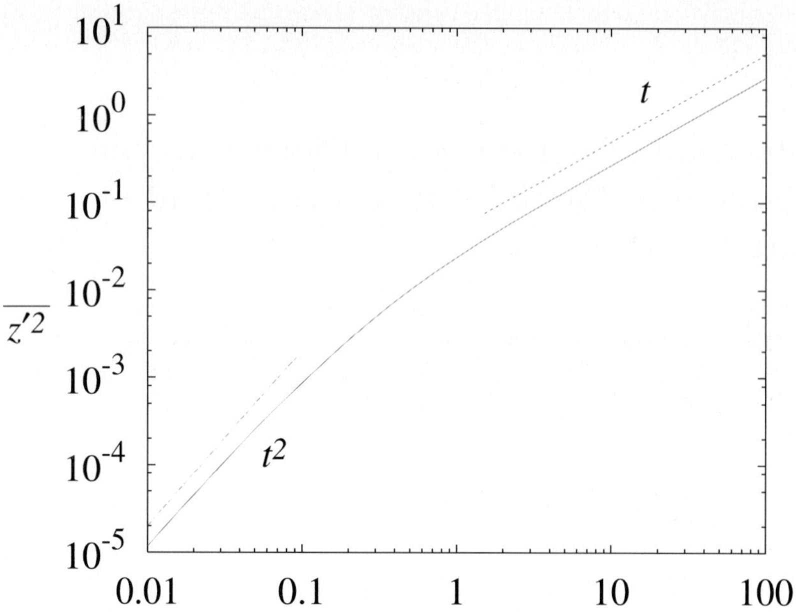

Fig. 1. *Ballistic and Brownian regimes of single particle dispersion*

$$\frac{\partial \overline{z'^2}}{\partial t} = 2\overline{wz'} = -2\Gamma^{-1}\overline{w\theta}, \tag{3}$$

in which the only unknown quantity is the scalar-velocity cross-correlation $\overline{w\theta}$. Using the EDQNM closure for the scalar-velocity cross-correlation spectrum [4, 5], $\overline{w\theta}$ can be evaluated. The resulting predictions for the mean squared displacement are found in good agreement with the classical dispersion theory of Taylor [6]. The results are shown in Fig. 1.

3 Dispersion in Homogeneous Shear Flow

Taylor predicted a t^3 behaviour for the horizontal dispersion in stationary homogeneous shear flow. While DNS is very limited to make large time or large Reynolds number predictions, a recent extension of the EDQNM theory to scalar mixing in homogeneous shear flow [7] allows to study the dispersion in horizontal and vertical direction for long times and virtually without limit in Reynolds number.

We consider homogeneous shear flow with a mean velocity gradient $S = \partial U/\partial z$. The vertical displacement obeys the same analogy with a non-diffusive scalar as in isotropic turbulence. The horizontal displacement is however explicitly affected by the mean velocity:

$$\frac{dx'}{dt} = u' + z'S, \tag{4}$$

with u' the x-component of the fluctuating velocity field. An additional term has to be added to the scalar equation to obtain a direct analogy. The equations for the scalars θ and χ corresponding to the vertical and horizontal displacements are now respectively:

$$\frac{d\theta}{dt} = -\Gamma_\theta w' \tag{5}$$

$$\frac{d\chi}{dt} = -\Gamma_\chi u' - \theta S \tag{6}$$

In these equations Γ_θ is a mean scalar gradient in the z-direction and Γ_χ is a mean scalar gradient in the x-direction. The equations for the mean squared displacement distance read:

$$\frac{d\overline{z'^2}}{dt} = 2\overline{w'z'} \tag{7}$$

$$\frac{d\overline{x'z'}}{dt} = \overline{u'z'} + \overline{w'x'} + \overline{z'^2}S \tag{8}$$

$$\frac{d\overline{x'^2}}{dt} = 2\overline{u'x'} + 2\overline{z'x'}S \tag{9}$$

(10)

The cross-correlations $\overline{u'z'}$ and $\overline{w'z'}$ are again treated as scalar fluxes, in the case of a vertical mean scalar gradient with value -1. The cross-correlations $\overline{u'x'}$ and $\overline{w'x'}$ are treated as scalar fluxes in the presence of a horizontal mean scalar gradient, taking into account the coupling term $(-\theta S)$ in (6). The

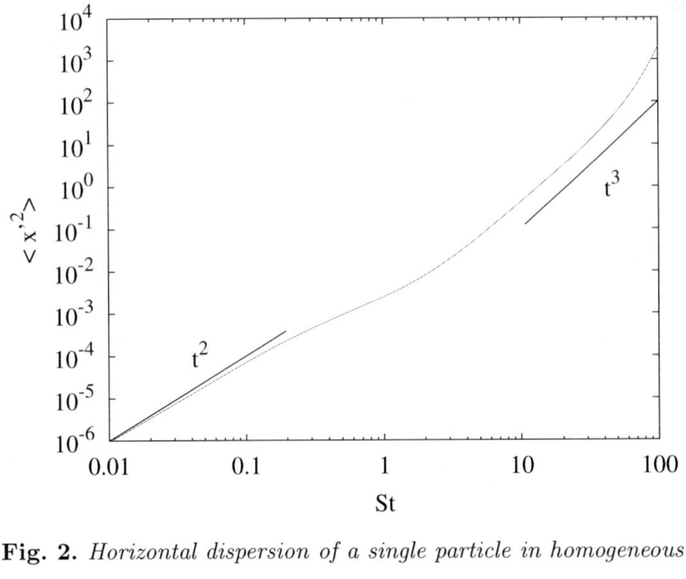

Fig. 2. *Horizontal dispersion of a single particle in homogeneous shear flow*

influence of this coupling term is neglected in the equation for the triple correlations. The details of the closure used to obtain quantitative results, based on the EDQNM theory, can be found in Bos [7].

The result is shown in Fig. 2. The ballistic regime is present again and after a transitional part a t^3 behaviour is observed. At the longest times the exponential growth of the kinetic energy in homogenous shear flow causes the dispersion to deviate from this scaling and to grow faster than t^3.

4 Conclusion

The analogy of the equation for the scalar fluctuation in a turbulent flow with a mean passive scalar gradient and the deplacement of a fluid particle allows to use closures developed for passive scalar mixing to obtain results for fluid particle dispersion. The EDQNM closure, developed for the scalar flux, is used without modification in this work to predict single particle dispersion in isotropic turbulence and homogeneous shear flow. The results are in agreement with classical scaling laws.

References

1. G.K. Batchelor. Diffusion in a field of homogeneous turbulence. i. eulerian analysis. *Aust. J. Sci. Res. Ser. A*, 2:437–450, 1949.
2. M. Larchevêque and M. Lesieur. The application of eddy-damped markovian closures to the problem of dispersion of particle pairs. *J. Mécanique*, 20:113–134, 1981.
3. G. Falkovich, K. Gawedzki, M. Vergassola. Particles and fields in fluid turbulence. *Rev. Mod. Phys.*, 73(4):913–975, 2001.
4. S. Herr, L.P. Wang, and L.R. Collins. EDQNM model of a passive scalar with a uniform mean gradient. *Phys. Fluids*, 8:1588–1608, 1996.
5. W.J.T. Bos, H. Touil, and J.-P. Bertoglio. Reynolds number dependency of the scalar flux spectrum in isotropic turbulence with a uniform scalar gradient. *Phys. Fluids*, 17(12), 2005.
6. G.I Taylor. Diffusion by continuous movements. *Proc. London Math. Soc.*, 2(20):196–212, 1921.
7. W.J.T. Bos. *Passive Scalar mixing in turbulent flow.* PhD thesis, Ecole Centrale de Lyon, 2005.

Clustering of Inertial Particles

L. Chen[1], S. Goto[2], and J.C. Vassilicos[3]

[1] Department of Aeronautics, Imperial College London, SW7 2AZ, UK
chen.lu@imperial.ac.uk
[2] Department of Mechanical Engineering and Science, Kyoto University, 606-8501, Japan goto@mech.kyoto-u.ac.jp
[3] Department of Aeronautics, Imperial College London, SW7 2AZ, UK
j.c.vassilicos@imperial.ac.uk

1 Introduction and Particle Motion Equation and Brief Details of Simulation

The clustering of inertial particles in turbulent flows has important applications from precipitation to chemical and pharmaceutical industries (see [2]). The mechanism in low Reynolds number turbulence was first proposed by [6] and believed to be the joint effect of vorticity and strain-rate(see [7]). But the mechanism in high Reynolds number turbulence remains unclear.

The equation of motion for small particles subjected to linear Stokes drag is approximated by (see [8])

$$\frac{d}{dt}\mathbf{v} = \frac{1}{\tau_p}[\mathbf{u}(\mathbf{x}_p, t) - \mathbf{v}(t)] \tag{1}$$

where $\mathbf{v}(t)$ is the velocity of the particle at its position $\mathbf{x}_p(t)$ at time t, $\mathbf{u}(\mathbf{x}, t)$ the fluid velocity field, a the radius, ρ_p the density, μ the surrounding fluid's dynamic viscosity and $\tau_p = \frac{2\rho_p a^2}{9\mu}$ the particle relaxation time.

We simulate two-dimensional (2D) homogeneous turbulent velocity fields $\mathbf{u}(\mathbf{x}, t)$ by Direct Numerical Simulation (DNS) following the method detailed in [5]. It has a broad range of scales as $L/\eta = 30$ ($T/\tau_\eta = 25$) by using 4096^2 grid points where L and η are integral and forcing length scale respectively. We also simulate 2D statistically homogeneous turbulent-like velocity fields with a $-5/3$ power law energy spectrum by Kinematic Simulation (KS) (see [3]). KS has $L/\eta = 10^3$ and $T/\tau_\eta = O(100)$ where L and η are outer and inner length scale respectively. Inertial particles are characterised by their Stokes number $St = \tau_p/\tau_\eta$. In all our simulations, $St \leq T/\tau_\eta$.

2 Acceleration Stagnation Points

Generalising the approach of [4] to zero-acceleration points, we define these points $\mathbf{s}_a(t)$ by $\mathbf{a}(\mathbf{s}_a(t), t) = 0$, their velocity by $\mathbf{V}_a \equiv \frac{d}{dt}\mathbf{s}_a(t)$ and write

$$\frac{D}{Dt}\mathbf{a} + (\mathbf{V}_a - \mathbf{u}) \cdot \nabla \mathbf{a} = 0 \tag{2}$$

Assuming that $(\mathbf{V}_a - \mathbf{u})$ and $\nabla \mathbf{a}$ are statistically uncorrelated, we deduce that

$$\langle(\mathbf{V}_a - \mathbf{u})^2\rangle^{1/2} \sim u'(L/\eta)^{-1/3} \tag{3}$$

where we reasonably assume Kolmogorov scaling for acceleration gradients. The important consequence is that $\langle(\mathbf{V}_a - \mathbf{u})^2\rangle^{1/2}/u' \to 0$ as $L/\eta \to \infty$, meaning that in the limit of high Reynolds numbers, zero-acceleration points move with their local fluid velocity $\mathbf{u}$. This conclusion can be seen as a quantitative formulation of the Tennekes sweeping hypothesis which states that energy containing turbulent eddies advect small-sale dissipative turbulent eddies [10]. The scaling relation (3) is validated by our 2D DNS (see Fig. 2(a)).

For $St \leq 0.3$, $\mathbf{v} \approx \mathbf{u} - \tau_p \mathbf{a}$ and $\frac{d}{dt}\mathbf{v} = \frac{1}{\tau_p}(\mathbf{u} - \mathbf{v})$ (see [2]) so that an inertial particle at a zero-acceleration point moves, on average, *with* this zero-acceleration point when $L/\eta \gg 1$ because they both move, statistically, approximately with the same velocity $\mathbf{u}$. Furthermore, the acceleration $\frac{d}{dt}\mathbf{v}$ of an inertial particle at a zero-acceleration point is zero, thus reducing the particle's ability to escape from the zero-acceleration point. Zero-acceleration points are therefore "sticky" for inertial particles, whereas high-acceleration points are repellent for much longer than particles remain in the vicinity of a given high acceleration point. This conclusion has been reached for small Stokes numbers but computational results in Fig. 1 suggest that it might be valid for a wide range of Stokes numbers all the way up to T/τ_η.

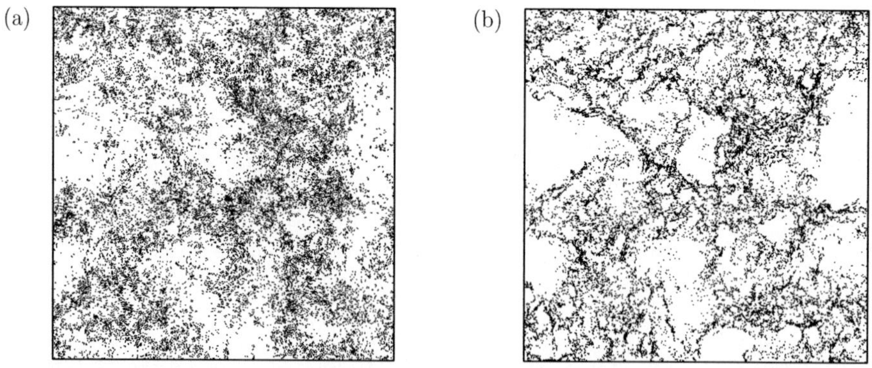

Fig. 1. Distribution of (a) inertial particles ($St = 1.9$) (black dots) and (b) zero-acceleration points (black dots). 2D DNS. Box size is about $4L$

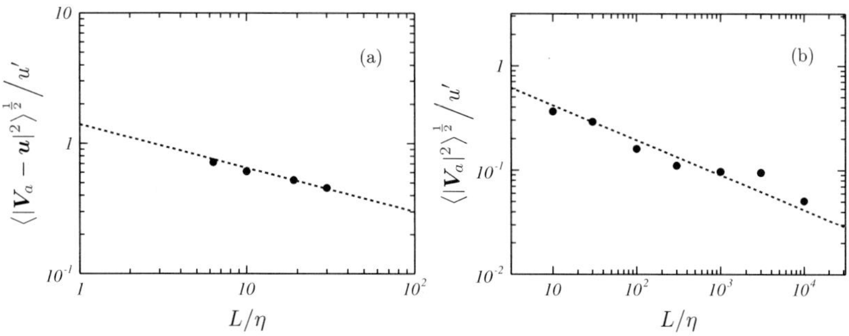

Fig. 2. (*a*) $(\mathbf{V}_a - \mathbf{u})'/u'$ as function of L/η in DNS. (*b*) $\mathbf{V}'_a/u'$ as function of L/η in KS with $\lambda = 0.5$. Dotted lines indicate -1/3 power law

3 Removing the Large-scale Sweeping of Small-scales

It is impossible to remove from the DNS turbulence the large-scale sweeping of small scales and the clustering of zero-acceleration points, so we now use a KS where the flow is synthesised in such a way that the sweeping and the acceleration clustering are absent by construction (see [3]).

Firstly, we confirm that the sweeping is indeed absent in KS because the scaling (3) does not hold. Instead $V_a'/u' \sim (L/\eta)^{-1/3}$ (see Fig. 2(b)). Secondly, we find that KS zero-acceleration points are uniformly distributed.

Without large-scale sweeping and without clustering of zero-acceleration points, the arguments of Sect 2 would suggest that KS cannot produce clustering of inertial particles. However inertial particle position fields do nevertheless develop well-defined near-empty spaces as time progresses.

4 Persistent Velocity Stagnation Points

In Fig. 3 we plot velocity stagnation points and inertial particles at a same time chosen long enough compared to τ_p in two cases.It is striking how inertial particles cluster in such a way as to avoid velocity stagnation points. This is consistent with inertial particles being centrifugally flung out of regions of high streamline curvature near velocity stagnation points.

We would expect the centrifugal effect around velocity stagnation points to exist in real high Reynolds number turbulence as it does in KS. However, our DNS results do not reveal a clear imprint of this effect on particle clustering. Instead, our DNS results strongly suggest that inertial particle clustering reflects the clustering of acceleration stagnation points.

Reference [5] has shown that the number-density of zero-acceleration points scales as $(L/\eta)^d$ ($d = 2, 3$ for 2D and 3D respectively) whereas the number-density of zero-velocity points scales as $(L/\eta)^{D_s}$ where $p + 2D_s/d = 3$ (see also [1]). It is natural to expect the effect of zero-acceleration points

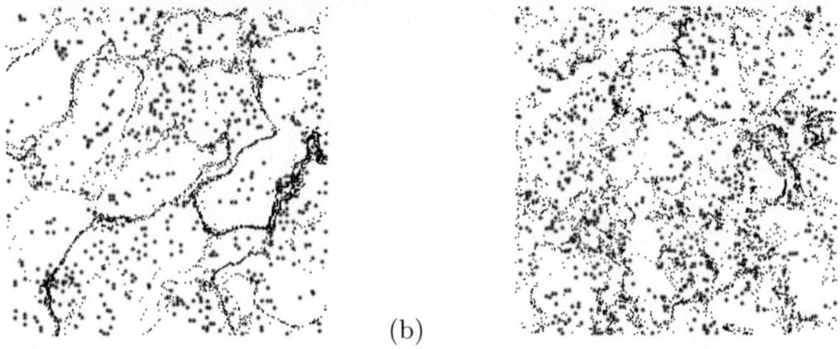

(a) (b)

Fig. 3. Spatial distribution of inertial particles ($St = 1.6$) (dot) and zero-velocity points (star) in 2D KS. $L/\eta = 1000$. Box size is $2L$. (*a*) $\lambda = 0$, (*b*) $\lambda = 0.5$

clustering on inertial particle clustering to dominate over any effect that the persistent zero-velocity points may have on inertial particle clustering.

5 Conclusion

In high Reynolds number 2D turbulence, inertial particles cluster in both DNS and KS. In DNS acceleration stagnation points and small inertial particles are swept by large-scale motions in such a way that inertial particles move towards acceleration stagnation points. In KS where large-scale sweeping is absent and where acceleration points do not cluster, inertial particle do nevertheless cluster as a result of the repelling action of velocity stagnation point clusters. This repelling action has a negligible effect on inertial particle clustering in high Reynolds number 2D DNS turbulence.

References

1. Davila, J. & Vassilicos, J.C. 2003 *Phys. Rev. Lett.* **91**, 144501.
2. Falkovich, G. & Pumir, A. 2004 *Phys. Fluids* **16**, L47.
3. Fung, J.C.H. & Vassilicos, J.C. 1998 *Phys. Rev. E* **57**, 1677.
4. Goto, S., Osborne, D.R., Vassilicos, J.C. & Haigh, J.D. 2005 *Phys. Rev. E* **71**, 015301.
5. Goto, S. & Vassilicos, J.C. 2004 *New J. Phys.* **6**, 65.
6. Maxey, M.R. & Corrsin, S. 1986 *J. Atmos. Sci.* **43**, 1112.
7. Maxey, M.R. 1987 *J. Fluid Mech.* **174**, 441.
8. Maxey, M.R. & Riley, J.J. 1983 *Phys. Fluids* **26**(4), 883.
9. Osborne, D.R., Vassilicos, J.C. & Haigh, J.D. 2005 *Phys. Fluids* **17**, 035104.
10. Tennekes, H. 1975 *J. Fluid Mech.* **67**, 561.

Self Similar Two Particle Separation Model

Beat Lüthi, Jacob Berg, Søren Ott and Jakob Mann

Risø National Laboratory, 4000 Roskilde, Denmark

1 Introduction

We present a new model for relative two particle separation in turbulence. Unlike models that treat the longitudinal relative velocity, dr/dt, of particle pairs, $r = |r_2 - r_1|$, by prescribing Eulerian statistics with K41 structure functions, like e.g. [1], we focus on the essential Lagrangian nature of the two particle separation process. The core of our approach is that we assume that the process of material line stretching [2, 3],

$$\frac{1}{2}\frac{dr^2}{dt} = r_i r_j s_{ij} = \frac{L(t)}{t^*} \cdot r^2, \qquad s_{ij} = \left(\frac{\partial u_i}{\partial x_j} + \frac{\partial u_j}{\partial x_i}\right), \quad t^* = \tau_\eta, \tag{1}$$

which in DNS and experiments has been observed to behave seemingly 'universal' [2, 4, 5], occurs in a similar way also in the inertial subrange. After the transition of the separation scale r from the viscous subrange to inertial subrange at $r \approx \alpha\eta$, where $\eta = \left(\nu^3/\epsilon\right)^{1/4}$ is the Kolmogorov scale of the flow and α specifies where precisely that transition occurs, the involved time, t^*, is scaled as $r \propto r^{2/3}$. In other words, we suggest that the process of relative two particle separation can be treated like a material line stretching process also in the inertial subrange, where the particle separation vector, r, becomes the coarse grained representative of the actual material line and where the separation-relevant part of the flow field is comprised by increasingly slower but larger eddies - by the coarse grained velocity derivative field, $\widetilde{\partial u_i/\partial x_j}$.

The model predictions for particle separations and separation velocities are in good agreement with numerical [1, 7] and experimental results [6, 8]. The model captures the influences of different initial particle separations and various Reynolds numbers both, in terms of fixed times and and in terms of fixed scales statistics. For particles with an initial separation of $0.1 < r_0/\eta < 100$ evolution times of $\mathcal{O}(100\tau_\eta)$ are required to reach the asymptotic solution of the Richardson-Obukhov law $\langle r^2 \rangle = C\varepsilon t^3$.

2 Model

Material line stretching is governed by the magnitude and distribution of the strain field and by the alignment of material line elements relative to that strain field. The overall outcome, the material line stretching rate, is in its nondimensional form equal to $L(t)$ of expression 1.

As can be seen from Fig. 1(a) the shape of the PDFs of $L(t)$ are reasonably close to being Gaussian which allows for a simple modelling. As was first shown in [2] and as can be seen from Fig. 1(a) both the mean, $L_M(t)$, and the variance, $\sigma^2(t)$, of $L(t)$ starting from $L_M(0) = 0$ and $\sigma^2(0) = 0.08$ reach asymptotic values of 0.12 and 0.04 respectively, after a short transition period of a few τ_η. These features are captured by the model treating $L(t)$ as a stochastic variable as

$$L(t) = L_M(t) + \sigma(t) \cdot S(t) \tag{2}$$

with

$$L_M(t) = L_\infty(1 - e^{-c_1 \cdot t^*}), \tag{3}$$

$$\sigma(t) = \sigma_\infty + \sigma_0 \cdot e^{-c_1 \cdot t^*}, \tag{4}$$

$$t^* = \begin{cases} \tau_\eta & , r \leq \alpha\eta \\ \left(\frac{r}{\alpha\eta}\right)^{2/3} & , r > \alpha\eta \end{cases} \tag{5}$$

and with $S(t)$ simulated as a 1D Ornstein-Uhlenbeck process. The variance of $S(t)$ is set to unity and - as the fourth free parameter - the integral time scale of $S(t)$ is set to $c_2 \cdot t^*$, e.g. the integral time scales also as $r^{2/3}$ once r is in the inertial subrange. With the free parameters set to $L_\infty = 0.115$, $\sigma_0 = 0.12$, $\sigma_\infty = 0.17$, $c_1 = 0.3$, and $c_2 = 1.8$ the behaviour of $L(t)$ is mimicked well by expression (2) as can be seen from Fig. 1(b). The remaining free parameter α that specifies the dimensional estimate for the Kolmogorov scale η and that defines where the model switches from a viscous process to an inertial range process is set to $\alpha = 25$. For inertial subrange separations with $\alpha\eta \leq r \leq \mathcal{L}$ the expression

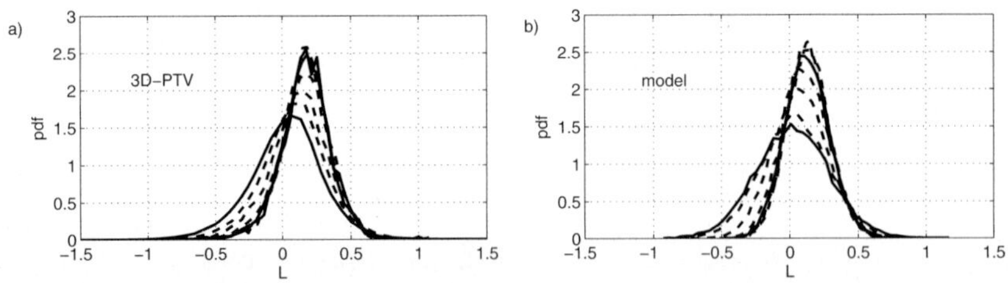

Fig. 1. Evolution in time of the nondimensional material line stretching rate as obtained from 3D-PTV (a) and as obtained from our model (b)

$$\frac{dr^2}{dt} = 2 \cdot \alpha^{\frac{2}{3}} \cdot L(t) \cdot \varepsilon^{\frac{1}{3}} \cdot r^{\frac{4}{3}} \tag{6}$$

is thus used to model the evolution of r. As the separations become larger than the integral length scale, $r > \mathcal{L}$, the governing scales are 'frozen' and r of eqn. 6 is replaced by $\mathcal{L}$ and the time scale is changed to $t^* = (\mathcal{L}/\alpha\eta)^{2/3}$.

3 Results

As part of the model verification we show in Fig. 2 how well the model predictions agree with DNS data from [7] and experimental data obtained from particle tracking [8].

We now ask the question: How does the influence of a finite inertial range (for $Re_\lambda \approx 280$ we have $\mathcal{L}/\eta \approx 600$) affect the relative particle dispersion? The Richardson constant, C, can be measured both with respect to fixed times, or with respect to fixed scales. From model results in Fig. 3 we see that for the former case (a) C roughly doubles its value from $C \approx 0.5$ to $C \approx 1$, while for fixed scales (b) the asymptotic value remains unchanged.

The separation distribution is also significantly affected by the extent of the inertial range. In Fig. 4 the modelled evolution in time of distributions for two different inertial ranges are shown. Distant neighbour functions, $q = p\left(r/\left\langle r^2\right\rangle^{1/2}\right)\left\langle r^2\right\rangle/4\pi r^2$, are shown for different times, (a): $Re_\lambda = 280$ and (b): $Re_\lambda = \infty$. For $Re_\lambda = 280$ the distribution starts to again approach a Gaussian shape after $t > 100\tau_\eta$. In Fig. 4(c) and (d) the skewness and flatness coefficients of the separation distributions are plotted over time for different initial separations. Again, we see how in the $Re_\lambda = 280$ case the distribution becomes Gaussian, while for $Re_\lambda = \infty$ the distribution takes on a constant shape after $t > 100\tau_\eta$, as predicted by the Richardson diffusion equation.

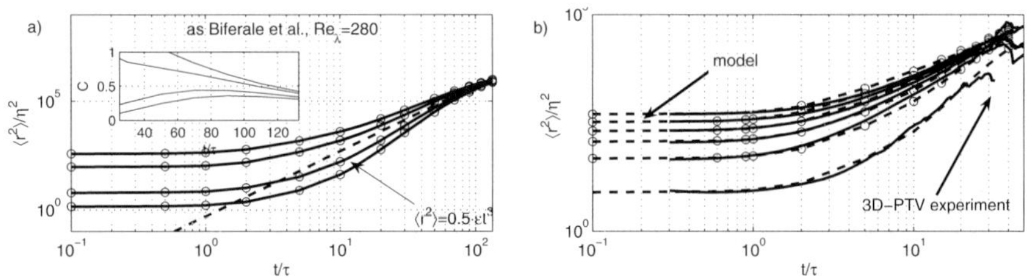

Fig. 2. Evolution in time of mean particle separations for different initial separations. In (a) the model is mimicking the DNS results of [7] and in (b) the model results are compared to results from a particle tracking experiment [8]

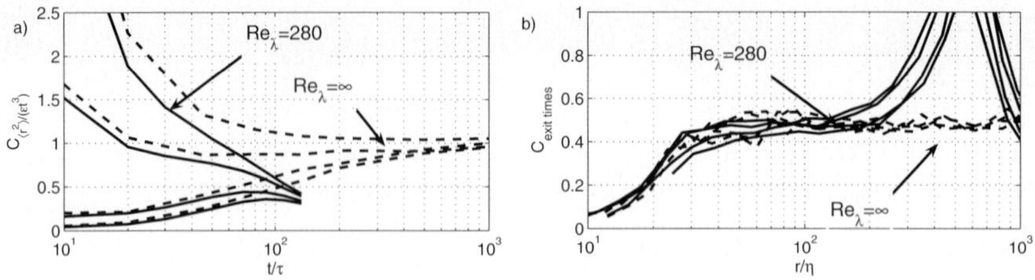

Fig. 3. The Richardson constant C as obtained for different methods and different inertial ranges. $C = \langle r^2 \rangle / \epsilon t^3$ (a), C is obtained from fixed scales statistics (b) as described in [7]

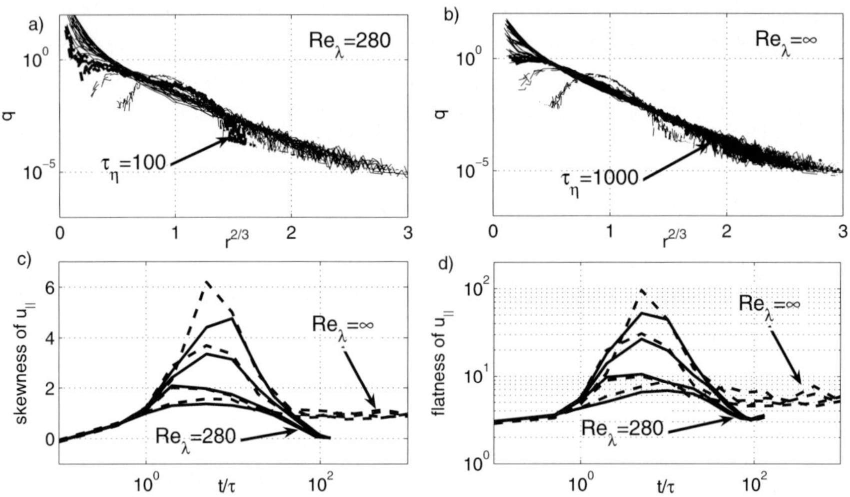

Fig. 4. Distant neighbour functions for different times for $Re_\lambda = 280$ (a) and for $Re_\lambda = \infty$ (b). Skewness (c) and flatness (d) coefficients for four initial separations

References

1. Borgas MS and Yeung PK (2004) J. Fluid Mech. 503:125-160.
2. Girimaji SS and Pope SB (1990) J. Fluid Mech. 220:427-458.
3. Tsinober A (2001) Kluwer, p. 88, p. 247.
4. Lüthi B, Tsinober A and Kinzelbach W (2005) J. Fluid Mech. 528:87-118.
5. Guala M, Lüthi B, Liberzon A, Tsinober A and Kinzelbach W (2005) J. Fluid Mech. 533:339-359.
6. Ott S and Mann J (2000) J. Fluid Mech. 422:207-223.
7. Biferale L, Boffetta G, Celani A, Devenish BJ, Lanotte A and Toschi F (2005) Phys. Fluids 17(11):111701.
8. Berg J, Lüthi B, Mann J, Ott S (2005) iTi 2005.

Coarse-grained Scalar Transport: Closures and Large-eddy Simulations

A. Celani[1], M. Martins Afonso[2], and A. Mazzino[2]

[1] CNRS, INLN, 1361 Route des Lucioles, 06560 Valbonne, France
[2] INFM-Department of Physics, University of Genova & INFN, Genova Section, via Dodecaneso 33, 16146 Genova, Italy, marcomar@fisica.unige.it

1 Introduction

A well-known problem faced in the numerical simulation of turbulent systems lies in the fact that the number of degrees of freedom necessary to correctly describe all active scales of motion rapidly grows by increasing the Reynolds number. To overcome such a limitation of numerical approach to turbulence, the large-eddy simulation (LES) technique has been proposed. The basic idea of such a strategy is to describe deterministically only a subset of active scales (typically the large scales, belonging to the inertial range) and parameterize the dynamical effect of the small, non-resolved, scales on the large ones. The problem arising from this method is that the cumulative effect of small scales on large scales is unknown: this is the renowned closure problem. The main aim of our presentation is to address the issue of small-scale parameterization in the framework of passive scalar turbulence where both analytical results and accurate direct numerical simulations (DNS) can be obtained. We focus on the Kraichnan advection model, which allows us to carry out exact analytical computations and derive systematically the dynamical equations describing the evolution of the large-scale component of the passive scalar field [1].

2 The LES Problem in Passive-scalar Turbulence

Our starting point is the well-known advection-diffusion forced equation ruling the dynamics of a passive scalar θ, advected by a given velocity field $\boldsymbol{v}$ and forced (at large scale) by an external source f:

$$\partial_t \theta + \boldsymbol{v} \cdot \boldsymbol{\partial} \theta = \kappa_0 \partial^2 \theta + f \ , \tag{1}$$

where κ_0 is the molecular diffusivity, acting at small scale η.
A strong analogy with the Navier–Stokes turbulence holds: the number of active spatial degrees of freedom rapidly increases with the Péclet number

(the analogous of the Reynolds number). Exploitation of LES to focus the attention on the sole large scales faces the problem of how to close the equation for the large-scale degrees of freedom.
More specifically, the LES strategy is carried out by convolving (1) with a low-pass filter P_l having a typical width l lying in the inertial range of scales. Doing this, we would like to obtain an equation for the coarse-grained fields (here denoted with a tilde) $\tilde{\theta} \equiv P_l * \theta$ and $\tilde{\boldsymbol{v}} \equiv P_l * \boldsymbol{v}$. Actually, this is not completely accomplished, because the resulting expression,

$$\partial_t \tilde{\theta} + \widetilde{\boldsymbol{v} \cdot \boldsymbol{\partial} \theta} = \kappa_0 \partial^2 \tilde{\theta} + \tilde{f} \ , \tag{2}$$

is not a closed equation for the large scale, as it still involves the original fields θ and $\boldsymbol{v}$ in the advective term. Therefore, we recast it in the following form:

$$\partial_t \tilde{\theta} + \tilde{\boldsymbol{v}} \cdot \boldsymbol{\partial} \tilde{\theta} = \kappa_0 \partial^2 \tilde{\theta} + \tilde{f} + \tau_\theta \ , \tag{3}$$

where $\tau_\theta \equiv \tilde{\boldsymbol{v}} \cdot \boldsymbol{\partial} \tilde{\theta} - \widetilde{\boldsymbol{v} \cdot \boldsymbol{\partial} \theta}$ can be interpreted as the subgrid scalar flux.
The LES challenge is to express τ_θ in terms of fields constructed from $\tilde{\theta}$ and $\tilde{\boldsymbol{v}}$, in order to have a closed equation for these quantities.

3 A New Closure for Passive-scalar Turbulence

Let us start from an exact result in passive-scalar advection. For times t larger than the largest velocity time scale, the averaged field $\langle\theta\rangle$ experiences the cumulative effect of velocity via a renormalized diffusion coefficient, the so-called eddy diffusivity (here supposed to be isotropic, as is indeed in presence of isotropic velocity fields), whose expression follows from the well-known Taylor formula. Having accounted for the nonlinear advective term through a linear diffusive one, in this particular case the problem faced while passing from (1) to (2) does not exist any more and the same eddy-diffusivity equation must then hold also for the averaged coarse-grained field $\langle\tilde{\theta}\rangle$.
We now look for an eddy-diffusivity-type closure in the equation for $\tilde{\theta}$ such that, starting from it, the same result is recovered. i.e. we ask that the closure is able to reproduce the long-time behavior of $\langle\tilde{\theta}\rangle$. The searched equation is

$$\partial_t \tilde{\theta} + \tilde{\boldsymbol{v}} \cdot \boldsymbol{\partial} \tilde{\theta} = \kappa_{\text{eff}} \partial^2 \tilde{\theta} + \tilde{f} \ , \tag{4}$$

where κ_{eff} can be determined using Taylor's formula again to obtain

$$\kappa_{\text{eff}} = \kappa_0 + \frac{1}{2d} \int_0^\infty \mathrm{d}\tau \ \left[\langle \boldsymbol{v}(\tau) \cdot \boldsymbol{v}(0) \rangle - \langle \tilde{\boldsymbol{v}}(\tau) \cdot \tilde{\boldsymbol{v}}(0) \rangle \right] \ . \tag{5}$$

In case of white-in-time velocity fields, the integral in (6) substantially reduces to the difference between the total and the large-scale kinetic energies.
It is worth noticing that a similar closure (i.e. exact in the same approximation) holds in presence of scale separation between velocity and scalar [2].

However, in what follows, we will focus on a situation where the velocity field is active at all scales: more specifically, we will consider the Kraichnan ensemble, where both $\boldsymbol{v}$ and f are Gaussian and δ-correlated-in-time stochastic fields, zero-mean and statistically stationary, homogeneous and isotropic.
Within this framework, the most natural quantity to be investigated is the two-point equal-time correlation function $C_2(r) = \langle \theta(\boldsymbol{r},t)\theta(\boldsymbol{0},t)\rangle$ in its steady state, with r lying in the inertial range. From the original equation (1), we can rigorously prove that, at the lowest (positive) order in the "natural" development parameter l/r, (4) still holds, in the sense that, starting from it, we are able to find an equation for the correlation of the coarse-grained scalar, $\langle \tilde{\theta}(\boldsymbol{r})\tilde{\theta}(\boldsymbol{0})\rangle$, whose solution reproduces its "exact" value, i.e. the expression found (by definition) by filtering $C_2(r)$ twice. In particular, for $d = 3$ and spherical top-hat filter P_l, the eddy-diffusivity coefficient (6) turns out to be

$$\kappa_{\text{eff}} = \kappa_0 \left[1 + \frac{2^{\xi} 24}{(4+\xi)(6+\xi)} \left(\frac{l}{\eta}\right)^{\xi} \right] , \tag{6}$$

ξ being the roughness exponent of the velocity field ($\xi = 4/3$ corresponds to the Kolmogorov–Obukhov–Corrsin (KOC) scaling). In the limit case $\xi = 0$, $\boldsymbol{v}$ plays the role of a white noise and this closure turns out to be exact.
Leaving aside the latter situation and the "opposite" limit $\xi = 2$ (corresponding to a smooth flow), we now want to improve the eddy-diffusivity closure, which indeed emerges from asymptotic arguments and does not seem appropriate to describe scales of motion r in the vicinity of the filter cut-off l. No exact closure exists in this framework. Anyway, we propose to add on the right-hand side of (4) a new contribution, which looks *ad-hoc* for the present situation (its exportability seems ruled out by the absence of a Lagrangian counterpart of this Eulerian analysis), but whose functional form and l-dependence can nevertheless be derived by first principles. Namely,

$$\partial_t \tilde{\theta} + \tilde{\boldsymbol{v}} \cdot \boldsymbol{\partial} \tilde{\theta} = \kappa_{\text{eff}} \partial^2 \tilde{\theta} + \tilde{f} - \frac{\xi}{15} l^2 \partial_\mu \partial_\nu (\tilde{v}_\mu \partial_\nu \tilde{\theta}) \, . \tag{7}$$

Terms proportional to l are ruled out by symmetry considerations and higher-order contributions in l have been neglected. The specific form of the new term is the most general which preserves tensorial structure, does not alter the value (7) already capturing the lowest order in l/r, and meanwhile correctly describes also the second order. Moreover, it is interpretable as a flux term and thus does not affect dissipation, i.e. it conserves $\langle \tilde{\theta}^2 \rangle$. In the next section the LES model (7) will be tested against DNS of passive scalars.
We would like to conclude this fully-analytical study by showing an example of what happens if a diversification in the filtering length between scalar and velocity is allowed, keeping however the same order of magnitude. In particular, if we define a new coarse-grained velocity as $\hat{\boldsymbol{v}} \equiv P_{\alpha l} * \boldsymbol{v}$ (with $\alpha = \sqrt{(3-\xi)/(3+\xi)}$, i.e. slightly less than unity), then we assert that an equation with the same structure as in (4),

$$\partial_t \tilde{\theta} + \hat{\boldsymbol{v}} \cdot \boldsymbol{\partial} \tilde{\theta} = \kappa_{\text{eff}} \partial^2 \tilde{\theta} + \tilde{f} ,$$

is able to reproduce for the correlation the same degree of approximation as in (7), without introducing any additive term.

4 Testing the LES Closure

In order to test our LES closure we have focused on two dimensions and we have simultaneously integrated (on a biperiodic lattice) the Navier–Stokes equation together with (1) for the "true" field θ, and both (4) and (7) for the coarse-grained field $\tilde{\theta}$ [3]. The lattice for the equation for $\boldsymbol{v}$ and θ is formed by 1024×1024 grid points, while $\tilde{\theta}$ is solved with solely 64×64 grid points. The field $\tilde{\boldsymbol{v}}$ appearing in the LES equation has been obtained by filtering the actual DNS field $\boldsymbol{v}$ with a Gaussian filter. In Fig. 1, the actual second-order structure function (dashed line) $S_2(r) = \langle [\theta(\boldsymbol{r}) - \theta(\boldsymbol{0})]^2 \rangle$ and the LES counterparts $\langle [\tilde{\theta}(\boldsymbol{r}) - \tilde{\theta}(\boldsymbol{0})]^2 \rangle$ (full lines) are shown for both the improved and

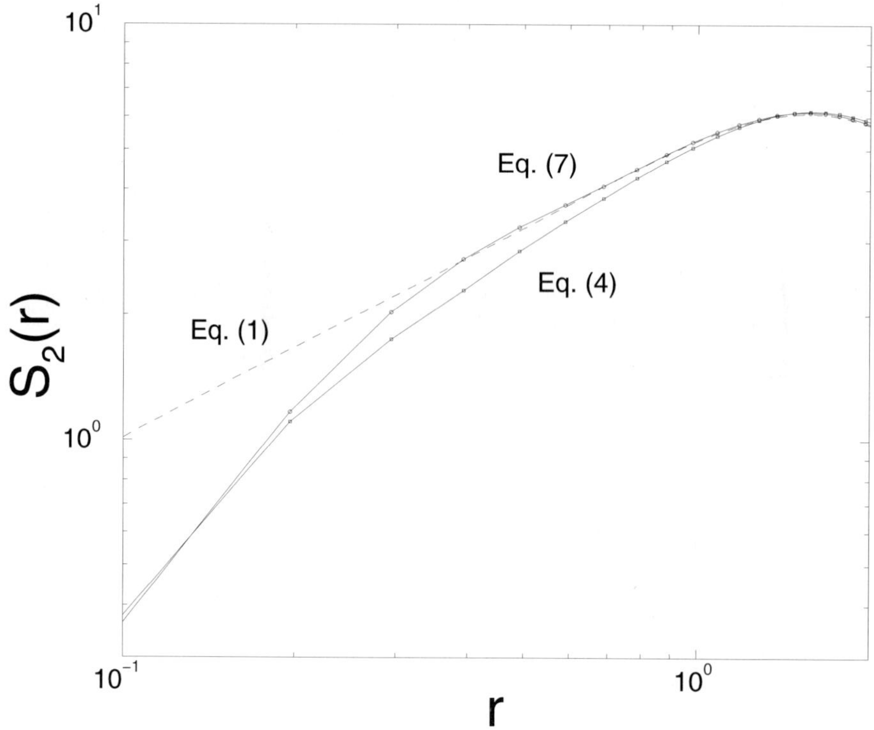

Fig. 1. DNS *vs* LES for $S_2(r)$. Notice how accurately the improved closure (7) reproduces the DNS prediction (dashed curve) from (1). Constant-eddy-diffusivity closure given by (4) is clearly insufficient to capture the correct inertial range scaling

the standard eddy-diffusivity closure. DNS simulation accurately reproduces the expected KOC scaling $r^{2/3}$. It is quite impressive as, despite the small resolution, our closure is able to accurately reproduce both the actual 2/3 exponent and the relative prefactor. This model outperforms the classical one and is therefore a natural candidate for subgrid modelling of scalar transport in generic turbulent flows since, on the contrary, there is no signature of inertial range for the constant-eddy-diffusivity closure.

References

1. Martins Afonso M, Celani A, Festa R, Mazzino A (2003) J Fluid Mech 496:355–364
2. Mazzino A (1997) Phys Rev E 56:5500–5510
3. Martins Afonso M, Celani A, Mazzino A (2004) Closures for Large-eddy simulations of Passive Scalars. In: Andersson H I, Krogstad P A (eds) Advances in Turbulence X, Proc. X European Turbulence Conference. CIMNE, Barcelona

Statistics and Geometry in High-Schmidt Number Scalar Mixing

Jörg Schumacher[1], Dan Kushnir[2], Achi Brandt[2], Katepalli R. Sreenivasan[3] and Herwig Zilken[4]

[1] Dept. of Physics, Philipps University, D-35032 Marburg, Germany
[2] Dept. of Computer Science and Applied Mathematics, The Weizmann Institute of Science, 76100 Rehovot, Israel
[3] International Centre for Theoretical Physics, 34014 Trieste, Italy
[4] Visualization Laboratory, Central Institute for Applied Mathematics, Research Centre Jülich, D-52425 Jülich, Germany

The mixing of substances occurs in various turbulent systems. Examples arise in reacting flows and combustion, mixing of salt and plankton in oceans and of chemical pollutants in the stratosphere [1]. The physics of scalar mixing depends strongly on the ratio of the kinematic viscosity ν of the fluid to the diffusivity κ of the scalar. This ratio is the Schmidt number $Sc = \nu/\kappa$. In the following, we focus to the so-called Batchelor regime of scalar mixing [2], i.e. $Sc > 1$. High-resolution simulations are used to explore some geometrical and statistical properties of the gradients of passive scalar fields, $\nabla\theta(\mathbf{x}, t)$ (for more details, see also [3, 4]). In order to resolve the fine scales very well, a larger than usual spectral resolution measure $k_{max}\eta_B$ is adopted here (see also caption of Fig.1). The Schmidt numbers throughout this work are 8 and 32. Both aspects, the fine resolution and $Sc > 1$, limit the accessible Taylor microscale Reynolds numbers of the advecting turbulent flow to $R_\lambda \leq 63$.

Regions with large scalar gradients can cause strong local mixing and are assigned to local maxima of the scalar dissipation rate. This field will be of interest for the following and is defined as

$$\epsilon_\theta(\mathbf{x}, t) = \kappa(\nabla\theta(\mathbf{x}, t))^2 \,. \qquad (1)$$

Figure 1 illustrates the shape and spatial distribution of its largest amplitude events. We see that regions with large dissipation rate are organized in thin extended sheets for cases $Sc > 1$, in contrast to the maxima of the energy dissipation rate which are plotted in the same panels. The figure indicates also a Reynolds number dependence of the mixing. With increasing Reynolds number the sheets become smaller, but more numerous. This is attributed to the local flow patterns which are responsible for the sheet formation. A larger range of scales in space and time is excited with growing Reynolds number.

Fig. 1. Joint isovolume plots of the energy dissipation field $\epsilon(\mathbf{x}, t)$ (light) and the scalar dissipation rate $\epsilon_\theta(\mathbf{x}, t)$(dark). The data are obtained from very well resolved pseudospectral simulations in a periodic box of sidelength 2π resolved with a grid of $N^3 = 1024^3$ points. The spectral resolution measure is $k_{max}\eta_B = 11.84$ (upper panel) and $k_{max}\eta_B = 3.39$ (lower panel) with $k_{max} = \sqrt{2}N/3$ and the Batchelor scale η_B. Usually $k_{max}\eta_B \approx 1.5$ is taken for pseudospectral simulations. The advecting turbulence is homogeneous and isotropic and is maintained stationary by stochastic forcing at low wavenumbers. The passive scalar fluctuations are kept stationary by a constant mean scalar gradient in y direction. The isovolume levels for both pictures are $5 \times \langle\epsilon\rangle$ for the energy dissipation rate and $7 \times \langle\epsilon_\theta\rangle$ for the scalar dissipation rate. The Schmidt number is $Sc = 8$ in both cases. Upper picture: $R_\lambda = 24$. Lower picture: $R_\lambda = 63$

The tail of the probability density function (PDF) of the scalar dissipation rate determines the statistical distribution of the maxima. The PDF is plotted in Fig. 2 for two Reynolds numbers. We find considerable deviations from log-normality, which exceed those previously reported (see [4] for a more detailed discussion). Deviations are detected for all Reynolds and Schmidt numbers studied here. In the figure, the tails of PDF were fitted with a stretched exponential

$$p(\epsilon_\theta \gg \langle\epsilon_\theta\rangle) \sim \epsilon_\theta^{-1/2} \exp\left(-C_2 \epsilon_\theta^{\alpha/2}\right) , \tag{2}$$

Such statistics were derived analytically for scalar advection in smooth and white-in-time flows in the limit of an infinite Péclet number. An exponent $\alpha = 2/3$ was found [5]. The tails remain below that limit, but above $\alpha = 1$ which corresponds to an exponential distribution of $|\nabla\theta|$.

Interesting for the small-scale modeling of mixing, e.g. for flamelets in combustion, is the cross-section thickness scale of the dissipation sheets. The thickness determines the scale across which the most intensive mixing events are present. The thickness is analysed here by a fast multiscale clustering algorithm [6], applied to two-dimensional planar cuts through snapshots of the scalar dissipation field [7]. The sheets appear in the cuts as filaments. A local principal component analysis is applied to subpieces of each separate filament,

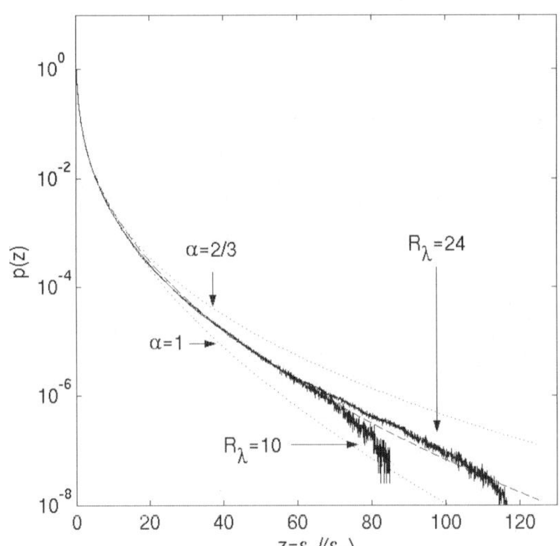

Fig. 2. Log-linear plot of the probability density function of the scalar dissipation rate, normalized to the mean value. Data are for two different Taylor microscale Reynolds numbers at $Sc = 32$. Fits to the data for $z \geq 10$ with the stretched exponential term of (2) are also plotted and the corresponding exponents α are shown. The dashed line is the optimum of a least square fit resulting in $\alpha = 0.86$

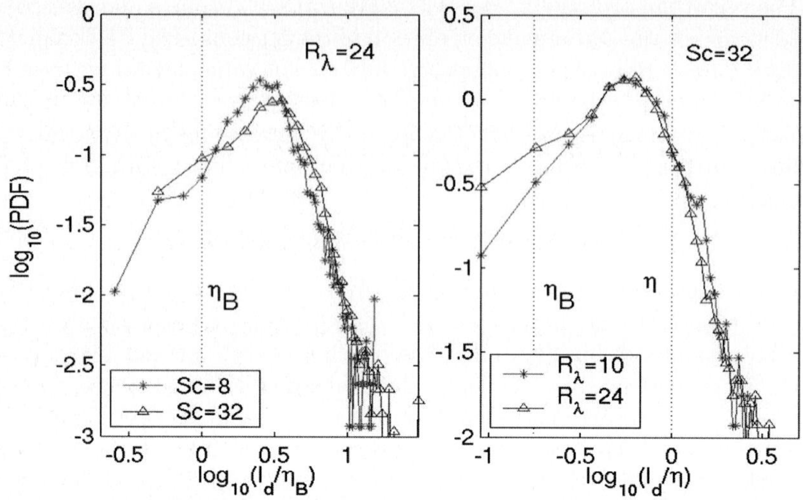

Fig. 3. Distribution of the local cross-section thickness l_d of the scalar dissipation rate filaments for $\epsilon_\theta \geq 4\langle\epsilon_\theta\rangle$. Left panel: Probability density function (PDF) $p(l_d/\eta_B)$ for two different Schmidt numbers at $R_\lambda = 24$. Right panel: PDF $p(l_d/\eta)$ for two different Reynolds numbers at $Sc = 32$

and the smaller eigenvalue is then taken as the local filament thickness, l_d. The thickness distribution is shown in Fig. 3. The PDF is supported by all scales within the viscous-convective range. Only a small number of the sheets have a thickness close to the Batchelor scale η_B, which is the finest scale in the turbulent mixing process. The collapse of the distributions in each of the panels suggests that the most probable thickness – the maximum of the PDF – varies as the Batchelor scale η_B with Sc at fixed Reynolds number (left) and as the Kolmogorov scale η with R_λ at fixed Schmidt number (right).

The numerical simulations were carried out on the IBM JUMP cluster at the John von Neumann Institute for Computing and we are grateful for their steady support. JS wishes to thank the Deutsche Forschungsgemeinschaft for support. Four of us (JS, DK, AB and KRS) were also supported by the US National Science Foundation.

References

1. Warhaft Z, (2000), Annu. Rev. Fluid Mech., 32:203-240
2. Batchelor GK, (1959), J. Fluid Mech., 5:113-133
3. Schumacher J, Sreenivasan KR and Yeung PK, (2005), J. Fluid Mech., 531:113-122

4. Schumacher J and Sreenivasan KR, (2005), Phys. Fluids, 17:125107
5. Chertkov M, Falkovich G and Kolokolov I, (1998), Phys. Rev. Lett., 80:2121-2124
6. Kushnir D, Galun M and Brandt A, (2006) Pattern Recognition, in press
7. Kushnir D, Schumacher J and Brandt A, (2006) Phys. Rev. Lett., submitted

Part VII

Polymer and Scalars

Scale by Scale Budget in Viscoelastic Wall Turbulence

E. De Angelis[1], N. Marati[2], C.M. Casciola[3], R. Piva[4]

[1] Dip. di Meccanica e Aeronautica – Università di Roma "La Sapienza"
elisabetta.deangelis@uniroma1.it
[2] Dip. di Meccanica e Aeronautica – Università di Roma "La Sapienza"
nicoletta.marati@uniroma1.it
[3] Dip. di Meccanica e Aeronautica – Università di Roma "La Sapienza"
carlomassimo.casciola@uniroma1.it
[4] Dip. di Meccanica e Aeronautica – Università di Roma "La Sapienza"
renzo.piva@uniroma1.it

Despite the growing interest on the subject, many of the efforts to the comprehension of practical aspects of turbulence in dilute polymer solutions, such as drag reduction, have proven inconclusive, suggesting that, even from the point of view of applications, a more fundamental approach is required. Recently the influence of polymers on the turbulent cascade has been studied in the contest of homogeneous and isotropic turbulence [2]. Those results allowed for the identification of a scale below which the flux towards polymers is larger than the nonlinear transfer. More relevant for the real problems is the study of wall turbulence where the alteration of the energy cascade might have a large effect. On the basis of the DNS data for the turbulent channel flow, an analysis of the interaction between polymers and turbulence at different distances from the wall is provided. A detailed balance of the scale by scale budget of the turbulent kinetic energy is shown.

1 Mathematical Model

As customary in dilute solutions, a polymeric stress tensor, T_{ij}, is superposed to the Newtonian contribution. Thus the momentum equation reads

$$\frac{\partial u_i}{\partial t} + u_j \frac{\partial u_i}{\partial x_j} = -\frac{\partial p}{\partial x_i} + \nu \left(\frac{\partial^2 u_i}{\partial x_j \partial x_j} \right) + \frac{\partial T_{ij}}{\partial x_j} . \tag{1}$$

In this approach each molecule is represented by a dumbbell, i.e. two massless beads connected by a nonlinear spring, whose configuration is given in terms of the end-to-end vector R_i. At each point the evolution of an ensemble of dumbbells, forced by the fluid dynamic field and by a Brownian noise, is considered and the following form of the extra-stress tensor is obtained [1]

$$T^p_{ij} = \frac{\eta_p}{De}\left(f(\mathcal{R}_{ii})\mathcal{R}_{ij} - \delta_{ij}\right). \tag{2}$$

In (2) the conformation tensor, $\mathcal{R}_{ij}$, is the covariance matrix of the distribution function for the vectors R_i of the ensemble and its equation reads

$$\frac{\partial \mathcal{R}_{ij}}{\partial t} + u_k \frac{\partial \mathcal{R}_{ij}}{\partial x_k} = -\frac{1}{De} f(\mathcal{R}_{ii})\mathcal{R}_{ij} + K_{ir}\mathcal{R}_{rj} + \mathcal{R}_{ir}K_{jr} + \frac{1}{De}\delta_{ij}\,. \tag{3}$$

The present model (FENE-P), though quite simple, is already realistic since is characterized by the nonlinearity of the spring coefficient and by a finite extension of the molecule. The relevant parameters are the Deborah number, De, ratio between the polymer relaxation time and the flow time scale and the relative viscosity η_p, ratio between the contribution of the polymers to the solution viscosity and the solvent viscosity. Another parameter is the maximal allowed elongation R^2_{max}, appearing in the nonlinear spring coefficient $f(\mathcal{R}_{ii})$.

2 Mean Quantities

For the present DNS, the dimensions of the domain are $2\pi h \times 2h \times 1.2\pi h$, where h is half the channel height. The simulations have been performed on a $128 \times 129 \times 64$ grid at a nominal Reynolds number of 4320, for both the cases, Newtonian and viscoelastic. The flow has been forced with the same pressure drop, so the resulting friction Reynolds number is the same and equal to $Re_\tau = 147$. The values for the polymers parameters are: $De = 10$, $\eta_p = 0.1$. and $R^2_{max} = 5000$. In this framework drag reduction corresponds to an increased flow rate[5] (see Fig. 1a), let's notice that the introduction of polymers alters also the turbulent kinetic energy κ (see Fig. 1b), whose equation for the viscoelastic case, if $g_i = \partial T'_{ij}/\partial x_j$ is

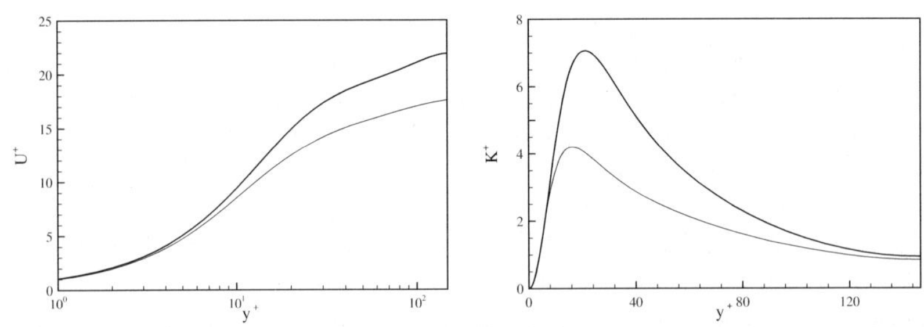

Fig. 1. Mean velocity and turbulent kinetic energy profiles for a Newtonian and viscoelastic simulation, thin and thick lines respectively, at $Re_\tau = 147$

$$-\langle uv \rangle \frac{dU}{dy} - \frac{d}{dy}\langle \kappa v \rangle - \frac{d}{dy}\langle pv \rangle + \frac{1}{Re}\frac{d^2}{dy^2}\langle \kappa \rangle - \langle \epsilon \rangle + \langle g_i u_i \rangle = 0. \quad (4)$$

3 Scale by Scale Budget

Defining as $U(y)$ the mean velocity, the equation for $\delta q^2 = \delta u_i \delta u_i$ reads

$$\frac{\partial \langle \delta q^2 \delta u_i \rangle}{\partial r_i} + \frac{\partial \langle \delta q^2 \delta U \rangle}{\partial r_x} + 2\langle \delta u \delta v \rangle \left(\frac{dU}{dy}\right)^* + \frac{\partial \langle v^* \delta u^2 \rangle}{\partial Y_c} = \quad (5)$$
$$- 4\langle \epsilon_N^* \rangle + 2\nu \frac{\partial^2 \langle \delta q^2 \rangle}{\partial r_i \partial r_i} - \frac{2}{\rho}\frac{\partial \langle \delta p \, \delta v \rangle}{\partial Y_c} + \frac{\nu}{2}\frac{\partial^2 \langle \delta q^2 \rangle}{\partial {Y_c}^2} + 2\langle \delta g_i \delta u_i \rangle$$

where $\phi^* = (\phi(x_i) + \phi'(x_i + r_i)/2$ and X_i is the middle point between x_i and $x_i + r_i$. ϵ_{N} is the Newtonian dissipation. Terms with r-derivate are related to the flux through scales while the ones with Y-derivate arise due to inhomogeneity. The term $2\langle \delta u \delta v \rangle \, (dU/dy)^*$ represents the production at scale r. It is easy to show that the large scale limit of (5) is equal to four times the balance of turbulent kinetic energy (4). It is possible to define an r-average of the terms in (5) as

$$Q_{\mathrm{r}}(r, Y_{\mathrm{c}}) = \frac{1}{r^2}\int_{-\frac{r}{2}}^{\frac{r}{2}}\int_{-\frac{r}{2}}^{\frac{r}{2}} q(r_{\mathrm{x}}, 0, r_{\mathrm{z}} | Y_{\mathrm{c}}) \, dr_{\mathrm{x}} dr_{\mathrm{z}} \,, \quad (6)$$

where $\boldsymbol{r} = (r_{\mathrm{x}}, 0, r_{\mathrm{z}})$ and $r_{\mathrm{x}} = r_{\mathrm{y}}$. Applying this average operator and casting together some terms, the (5) reads, in a compact form,

$$T_{\mathrm{r}}(r, Y_{\mathrm{c}}) + \Pi_{\mathrm{e}}(r, Y_{\mathrm{c}}) = D_{\mathrm{e}}(r, Y_{\mathrm{c}}) + E(r, Y_{\mathrm{c}}) + G_{\mathrm{e}}(r, Y_{\mathrm{c}}) + E_P(r, Y_{\mathrm{c}}) \,. \quad (7)$$

In analogy with an homogeneous and isotropic case, (7) can be interpreted as the sum of transfer through the scales and effective production balanced by the dissipative contributions due both to the solvent and the polymers. Hence in this framework the respective role of the nonlinear transfer and the flux to the polymers can be evaluated in different regions of the boundary layer. Namely, to write (7), an effective production has been defined as the sum between the turbulent transport and real production $\Pi_{\mathrm{e}}(r, Y_{\mathrm{c}}) = \Pi(r, Y_{\mathrm{c}}) + T_{\mathrm{c}}(r, Y_{\mathrm{c}}) - P(r, Y_{\mathrm{c}})$ and analogously also the viscous and polymeric contributions have been expressed as the sum of a transport term, $D_{\mathrm{e}}(r, Y_{\mathrm{c}})$ and $G_{\mathrm{e}}(r, Y_{\mathrm{c}})$, grouping together the r-fluxes and the Y-fluxes, and a local term, local dissipation, $E(r, Y_{\mathrm{c}})$, and local transfer to the microstructure, $E_P(r, Y_{\mathrm{c}})$, respectively. In other words the energy avaliable at given position Y_{c} and scale r is partly produced locally by the production term Π and partly arrives or leaves thanks to the spatial transfer, the remaining part is either dissipated or transferred to the microstructure. Here we propose an extention to a viscoelastic case of the scale by scale analysis, already studied for a Newtonian channel flow in [3].

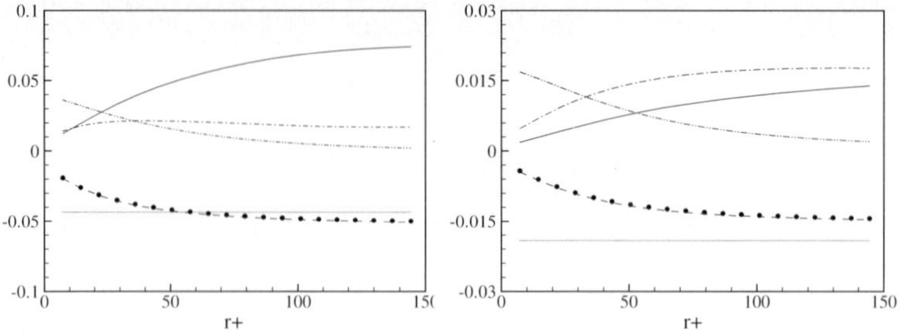

Fig. 2. Detailed balance (7) at $y^+ = 80, 147$ for the viscoelastic case. The sum $(T_r + \Pi_e + E_P + G_e)$ is represented by the filled circles, $E + D_e$ is given by the dashed line. The solid line is $-\Pi_e$, the dash-dotted line corresponds to $-T_r$. For the polymers the dotted line is E_P and the dash-double-dotted is G_e

From the inspection of the results the most interesting information seem to come from the region above $y^+ = 80$ where the polymer contribution becomes relevant. In Fig. 2 the detailed balances at two different distances from the wall are shown. In the center of the channel the behaviour of the fluctuations resembles the one observed in the homogeneous and isotropic case, at large scale (larger than the plot shows) the dominant term is the effective production, then the energy is transferred through the scales until it reaches a point where the transfer to the polymers becomes larger, thus the dynamics seem affected by the polymers only at the very small scales. Instead, in the left plot it has been singled out a region where we observe what can be defined an *elastic* behaviour, i.e. at large scales the dominant term is still the production, however there is an intermediate range of scales where the transfer to polymers becomes relevant. There is no range of scale where the cascade is not affected by the polymers. We argue that at larger values of drag reduction the region in which such a behaviour is observed should increase considerably.

References

1. Bird, R.B., Curtiss, C.F., Armstrong, R.C., Ole Hassager,(1987) Dynamics of polymeric liquids. Wyley-Interscience.
2. De Angelis E., Casciola C. M., Benzi R., Piva R. (2004), Homogeneous isotropic turbulence in dilute in dilute polymers, Journal Fluid Mech 531:1 - 10.
3. Marati N., Casciola C. M., Piva R. (2004), Energy cascade and spatial fluxes in wall turbulence Journal Fluid Mech 521:191 - 215.
4. De Angelis, E., Casciola, C.M., Piva, R., DNS of wall turbulence: dilute polymers and self-sustaining mechanisms, Computers and Fluids 31:495-507.
5. Tiederman, W.G., Luchik, T.S., Bogard, D.G., *Wall-layer structure and drag reduction.*, J.Fluid Mech. **156**: 419-37, 1985.

Alignment Statistics of a Passive Scalar Gradient in Nonstationary Flow Topology

A. Garcia, M. Gonzalez, L. Danaila and P. Paranthoën

CNRS, UMR 6614, Laboratoire de Thermodynamique
CORIA, Site universitaire du Madrillet, 76801 Saint-Etienne du Rouvray, France
Aurelien.Garcia@coria.fr, Michel.Gonzalez@coria.fr

1 Introduction

The present study is mostly focused on the response of a scalar gradient to time-varying velocity derivatives. The influence of Lagrangian fluctuations of strain persistence upon the alignment properties of a passive scalar gradient has been studied using measurements in a two-dimensional, low-Reynolds number experiment and a simple, numerical analysis.

In a two-dimensional, incompressible flow, the equation for the orientation of the scalar gradient, $|\mathbf{G}|(\cos\theta, \sin\theta)$, neglecting molecular diffusivity is [1]:

$$\frac{d\zeta}{dt} = \sigma(r - \cos\zeta) \tag{1}$$

where σ is the strain intensity and $\zeta = 2(\theta + \Phi)$. Angle Φ gives the orientation of the strain principal axes. Parameter $r = (\omega + 2d\Phi/dt)/\sigma$ defines strain persistence by comparing the influence of effective rotation to that of strain.

Analysis of (1) assuming slow Lagrangian variations of σ and r [1] shows that in strain-dominated regions ($r^2 < 1$), the gradient has an equilibrium orientation, $\zeta_{eq} = -\arccos(r)$, which coincides with the compressional direction, $\zeta_c = -\pi/2$, for $r = 0$. In rotation-dominated regions ($r^2 > 1$), the scalar gradient has a most probable direction corresponding to its minimal rotation rate. This picture has been confirmed by numerical simulations [1, 2].

2 Temperature Gradient Orientation in a Bénard - von Kármán Experiment

The temperature field downstream of a thin heated line located on the centerline of a Bénard - von Kármán street (Fig. 1) has been studied experimentally [3, 4]. The cylinder Reynolds number based on diameter is equal to 63 and the source Reynolds number is smaller than unity. The thermal

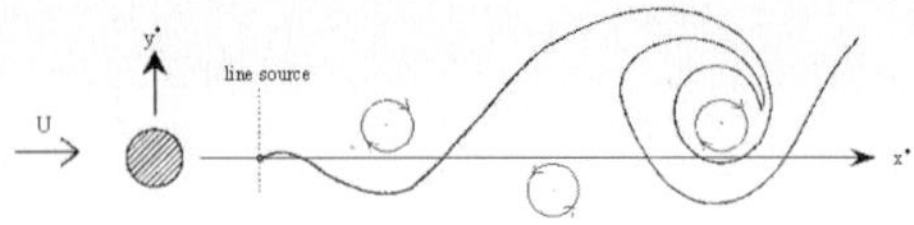

Fig. 1. Experimental set-up

field is passive and the velocity field is two-dimensional. Velocity components and temperature have been measured simultaneously using a two-component Laser Doppler Anemometry system combined with a cold wire probe. A phase reference for the velocity and temperature signals was taken from a second cold wire.

By numerically tracking fictitious particles assumed to move at the measured velocity, we have derived the Lagrangian history of the temperature gradient and computed its orientation properties. Alignment of the temperature gradient with the compressional direction is initially weak (Fig. 2a), but is the preferential orientation further from the source (Fig. 2b). This suggests that the compressional direction is gradually promoted at the expense of the theoretical, equilibrium direction predicted by Lapeyre et al. model. Note that the initial bimodal p.d.f of $\zeta - \zeta_c$ results from the special condition of heat injection at the line source.

The apparent discrepancy between Lapeyre et al. analysis and experiment is not due to some effect of molecular diffusion. Solving (1) (which is non-diffusive) with the measured σ and r leads, as experimental data, to better alignment with the compressional direction than with the theoretical one.

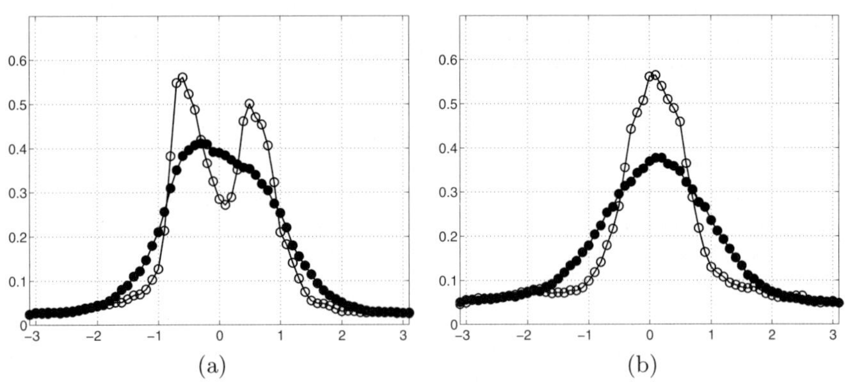

Fig. 2. Alignment pdf in the hyperbolic regions ($|r| < 1$) for $|\mathbf{G}| \geq 100$; (a) "close" to the heat source ($\Delta x^* < 4$); (b) "far" from the heat source ($\Delta x^* > 4$); $\circ$: pdf of $\zeta - \zeta_c$; $\bullet$: pdf of $\zeta - \zeta_{eq}$

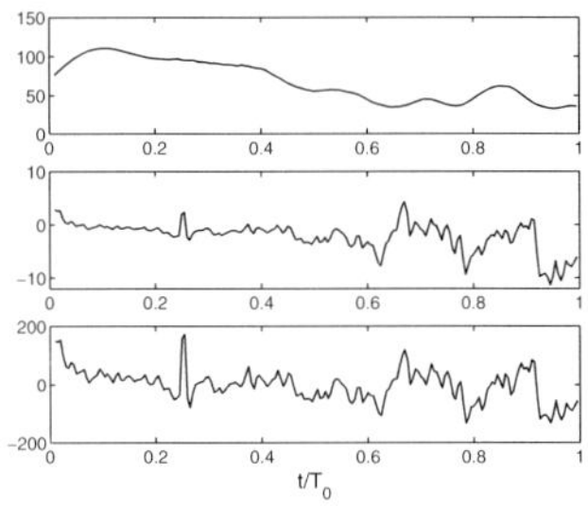

Fig. 3. Lagrangian evolution of σ (top), r (middle) and $d\Phi/dt$ (bottom)

Besides, Lagrangian tracking along a typical, experimental trajectory reveals that r varies as $d\Phi/dt$, on a shorter time scale than σ does (Fig. 3). In this experiment, the assumption made by Lapeyre et al. of slowly-varying σ and r is therefore questionable and is at best valid for σ.

Equation (1) describes the dynamic response of gradient orientation to the time-varying forcing r with a relaxation time scale of the order of $1/\sigma$. Thus quantity σT (where T is the fluctuation time scale of r) arises as a key parameter. Large σT values imply that ζ is always close to the equilibrium imposed by the instantaneous value of r whereas low σT bring about a poor response of the gradient.

An experimental estimate of σT derived from $\langle\sigma\rangle \simeq 70s^{-1}$ and the Lagrangian autocorrelation time scale of r, $T \simeq 2ms$, is $\sigma T \simeq 0.15$. This implies significant nonstationary effects through Lagrangian fluctuations of r and a weak response of the gradient that may explain misalignement with respect to Lapeyre et al. theoretical orientation.

3 Analysis of Alignment Statistics Using a Simple Numerical Model

Equation (1) with time normalized as $\tau = \sigma t$ has been solved with a given r signal. The statistics of r is certainly not Gaussian, but we have used a Langevin equation for analyzing the effect of varying r:

$$dr(\tau) = -r(\tau)\beta^* d\tau + \left(2r'^2\beta^*\right)^{1/2} dW(\tau) \tag{2}$$

where r'^2 is the variance, $dW(\tau)$, a Wiener process and $\beta^* = 1/\sigma T$ with T, the integral time scale of r signal. The mean value, $\langle r\rangle$, is zero, complying with experimental data. The $\langle r\rangle$-defined orientation, $\zeta_{\langle r\rangle} = -\arccos\langle r\rangle$, therefore coincides with the compressional one, ζ_c. Numerical results show that, for $|r| < 1$, the scalar gradient aligns preferentially with the asymptotic, theoretical orientation when β^* is small whereas the best alignment is the $\langle r\rangle$-based direction for large values of β^* (Fig. 4). Thus the influence of strain persistence

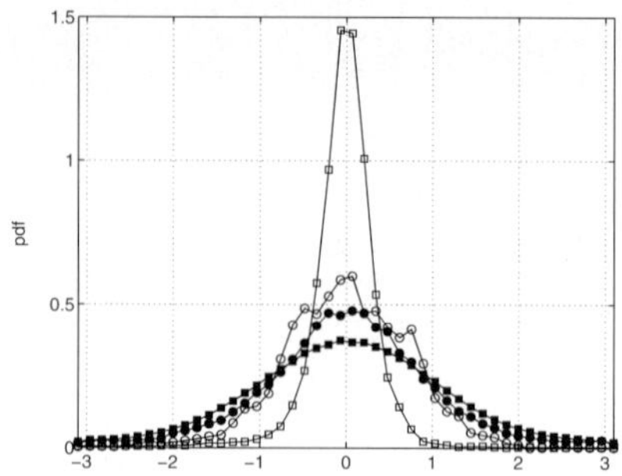

Fig. 4. Alignment pdf for $|r| < 1$; $r'^2 = 4$; with $\beta^* = 5$: • pdf of $\zeta - \zeta_{\langle r \rangle}$, ■ pdf of $\zeta - \zeta_{eq}$; with $\beta^* = 0.01$: ∘ pdf of $\zeta - \zeta_{\langle r \rangle}$, □ pdf of $\zeta - \zeta_{eq}$

fluctuations upon scalar gradient alignment conjectured from the experimental data is confirmed. It also appears that the Lapeyre et al. model is verified for fluctuating strain persistence r provided the gradient time response is short compared to r fluctuations time scale.

4 Conclusion

Measurements in a two-dimensional, low-Reynolds number experiment as well as a simple model show that the scalar gradient alignment is significantly influenced by Lagrangian fluctuations of the strain persistence parameter, r. In strain-dominated regions, the scalar gradient statistically aligns with the equilibrium orientation determined by the instantaneous value of r if its time response is short compared to the fluctuations time scale of r. Otherwise, the preferential alignment is determined by $\langle r \rangle$ and coincides with the compressional direction if $\langle r \rangle \simeq 0$. This suggests that better understanding of alignment properties in turbulent flows would require estimating the Lagrangian time scale of strain persistence from simulation or, possibly, experimental data.

References

1. Lapeyre G, Klein P, Hua B L (1999) Phys. Fluids 11:3729–3737
2. Lapeyre G, Hua B L, Klein P (2001) Phys. Fluids 13:251–264
3. Godard G (2001) Etude expérimentale de la structure du champ d'un scalaire passif dans une allée de Bénard - von Kármán. PhD Thesis, University of Rouen
4. Paranthoën P, Godard G, Weiss F, Gonzalez M (2004) Int. J. Heat Mass Transf. 47:819–825

An Improved Model for the Turbulent Atmospheric Boundary Layer Including Urban Canopy*

A.F.Kurbatskiy[1], A.V.Lonchakov[2], and L.I.Kurbatskaya[3]

[1] Institute of Theoretical and Applied Mechanics of Russian Academy of Sciences, Siberian Branch kurbat@nsu.ru
[2] Novosibirsk State University kurbat@nsu.ru
[3] Institute of Computational Mathematics and Mathematical Geophysics RAS, Siberin Branch L.Kurbatskaya@ommgp.sscc.ru

1 Introduction

This study attempts to formulate an improved model for the turbulent atmospheric boundary layer (ABL). The present model employs three new ingredients: 1) updated expression for the pressure-velocity correlation, 2) updated expression for the pressure-temperature/concentration correlation, and 3) deduced fully explicit anisotropic expressions for the turbulent fluxes of a momentum, heat/concentration. A parameterization scheme is used to represent the impact of urban buildings on airflow in mesoscale atmospheric models. The buildings are not explicitly resolved in this model, but their effects on the grid-averaged variables are parameterized [1] as the extra terms in the transport equations of the momentum, heat , turbulent kinetic energy and its dissipation rate. This improved model predicts more realistically the impact of the urban surfaces on the airflow as compared to the one-parametrical turbulence models.

2 Mesoscale RANS-model for the Turbulent ABL

For the 2D case the basic equations can be written in the form

$$U_x + W_z = 0, \tag{1}$$

$$U_t + UU_x + WU_z = -\frac{1}{\rho_0} P_x - \langle wu \rangle_z + fV + D_u, \tag{2}$$

* This work was supported financially by the Russian Foundation for Basic Research (grant No. 06-05-64002)

$$V_t + UV_x + WV_z = -\langle wv \rangle_z - fU + D_v, \tag{3}$$

$$W_t + UW_x + WW_z = -\frac{1}{\rho_0} P_z - \langle ww \rangle_z + \beta \Theta g, \tag{4}$$

$$\Theta_t + U\Theta_x + W\Theta_z = -\langle u\theta \rangle_x - \langle w\theta \rangle_z + D_\theta. \tag{5}$$

The dependent variables in (1)–(5) are the mean velocities U, V, and W along the x, y, and z axes, respectively; Θ is the mean deviation of the potential temperature from the standard value T_0 ; f is the Coriolis parameter, β is the volume expansion coefficient of the air ($3.53 \times 10^{-3} K^{-1}$); ρ_0 is the mean air density. The extra terms of D_u, D_v and D_θ describe the impact of the urban roughness on the air flow. The fully explicit algebraic models for the Reynolds shear stresses and the turbulent heat flux are formulated in the following subsection.

2.1 Fully Explicit Algebraic Reynolds Stress and Scalar Flux Models

The coupled algebraic equations system for the traceless Reynolds tensor and the turbulent scalar fluxes (temperature/concentratin) can be solved in the 2D case using symbolic algebra. Below we represent the equations for turbulent momentum and heat fluxes only, which are used in the numerical tests for solution of the equations system (1)–(5):

$$(\langle uw \rangle, \langle vw \rangle) = -K_M \left(\frac{\partial U}{\partial z}, \frac{\partial V}{\partial z} \right), \tag{6}$$

$$\langle w\theta \rangle = -K_H \frac{\partial \Theta}{\partial z} + \gamma_c, \tag{7}$$

$$K_M = E\tau S_M, \quad K_H = E\tau S_H, \tag{8}$$

$$S_M = \frac{1}{D} \left\{ s_0 \left[1 + s_1 G_H \left(s_2 - s_3 G_H \right) \right] + s_4 s_5 \left(1 + s_6 G_H \right) (\tau \beta g)^2 \frac{\langle \theta^2 \rangle}{E} \right\}, \tag{9}$$

$$S_H = \frac{1}{D} \left\{ \frac{2}{3} \frac{1}{c_{1\theta}} \left(1 + s_6 G_H \right) \right\}, \tag{10}$$

$$\gamma_c = \frac{1}{D} \left\{ 1 + \frac{2}{3} \alpha_2^2 G_M + s_6 G_H \right\} \alpha_5 (\tau \beta g) \langle \theta^2 \rangle \tag{11}$$

is the countergradient term, which is absent in the 2nd and 2.5–order closure models [3].

$$G_H \equiv (\tau N)^2, \quad G_M \equiv (\tau S)^2, \tag{12}$$

$$N^2 = \beta g \tfrac{\partial \Theta}{\partial z}, \quad S^2 \equiv \left(\tfrac{\partial U}{\partial z} \right)^2 + \left(\tfrac{\partial V}{\partial z} \right)^2, \tau = E/\varepsilon.$$

All parameters in the (6)–(12) are expressed though the three constants of modeling expressions for the pressure-velocity and pressure-temperature correlations that can be found in [4,5]. The three-parametric turbulence model $E - \varepsilon - <\theta^2>$ is used for closure of the expressions (6)–(7) for the turbulent fluxes.

2.2 Numerical Test

The new mesoscale model ABL is tested in the simple 2D case. The horizontal extension of the domain is $120km$ with a resolution $1km$. The vertical resolution is $10m$ in the first $50m$ above the ground and then it is stretched up to 1 km at the top ($6km$). The topography is flat with a $10 - km$ wide city (10 'urban' points) surrounded by a rural area. The ground temperature is the only unsteady boundary condition. This thermal boundary condition simulates a 24 hour cycle of heating by the sun on a land mass located from 45 km to 55 km. In our study the urban heat island effect is specified by an urban-rural temperature difference . The meteorological initial conditions are a geostrophic wind from the west of 3 and $5ms^{-1}$, and atmospheric thermal stratification is equal to 3.5 $^oKkm^{-1}$ in potential temperature.

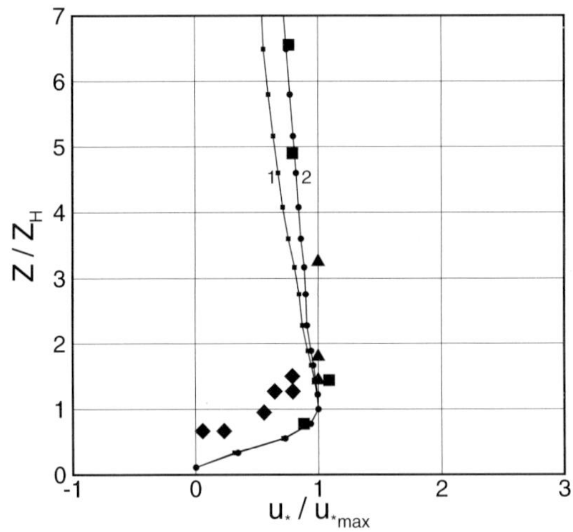

Fig. 1. Vertical profiles of "local"friction velocity u_* (defined as $(<uw>^2 + <vw>^2)^{1/4}$). Solid lines are results of the present model (**1** - $3ms^{-1}$ wind speed, **2** - $5ms^{-1}$ wind speed); squares are data of [7]; diamonds are data of [8–10]; triangles are data of [11]. Z_H is the average building height

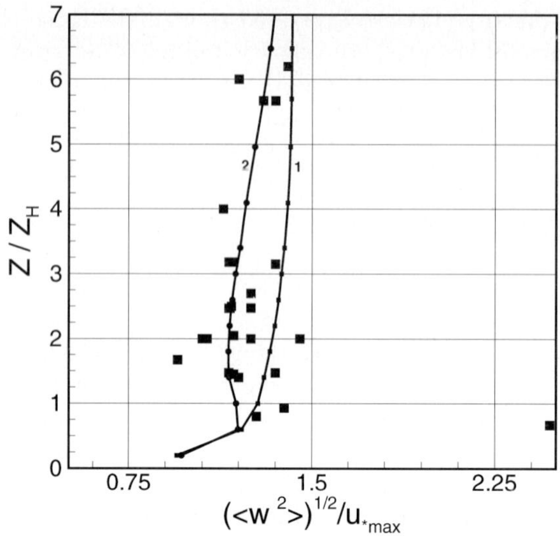

Fig. 2. Vertical profiles of normalized vertical velocity variance at the center of the urbanized area. Solid lines are results of the present model (**1** - $3ms^{-1}$ wind speed, **2** - $5ms^{-1}$ wind speed); the symbols are the measurement data of [12]

3 Results

Figure 1 illustrates the vertical distribution of the scaled frictional velocity, computed by the numerical model and the observational data of [7,8,9,10,11]. As seen in the figure, the computed frictional velocity increases with the height from the ground surface, reaches a maximum at the roof level and then decreases slightly with height. This is in reasonably agreement with observations [6, 11]. Figure 2 shows the vertical distribution of the vertical velocity variance. The important feature of this three parametric model is the improvement in evaluation of the the turbulent transport processes in the vertical and horizontal directions under different stratifications conditions, which are commonly observed in the urban canopy.

References

1. Raupach M (1992) Bound-Layer Meteor 60: 375–395
2. Martilli A (2002) Bound-Layer Meteor 104: 261–304
3. Cheng Y, Canuto V, Howard A (2002) J Atmos Sci 59: 1550–1565
4. Kurbatskii A (2005) Thermophysics and Aeromechanics (Russia) 12: 39–58
5. Kurbatskii A, Kurbatskaya L (2005) Atmospheric Oceanic Optik (Russia) 18: 399–408
6. Kurbatskii A (2001) J Appl Meteor 40: 1748–1761

7. Oikawa S, Meng Y (1995) Bound-Layer Meteor 74: 289–312
8. Rotach M (1991) Turbulence Within and Above an Urban Canopy. ETH Diss. 9439, ZGS, Heft 45, Verlag vdf, Zurich
9. Rotach M (1993) Bound-Layer Meteor 65: 1–28
10. Rotach M (1995) Atmos Environ 29: 1473–1486
11. Feigenwinter C (1999) The vertical Structure of Turbulence above an Urban Canopy. Ph.D.Thesis, University Basel
12. Roth (2000) Q J R Meteor Soc 126: 941–990

LES of Turbulent Low Mach Number Shear Layers with Active Scalars Using Explicit Filtering

Inga Mahle[1], Juan Pedro Mellado[2], Jörn Sesterhenn[1] and Rainer Friedrich[1]

[1] Fachgebiet Strömungsmechanik, Technische Universität München, Germany
mahle@aer.mw.tum.de
[2] Ingenieria Aeroespacial, Escuela Superior de Ingenieros, Sevilla, Spain

Performing Large Eddy Simulations (LES) at moderate Reynolds numbers with active scalars and detailed molecular transport mechanisms is quite complicated as they are fully coupled with the flow field and influence it directly for example via density gradients resulting from varying mass fractions. The use of the explicit filtering method (EFM) is especially attractive due to its simplicity: The Navier-Stokes and species transport equations in the original DNS formulation are solved on the LES grid and the resulting flow fields are explicitly low-pass filtered in a specific way after each time step. No heuristic or physical subgrid models are used. Up to now, the EFM has been successfully applied to compressible turbulent channel flows [2]. In this work, it is tested for turbulent shear layers in order to find out how it performs in the presence of active scalars.

The first section of the paper describes the inert shear layer configuration and the parameters of the simulations. In addition, the EFM is briefly explained. The method is tested qualitatively and quantitatively in the next section by comparing LES results with results from Direct Numerical Simulations (DNS). The last part of the paper is dedicated to the application of the EFM to reacting shear layers with infinitely fast chemistry.

1 Test Cases and Numerical Method

Figure 1 shows an instantaneous image of the temporally evolving shear layer. All test cases are 3D with x and y denoting the streamwise and spanwise directions and z denoting the transverse direction of the shear layer. The upper stream (index 1) is pure oxygen while the lower stream (index 2) is either nitrogen which results in a nearly constant density (case O_2-N_2) or a mixture of hydrogen and oxygen with mass fractions Y chosen in a way that the density ratio between this stream and the upper stream is 1:4 (case O_2-O_2/H_2). Temperature and pressure are initially constant. The convective

Mach number Ma_c, computed with the velocity difference between the two streams, Δu, and the sum of their sonic speeds, is 0.15 reflecting the low Mach number situation in combustion chambers. For the initialization of the LES, turbulent DNS fields are interpolated onto the LES grid and filtered by the EFM filter (see Sect. 3). DNS grids and domain sizes are comparable to those used by Pantano and Sarkar [4]. The LES grid of the case O_2-N_2 (O_2-O_2/H_2) has approximately 30 (14) times less points than the DNS grid.

The EFM, which is used in the LES, is derived from the Approximate Deconvolution Model (ADM) [5] based on the observation that the filtering and deconvolution steps can be combined into a single filtering step [2]. So, in the LES the same code as in the DNS can be used, which integrates the compressible Navier-Stokes and species equations. The effective cut-off wavenumber of the low-pass filter lies in the range where the sixth-order accurate numerical scheme allows for excellent reconstructions of the original unfiltered field.

2 Results

Figure 1 is a comparison between the instantaneous hydrogen mass fraction fields of the LES and the DNS (case O_2-O_2/H_2) at $\tau = t \cdot \Delta u / (\delta_{\omega,0}) = 62$ after the start of the LES. $\delta_{\omega,0}$ is the initial vorticity thickness of the DNS. One can see that the large structures are resolved, but the LES is somewhat blurred by the missing of the small scales. The agreement between the DNS data, which have been interpolated onto the coarser LES grid and low-pass filtered, and the LES data for the scalar variance $\langle Y''^2 \rangle_f$ is excellent (Fig. 2 for O_2-N_2). Averages over the two homogeneous directions and in time over the self-similar state are denoted by $\langle \bullet \rangle$, Favre averages by $\langle \bullet \rangle_f$. The corresponding fluctuations are $\bullet'$ and $\bullet''$. Before averaging in time, a non-dimensionalization by δ_θ, the momentum thickness, by $\langle \rho \rangle$ and Δu is done [4]. The computation of the scalar dissipation rate $\langle \epsilon_Y \rangle$ involves derivatives which are small scale quantities and are therefore not represented in the LES. So the interpolated and filtered DNS data and the LES data are not as close to the unfiltered ones (Fig. 3) as it is the case in Fig. 2. The scalar dissipation rate computed with filtered quantities is overestimated by the LES. The subgrid part of the scalar dissipation rate $\epsilon_{Y,sg}$, which has been taken away from the DNS by interpolation to the coarser grid and by filtering and which is the difference

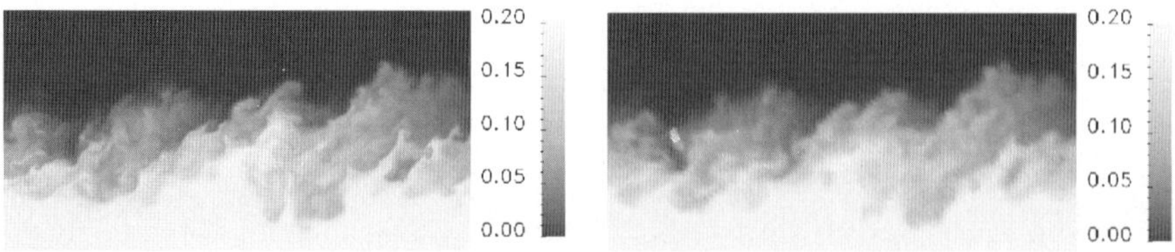

Fig. 1. Instantaneous mass fraction field of H_2 for O_2-O_2/H_2, left: DNS, right: LES, $\tau = 62$, x-z-plane in the middle of the computational domain

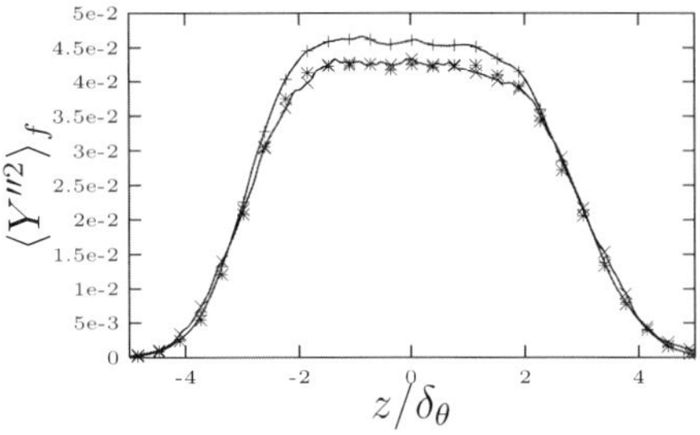

Fig. 2. Scalar variance for O_2-N_2, +: DNS, ×: LES, ∗: filtered DNS data

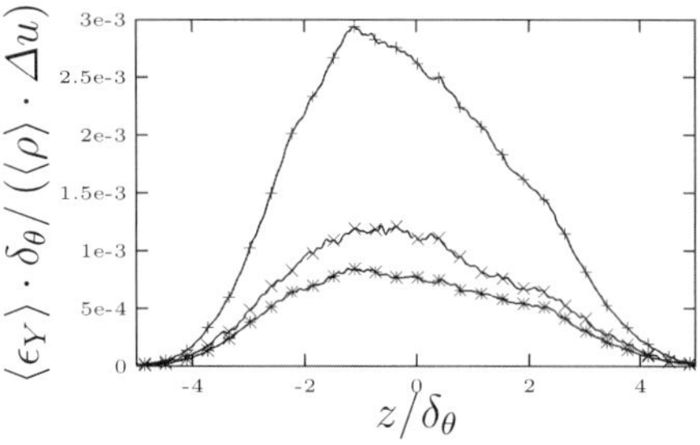

Fig. 3. Scalar dissipation rate for O_2-N_2, +: DNS, ×: LES, ∗: filtered DNS data

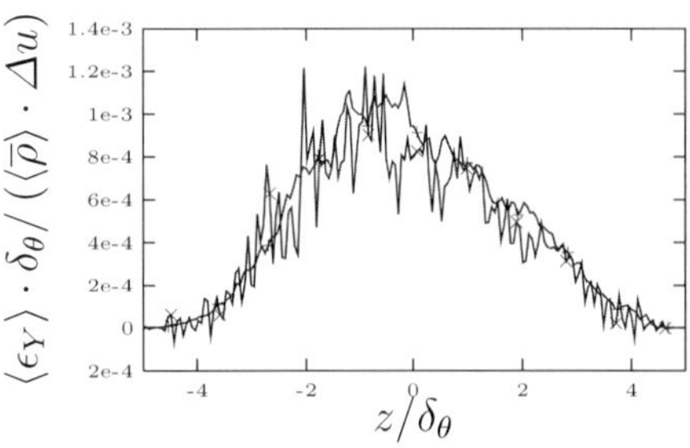

Fig. 4. Spatially averaged profiles at $\tau = 62$ for O_2-N_2, +: $\epsilon_{Y,sg}$, ×: $\epsilon_{Y,ef}$

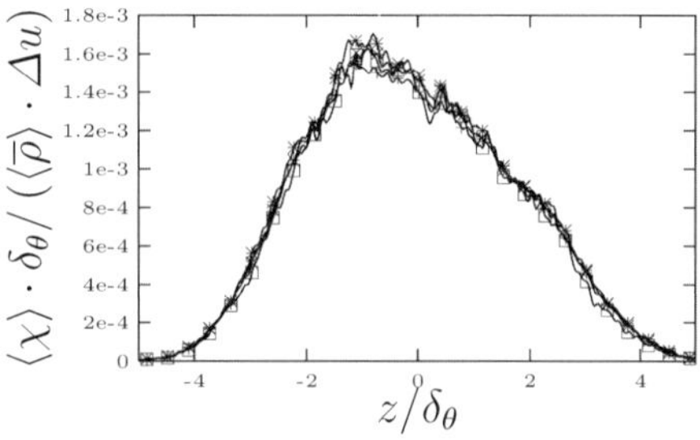

Fig. 5. +: $\widetilde{\chi}$, ×: $\chi_{ls} + \chi_{sg,turb}$, ∗: $\chi_{ls} + \chi_{sg,sm}$, □: $\chi_{ls} + \chi_{sg,ss}$ (data of O_2-N_2)

between the profiles for the DNS and the filtered DNS in Fig. 3, has to be replaced by the EFM during the LES in order to capture the development of the flow correctly. For comparison with $\epsilon_{Y,sg}$, a dissipation rate caused by the low-pass filter, $\epsilon_{Y,ef}$, can be computed: After performing one iteration with and one without filtering on the LES grid, this filter dissipation rate is the difference of the resulting $\frac{1}{2}\langle \rho Y'' Y'' \rangle$ divided by the time step. The agreement with $\epsilon_{Y,sg}$ in Fig. 4 is very good.

3 Application of the EFM to Reacting Shear Layers and Modeling of the Heat Release Term

In LES of reacting shear layers, filtered reaction rates have to be modeled. They arise in addition to the subgrid nonlinearities which are already taken care of by the EFM. However, as the chemical reactions are taking place in the unresolved subgrid scales, it is impossible to get a useful approximation of the source terms by explicit filtering.

As a first step, we concentrate on an infinitely fast, irreversible reaction: Instead of solving the species transport equations, a transport equation for a passive scalar, the mixture fraction Z, is computed to which the species are explicitly linked. However, simplifications e.g. concerning the diffusion have to be made. For the subgrid probability density function (pdf) $\widetilde{F}(Z)$, a β-pdf can be used. To model the necessary subgrid variance, scale-similarity is assumed which requires additional filter operations on the mixture fraction and density fields.

In order to retain a compressible formulation, the pressure equation is solved. The filtered heat release term in this equation takes the following form, which is similar to RANS [1]:

$$\overline{\omega}_p = (\gamma - 1) \, Q_e \widetilde{\chi}_s \widetilde{F} \, (Z_s) \tag{1}$$

The ratio of the specific heat capacities is γ, Q_e represents a heat release parameter, which is computed from chemical data (stoichiometric coefficients, enthalpies...). Z_s is the stoichiometric value of the mixture fraction. The crucial part in this formulation is the modeling of the conditionally filtered scalar dissipation rate[3] $\widetilde{\chi}_s$. Using a 1D model for mixing derived from laminar configurations [3], $\widetilde{\chi}_s$ can be linked to the filtered scalar dissipation rate $\widetilde{\chi}$. This quantity can be split up in a resolved large scale part χ_{ls} and an unresolved subgrid part which is of similar size as the turbulent subgrid dissipation rate $\chi_{sg,turb} = -\overline{\rho} \left(\widetilde{u_i Z} - \widetilde{u}_i \widetilde{Z} \right) \partial \widetilde{Z} / \partial x_i$. In Fig. 5, two models are investigated in an a priori test for this quantity, a Smagorinsky-type gradient model $\chi_{sg,sm}$ and a scale-similarity model $\chi_{sg,ss}$. With the agreement of the curves being good, the next step of our work will be to use the EFM and the source term modeling in a combustion LES.

References

1. Bilger R W (1980) Turbulent Flows with Nonpremixed Reactants. In: Libby P A, Williams F A (eds) Turbulent Reacting Flows. Springer, Berlin Heidelberg New York
2. Mathew J, Lechner R, Foysi H, Sesterhenn J, Friedrich R (2003) Physics of Fluids 15(8):2279–2289
3. Peters N (2000) Turbulent combustion. Cambridge University Press
4. Pantano C, Sarkar S (2002) Journal of Fluid Mechanics 451:329–371
5. Stolz S, Adams N A (1999) Physics of Fluids 11(7):1699–1701

[3] χ describes the scalar dissipation rate of a passive scalar Z while ϵ_Y is used to denote the scalar dissipation of the active scalar Y

Statistical Closures for Homogeneous Shear Flow Turbulence of Dilute Polymer Solutions

D. Vincenzi[1,2], S. Jin[3], T. Vaithianathan[3], L. R. Collins[3] and E. Bodenschatz[1,2]

[1] Max-Planck-Institut für Dynamik und Selbstorganisation, Busenstr. 10, 37073 Göttingen, Germany
[2] Laboratory for Atomic and Solid State Physics, Cornell University, Ithaca, NY 14853, USA
[3] Sibley School of Mechanical and Aerospace Engineering, Cornell University, Ithaca, NY 14853, USA

1 Introduction

Dilute polymer solutions exhibit macroscopic behaviors that distinguish them from ordinary Newtonian fluids. For example, minute concentrations of polymers (parts per million on a weight basis) can lead to impressive reductions in the drag on solid surfaces (by up to 80%) [1]. Numerical simulations of viscoelastic flows are generally based on an evolution equation for the conformation tensor of the polymer, $C_{ij} \equiv \langle r_i r_j \rangle$, where $\mathbf{r}$ is the separation vector between the ends of the molecule and angle brackets indicate the average over the Brownian configuration space of the molecule. Direct numerical simulations (DNS) of viscoelastic turbulence are able to reproduce the key phenomenology found in experiments; however, in view of the large number of degrees of freedom, they are limited to modest values of the Reynolds number [2–4]. An alternative approach is to seek a closed equation for the average configuration tensor $\overline{C}_{ij}$, where the average should now be intended over statistical realizations of the turbulent flow. That equation could then be coupled to a Reynolds averaged Navier-Stokes (RANS) solver [5]. Here, we present preliminary results based on a statistical analysis of the transport equation for C_{ij}.

Continuum models for dilute polymer solutions are based on a coarse-grained description of the polymer molecule. Within two-bead dumbbell models, the conformation tensor C_{ij} obeys the transport equation [6]:

$$\frac{\partial C_{ij}}{\partial t} + u_k \frac{\partial C_{ij}}{\partial x_k} = C_{ik} \frac{\partial u_j}{\partial x_k} + C_{jk} \frac{\partial u_i}{\partial x_k} - \frac{f(r)C_{ij} - r_0^2 \delta_{ij}}{\tau_p} , \qquad (1)$$

where r_0 is the equilibrium extension of the polymer, τ_p is its relaxation time, $\mathbf{I}$ is the identity tensor, and r^2 denotes the square of the mean separation between the ends of the molecule: $r^2 = C_{ii}$. The function f models entropic forces restoring stretched molecules to their equilibrium configuration. In the Oldroyd-B model, the restoration force is assumed to be linear ($f(r) = 1$); this assumption is however appropriate only for weakly stretched polymers. A more realistic model is the FENE-P model, which takes into account the finite extensibility of polymers by means of an nonlinear entropic force that prevents polymers to extend beyond the maximum extension L: $f(r) = (L^2 - 3r_0^2)/(L^2 - r^2)$. The relevant nondimensional number for (1) is the Weissenberg number $We \equiv \tau_p/\tau_\eta$, where τ_η denotes the Kolmogorov time scale. The Weissenberg number measures the strength of the turbulence relative to the restoration force of the polymer.

2 Mean Polymer Conformation Equation

The possibility of introducing a RANS solver for viscoelastic turbulence depends on our ability to deduce an effective evolution equation for $\overline{C}_{ij}$. Averaging (1) over the statistical realizations of the turbulent flow yields

$$\frac{\partial \overline{C}_{ij}}{\partial t} + \overline{u_k \frac{\partial C_{ij}}{\partial x_k}} = \overline{C}_{ik}\frac{\partial \overline{u}_j}{\partial x_k} + \overline{C}_{jk}\frac{\partial \overline{u}_i}{\partial x_k} + \overline{C_{ik}\frac{\partial u'_j}{\partial x_k} + C_{jk}\frac{\partial u'_i}{\partial x_k}} - \frac{\overline{f(r)C_{ij}} - r_0^2\delta_{ij}}{\tau_p}, \tag{2}$$

where $\mathbf{u} = \overline{\mathbf{u}} + \mathbf{u}'$. For notational convenience, we hereafter denote the velocity gradient by $\Gamma_{ij} \equiv \partial u_i/\partial x_j$ and decompose it into the sum of the mean gradient $\overline{\Gamma}_{ij}$ and zero-mean fluctuations Γ'_{ij}. When deriving a closed equation for $\overline{C}_{ij}$ from (2) under homogeneous conditions, we are faced with two closure problems. The first one deals with the contribution to the stretching term coming from fluctuations of the velocity gradient: $\Lambda_{ij} \equiv \tau_p \overline{C_{ik}\Gamma'_{jk} + C_{jk}\Gamma'_{ik}}$. The second one arises only within nonlinear models, such as the FENE-P model, and deals with the closure of the elastic term $\overline{f(r)C_{ij}}$. Here we focus on the first closure, which is more related to the description of turbulent fluctuations rather than to polymer rheology. Our goal is then to derive a closed expression for Λ_{ij} in terms of $\overline{C}_{ij}$.

The appropriate reference frame to consider the deformation of a polymer molecule is the Lagrangian frame. In applications, polymer extension is known to be smaller than the Kolmogorov scale of the flow. In this range of scales the velocity increment between two points in the flow can be assumed to depend linearly on the relative separation between these points. This means that the velocity gradient depends only on time. In order to make analytical progress, we describe the velocity gradient fluctuations $\Gamma'_{ij}(t)$ according to the model introduced by Brunk et al. [7, 8] and subsequently applied to inertial particle motion in turbulent flows [9, 10]. Velocity gradient fluctuations are assumed to be Gaussian, statistically isotropic and stationary in time. The random gradient Γ'_{ij} can be decomposed into the sum of the rate-of-strain and the rate-of-rotation. From DNS of isotropic turbulence, the autocorrelation times of strain and rotation are known to be different and proportional to τ_η: $\tau_s = 2.3\tau_\eta$ and $\tau_r = 7.2\tau_\eta$, respectively [11]. This range of time scales cannot be solved analytically. Our approach is then to compute Λ_{ij} in the limit of very small autocorrelation times and deduce from that limit the general structure of Λ_{ij} in terms of $\overline{C}_{ij}$. The turbulent regime ($\tau_s = 2.3\tau_\eta$, $\tau_r = 7.2\tau_\eta$) could then be described by preserving the same structure of Λ_{ij} and properly tuning the coefficients appearing in that expression to match DNS.

We treat the small autocorrelation-time expansion of Λ_{ij} by generalizing the method presented in [13] to tensorial stochastic differential equations. We then let $\Omega = \tau_s/\tau_\eta$ tend to zero, while keeping $\tau_r/\tau_s = 7.2/2.3$ and $\Omega\, We$ constant. The latter condition ensures that the level of polymer stretching stays constant as Ω tends to zero.

The first term of the small Ω expansion corresponds to the white-noise limit and reads

$$\Lambda_{ij} = \Omega\, We(\alpha_0\, \overline{C}_{kk}\, \delta_{ij} - \beta_0 \overline{C}_{ij}) \tag{3}$$

with $\alpha_0 = 0.62$ and $\beta_0 = 1.20$. The above expression depends linearly on $\overline{C}_{ij}$ and explicitly shows that small-scale isotropic turbulent fluctuations make the conformation tensor isotropic. It is worth noting that the limiting expression (3) is independent of $\overline{\Gamma}_{ij}$ and the form of the entropic force $f(r)$. When $\tau_r = \tau_s$, (3) reduces to the expression derived in [14]. However, with that restriction, the model does not capture some key aspects of the phenomenology. For instance, if $\tau_r = \tau_s$, then Λ_{11} is forced to be positive for all We, which is not the case in the presence of a large mean shear.

The next order in the small Ω expansion gives

$$\begin{aligned}\Lambda_{ij} = {} & [(0.62\Omega\, We - 0.15\Omega^3\, We)\overline{C}_{kk}\delta_{ij} - (1.20\Omega\, We + 0.007\Omega^3\, We)\overline{C}_{ij}] \\ & -0.87\Omega^2[2\overline{C}_{lk}\overline{\Gamma}_{kl}\delta_{ij} - \overline{C}_{kk}(\overline{\Gamma}_{ij} + \overline{\Gamma}_{ji}) - \overline{C}_{kj}\overline{\Gamma}_{ki} - \overline{C}_{ki}\overline{\Gamma}_{kj}] \\ & +1.43\Omega^2(\overline{C}_{kj}\overline{\Gamma}_{ik} + \overline{C}_{ki}\overline{\Gamma}_{jk}) \\ & -\tfrac{2}{15}\Omega^2\, \overline{(3C_{ik}C_{kj} - C_{kk}C_{ij})f'(r)} - \tfrac{2}{3}R_0^2\Omega^2\delta_{ij}\ ,\end{aligned} \tag{4}$$

where $f'(r) = df/dr$. Although (4) is closed only for the Oldroyd-B model, it suggests that a contribution similar to that coming from the white-noise limit is present also for finite autocorrelation times of strain and rotation. Moreover, the contribution from the mean velocity gradient consists of all the possible combinations of $\overline{C}_{ij}$ and $\overline{\Gamma}_{ij}$ that give a symmetric second-order tensor. For the isotropic case ($\overline{\Gamma} = 0$) the above predictions were validated by Brownian Dynamics simulations based on the flow introduced in [7, 8] (Fig. 1). Numerical simulations were conducted by means of a second-order numerical scheme and the averages were computed both as ensemble averages over independent Lagrangian trajectories and steady-state time averages.

Even if they have been derived in the asymptotic range of small autocorrelation times, we expect that our predictions will be useful for deducing a closed equation for the mean conformation tensor in real turbulent viscoelastic flows. This result may be achieved by direct comparison with DNS.

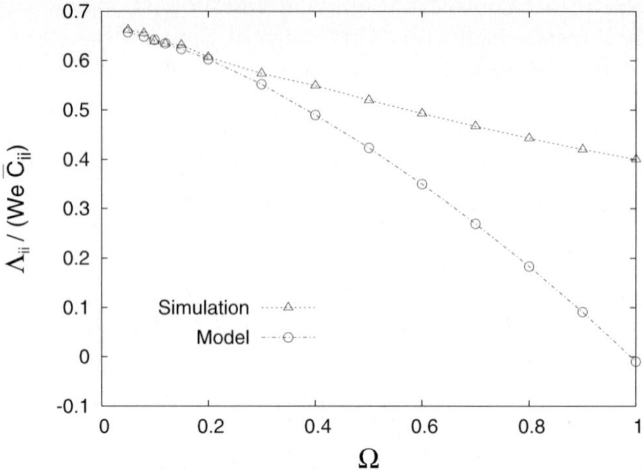

Fig. 1. Validation of the model prediction in the limit of small autocorrelation times of strain and rotation by means of Brownian Dynamics simulations ($\Omega\, We = 10$)

Acknowledgments

This work has been executed under the auspices of the International Collaboration for Turbulence Research. DV is grateful to A. Celani for useful suggestions.

References

1. Virk P S (1975) AIChE J 21:625-656
2. Sureshkumar R, Beris A N (1995) J Non-Newtonian Fluid Mech 60:53-80
3. Vaithianatan T, Collins L R (2003) J Comput Phys 187:1-21
4. Dubief Y, White C M, Terrapon V E, Shaqfeh E S G, Moin P, Lele S K (2004) J Fluid Mech 514:271-280
5. Leighton R, Walker D T, Stephens T, Garwood G (2003) Reynolds Stress Modeling for Drag Reducing Viscoelastic Flows. In: Proceedings of the 4th ASME/JSME Joint Fluids Engineering Conference, Honolulu, HI, USA, July 6-11 (FEDSM2003-45655), pp. 735-744
6. Bird R B, Hassager O, Armstrong R C, Curtiss C F (1977) Dynamics of Polymeric Liquids, Vol. 2. Wiley, New York
7. Brunk B K, Koch D L, Lion L W (1997) Phys Fluids 9:2670-2691
8. Brunk B K, Koch D L, Lion L W (1998) J Fluid Mech 364:81–113
9. Zaichik L I, Alipchenkov V M (2003) Phys Fluids 15:1776-1787
10. Chun J, Koch D L, Rani S L, Ahluwalia A, Collins L R (2005) J Fluid Mech 536:219–251
11. Girimaji S S, Pope S B (1990) Phys Fluids A 2:242-256
12. Jeffreys H (1931) Cartesian Tensors. Cambridge University Press, Cambridge
13. Dekker H (1982) Phys Lett A 90:26-30
14. L'vov V S, Pomyalov A, Procaccia I, Tiberkevich V (2005) Phys Rev E 71:016305

Part VIII

Large Eddy Simulation

Towards Wall Models for LES of Separated Flows

Michael Breuer[1], Boris Kniazev[1], and Markus Abel[2]

[1] Institute of Fluid Mechanics, University of Erlangen–Nürnberg, D–91058 Erlangen, Germany, breuer@lstm.uni-erlangen.de

[2] Institute of Physics, University of Potsdam, D–14415 Potsdam, Germany

1 Introduction

A key technology for the application of LES to high–Re flows of engineering interest is an appropriate wall modeling strategy bridging the near–wall region and thus avoiding the expensive DNS–like resolution. Standard wall models such as Schumann's model [1] work well for attached flows, e.g. plane channel flow, where the averaged quantities required can be calculated. The model reads: $\boldsymbol{\tau}_w = [\langle|\boldsymbol{\tau}_w|\rangle \, / \, \langle|\mathbf{v}_{tan}|\rangle] \, \mathbf{v}_{tan}$. Hence, a phase coincidence of the instantaneous wall shear stress $\boldsymbol{\tau}_w$ and the tangential velocity $\mathbf{v}_{tan}$ at the wall–nearest grid point is assumed. More complex 3–D flows with large pressure gradients or local separation are not well reproduced. For such flows the mean values required are not accessible and the validity of the law of the wall used is questionable. The model of Werner and Wengle [2] (WW) reads $\boldsymbol{\tau}_w = f(|\mathbf{v}_{tan}|) \, \mathbf{v}_{tan}$ where $f(|\mathbf{v}_{tan}|)$ is expressed by a law of the wall applied directly to the instantaneous quantities. Therefore, this model is also not able to predict separated flows correctly. The main reason is the dependence of $\boldsymbol{\tau}_w$ on additional quantities such as the pressure gradient $\boldsymbol{\nabla} p$. Thus, attempts were made to take $\boldsymbol{\nabla} p$ into account, e.g. by Hoffmann and Benocci [3].

2 Wall Modeling Using Statistical Data Analysis

The objective is to develop wall models which are more generally applicable, in particular for separated flows. For this purpose, generalizations of the classical wall models were tested which take only instantaneous quantities into account, i.e. $\boldsymbol{\tau}_w = f(\mathbf{v}, \boldsymbol{\nabla} p) \, \mathbf{v}_{tan}$. To find promising dependences $f(\mathbf{v}, \boldsymbol{\nabla} p)$ an *a–priori* data analysis was carried out, inferring conditional statistics from highly resolved data [4]. The procedure is: **1.** data generation on a fine grid, **2.** data analysis, **3.** model derivation, **4.** *a–posteriori* tests on a coarse grid.

As the first reasonably complex test case (besides channel flow) the flow over periodic hills at $Re = 10,595$ was considered. Data from a wall–resolved a difference of less than 1% between two continuous calculations was obtained. The y^+ value of the first grid point adjacent to the wall was less than 5. In addition, for present fully developed turbulent flow, periodic conditions were employed in the main flow direction and mass flow rate was specified to create a flow field. The calculations were terminated when the absolute residuals of all the variables became less than 10^{-6}.

$$\frac{\partial}{\partial x_j}(\rho U_j) = 0 \tag{1}$$

$$\frac{\partial}{\partial x_j}(\rho U_j U_i) = -\frac{\partial P}{\partial x_i} + \frac{\partial}{\partial x_j}[\mu(\frac{\partial U_i}{\partial x_j} + \frac{\partial U_j}{\partial x_i})] + \frac{\partial}{\partial x_j}(-\rho\overline{u_i u_j}) \tag{2}$$

$$\frac{\partial}{\partial x_j}(\rho U_j T) = \frac{\partial}{\partial x_j}(\frac{\mu}{Pr}\frac{\partial T}{\partial x_j} - \rho\overline{u_j t}) \tag{3}$$

3 Results and Discussions

Three planes right after the ribs in the triangular duct were investigated, as shown in Fig. 1. Here, U_0 stands for the bulk velocity, e for the rib size. Present predicted velocity field in Plane I was shown in Fig. 2(a). Two counter-rotating vortices were formed in each rib-formed corners of the triangular duct. The six vortices carried high momentum fluid towards the corners along the bisectors and then outwards to the middle plane along the walls. It agreed reasonably well with previous measured results in [7], see Fig. 2(b).

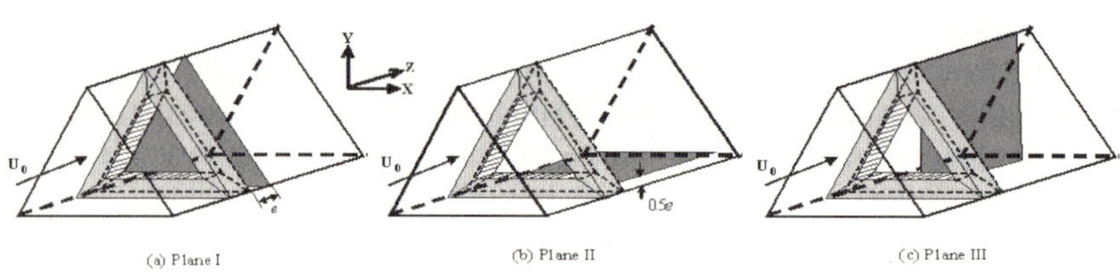

Fig. 1. Interested planes for the ribbed triangular duct

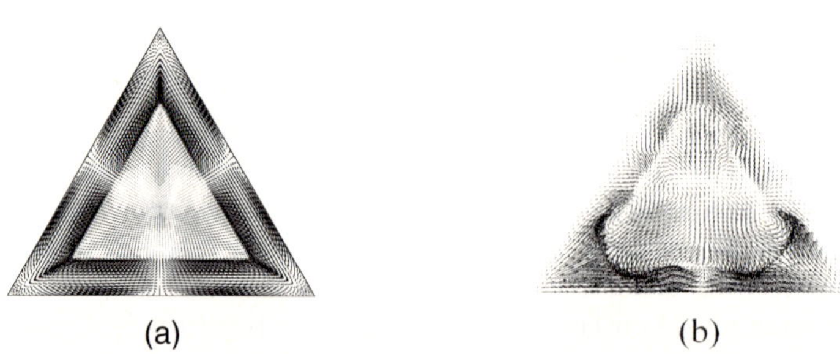

Fig. 2. (a) Predicted (b) Measured Velocity fields in Plane I

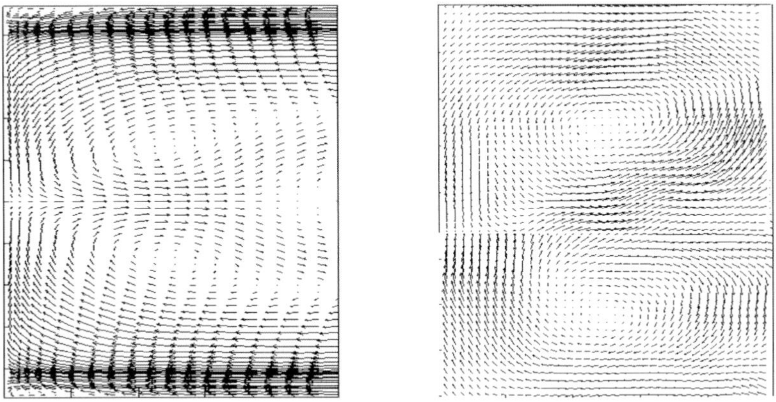

(a) (b)

Fig. 3. (a) Predicted (b) Measured Velocity fields in Plane *II*

In Plane *II*, one large vortex cell existed in each half part of the plane (Fig. 3), which suggested three-dimensional flow structures generated by the ribs. Again, good agreement was observed between predicted and measured results.

In the vertical Plane *III*, vortices were observed behind the ribs, both in predicted and measured work. For clear observation of the flow patterns, Plane *III* was divided into two parts: upper (a) and lower part (b), as shown in Figs. 4,5. The main flow reattached to the floor plane about 3–4 rib height away from the upstream rib. It was owed to the sudden expansion of the flow field and an acceleration of the flow. High turbulence intensities and velocity gradients in this region also contributed, which were brought in by the ribs which acted as turbulent promoters.

Compared to the Nusselt number distribution measured in [3], it could be concluded that rib-generated recirculation flows and main flow attachment on the side wall accounted for the enhancement of heat transfer of turbulent flow in the triangular duct.

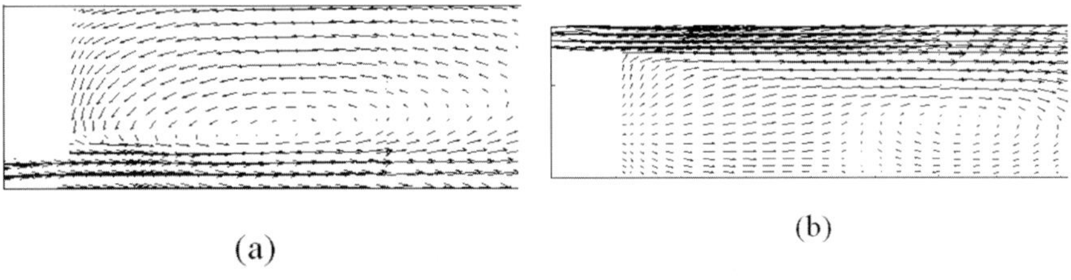

(a) (b)

Fig. 4. Predicted Velocity fields (a) upper part (b) lower part in Plane *III*

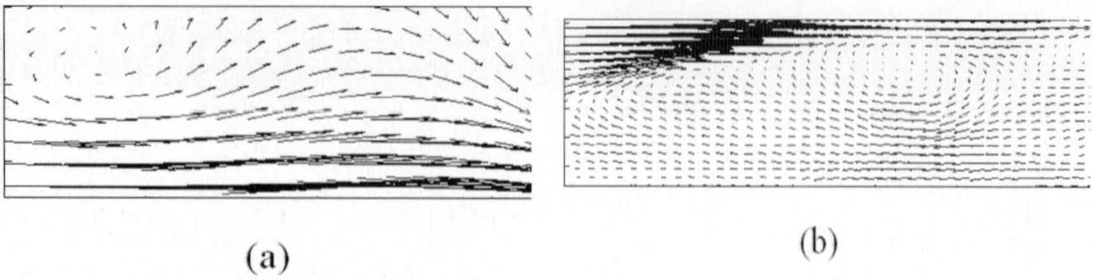

Fig. 5. Measured Velocity fields (a) upper part (b) lower part in Plane *III*

4 Conclusions

In present study, a well predication of the turbulent flow in a triangular duct, which was attached with square-sectional ribs on its internal surfaces, was obtained by RSM. Three-dimensional flow patterns were observed by the introduction of ribs. Vortices were developed after the ribs. Rib-induced recirculation flows around the ribs were detected and concluded to contribute most to the improvement of the thermal performance of present system.

References

1. Luo DD, Leung CW and Chan TL (2003) Optimum rib-to-rib spacing to enhance forced convection in the triangular duct with ribbed innner surfaces, ASME, Washington, DC, IMECE2003-41744
2. Acharya S, Dutta S, Myrun TA and Baker RS (1994) Turbulent flow past a surface-mounted two-dimensional rib, ASME Journal of Fluids Engineering, 116:238–246
3. Acharya S, Myrum TA, Qiu X and Sinha S (1996) Developed and periodically developed flow, temperature and heat transfer in a ribbed duct, International Journal of Heat and Mass Transfer, 40:461–479
4. Iacovides H (1996) Computation of flow and heat transfer through rotating ribbed passages, Biennial Colloqium on Computational Fluid Dynamics, UMIST, 3:19–24
5. Rigby DL (1998) Predication of heat and mass transfer in a rotating ribbed coolant passage with a 180 degree turn, ASME, 98-GT-329
6. Saidi A and Sunden B (2000) Numerical simulation of turbulent convective heat transfer in square ribbed ducts, Numerical Heat Transfer, Part A, 38:67–88
7. Luo DD, Leung CW, Chan TL and Wong WO (2005) PIV study of flow characteristics in a triangular duct with internal ribbed surfaces, 8th Flucome, Chengdu, China, Paper No. 375.

Multi-Cycle Simulations of In-Cylinder Flows on Unstructured Grids using a Hybrid VLES Model

F. Freikamp[1], J. Ewald, and N. Peters

Institut für Technische Verbrennung, RWTH Aachen
Templergraben 64, D-52056 Aachen, Germany
`freikamp@itv.rwth-aachen.de`

In an industrial context for the simulation of turbulent flows in reciprocating internal combustion engines typically RANS (Reynolds Averaged Navier Stokes) based modelling is deployed. Due to the growth of computational power and improvements in numerical techniques, the demands on the accuracy, reliability, and predictability of the numerical results increase. While RANS based modelling has been successfully employed for the prediction of mean quantities, transient effects like cycle-to-cycle variability cannot be captured accurately. An alternative to RANS-based models is the technique of Large Eddy Simulation (LES) where the governing equations are spatially filtered instead of being ensemble-averaged. Though numerically more demanding then RANS, LES promises more direct access to physical processes.

The moderate Reynolds numbers observed in IC engine flows render it an appealing candidate for LES. While LES mesh requirements are comparable to those of RANS [1] (except for near wall boundary layers), a sufficient grid resolution for LES, i.e. the mesh spacing and thus the filter width lies well within the inertial range of the turbulent spectra, cannot *pre se* be guaranteed in complex geometries, especially considering the problem of moving geometries and meshes. On the other hand the resolution of currently used meshes allows in principle to resolve turbulent structures well beyond the integral length scale, onto with the RANS models are usually based on, and thus additional information about the flow field and depending quantities are available.

In Fig. 1 the ratio of the filter width Δ, which is taken to be proportional to the cell size, to the integral length scale ℓ_t for a model engine is shown; the numerical grid used in the calculation is of a moderate mesh size of around 150 000 cells. The white line indicates the isosurface of $\Delta/\ell_t = 1$. Obviously the mesh resolution inside the core region sufficiently resolves the integral length scale which implies that in this part of the domain a large eddy simulation could be performed. The near-wall regions on the other hand are

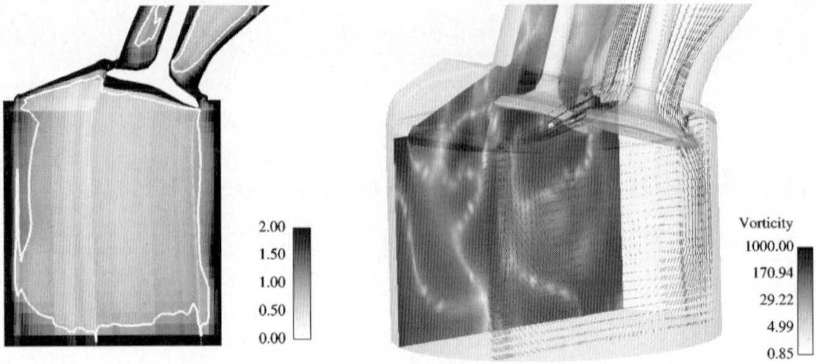

Fig. 1. Right: Ratio of filter width Δ to integral length scale ℓ_t as a measure for the mesh resolution; left: 3d view of the intake flow simulation with velocity vectors and the vorticity magnitude being depicted on cutplanes through the intake valves

under-resolved, therefore the usage of a LES model is not justifiable. To circumvent these problems, a hybrid RANS-LES model combining both LES and RANS modelling approaches will be presented in the next section. The model has been applied to a four-valve SI engine in order to study the phenomenon of cyclic variability in the fuel mixing and combustion process.

1 Model Formulation

The hybrid VLES model consists of two sub-models, a RANS and an LES model. The mode of the model is based on the ratio of the filter width Δ to the integral length scale ℓ_t. In regions of high mesh resolution a one-equation subgrid-k model [2] is used, otherwise the standard k-ε model [4] is employed.

The one-equation model by Yoshizawa et al. [2] for the subgrid kinetic energy k_r is given by

$$\frac{\partial \rho k_r}{\partial t} + \frac{\partial \rho u_j k_r}{\partial x_j} = \frac{\partial}{\partial x_j}\left[(\mu + \mu_t)\frac{\partial k_r}{\partial x_j}\right] + \mathcal{P}_r - \rho\varepsilon_r\,. \tag{1}$$

Here the subgrid dissipation ε_r and the subgrid viscosity μ_t are given by

$$\varepsilon_r = c_\varepsilon \frac{k_r^{3/2}}{\Delta} \qquad \text{and} \qquad \mu_t = C_k \bar{\rho}\sqrt{k_r}\,\Delta \tag{2}$$

and all field quantities are filtered[1].

The standard k-ε equations for the integral turbulent kinetic energy k and the dissipation rate ε read[2].

$$\frac{\partial \rho k}{\partial t} + \frac{\partial \rho u_j k}{\partial x_j} = \frac{\partial}{\partial x_j}\left([\mu + \frac{\mu_T}{\sigma_k}]\frac{\partial k}{\partial x_j}\right) + \mathcal{P} - \rho\varepsilon \tag{3}$$

[1] For the sake of brevity, the overbar indicating filtered quantities has been omitted.

[2] Here all field quantities are to be considered as Reynolds averaged.

$$\frac{\partial \rho\varepsilon}{\partial t} + \frac{\partial \rho u_j \varepsilon}{\partial x_j} = \frac{\partial}{\partial x_j}\left([\mu + \frac{\mu_T}{\sigma_\varepsilon}]\frac{\varepsilon}{\partial x_j}\right) + C_1 \mathcal{P}\frac{\varepsilon}{k} - C_2 \rho\frac{\varepsilon^2}{k} + C_3 \rho\varepsilon\frac{\partial u_j}{\partial x_j} \tag{4}$$

where the turbulent eddy viscosity μ_t is given by $\mu_T = C_\mu \rho \frac{k^2}{\varepsilon}$ and the turbulent production $\mathcal{P} = \tau_{t,ij}\frac{\partial u_j}{\partial x_i}$, with the Reynolds stress tensor $\tau_{t,ij}$ being modeled using the Boussinesq approximation; see [3, 4] for more details.

The model equations are solved in the whole domain. Using the integral value of the turbulent kinetic energy and the dissipation rate, the integral length scale ℓ_t can be estimated while the filter width Δ is related to the local cell size of the mesh, i.e. determined by means of the finite cell volume V_{cell}:

$$\ell_t = \frac{k^{3/2}}{\varepsilon} \quad , \quad \Delta = \beta \sqrt[3]{V_{cell}} \, ; \tag{5}$$

the factor β is usually taken to be $2 \ldots 3$. The model mode, i.e. whether the LES or the RANS model is used in determining the turbulent eddy viscosity, is determined by the ratio Δ/ℓ_t. For $\Delta/\ell_t < 1$ the mesh is fine enough to fully resolve the integral length scale thus the LES mode of the model is used (6), while for values larger than 1 the RANS mode is active (7).

$$\Delta < \ell_t : \quad \varepsilon_r = c_\varepsilon \frac{k_r^{3/2}}{\Delta} \quad \mu_t = C_k \rho \sqrt{k_r}\, \Delta \tag{6}$$

$$\Delta \geq \ell_t : \quad \varepsilon_r = \varepsilon \quad \mu_t = \mu_T \tag{7}$$

In the whole domain the ε equation is solved for, but in the LES region the algebraic model for the sub-grid dissipation (6) is employed. The standard model constants for the k-ε equations are used, while the constants for the LES model are chosen such, that the turbulent eddy viscosity changes continuously between the two regimes. The standard wall function approach is used as boundary conditions for the RANS equation.

2 Multi Cycle Simulation

The model has been applied to the simulation of multiple cycles a four-valve SI engine with a flat piston design; further specifications are given in Table 1. The mesh size used for the calculations were of low-moderate size of approximately 150 000 cells.

The history of the turbulent kinetic energy and the axial velocity near the spark plug are shown in Fig. 2 for ten full cycles employing the hybrid VLES model. The first three cycles of the simulation reveal significant variations of the flow pattern. After this initial transition phase stochastic variations of the flow field around an average value can be detected in the following cycles indicating the cycle-to-cycle variability phenomenon. The largest variations in the profiles can be detected during the compression stroke around $-100°$ to $-50°$ before top dead center, i.e. during the essential phase for the air-fuel mixing process.

Table 1. Engine characteristics

stroke	86 mm
bore	92 mm
engine speed	2000 rpm
intake valve closure	580°
	370° – 580°
simulation intervals	580° – 890°
	890° – 1090°

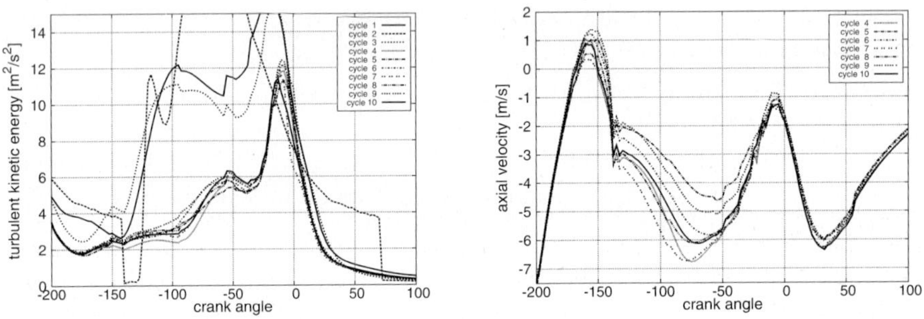

Fig. 2. History of the turbulent kinetic energy (left) and the axial velocity over multiple cycles indicating cycle-to-cycle variation of the in-cylinder flow structures

3 Summary

A new hybrid RANS-LES model has been presented and applied to the simulation of a four-stroke SI engine. The model has been shown to be capable of capturing the effects of cycle-to-cycle variability in gasoline engines.

This research work was funded by the Bundesministerium für Bildung und Forschung under contract 03 PEA 1 AC.

References

1. Haworth D C and Jansen K (2000) Large-Eddy Simulation on Unstructured Deforming Meshes: Towards Reciprocating IC Engines. Computers & Fluids, 19: 493-524
2. Yoshizawa A and Horiuti K (1985) A Statistically-Derived Subgrid-Scale Kinetic Energy Model for the Large-Eddy Simulation of Turbulent Flows. Journ. Phys. Soc. Japan, 54 (8):2834-2839
3. Launder B E and Spalding D B (1974), The Numerical Computation of Turbulent Flows, Comp. Meth. Appl. Mech. Eng., 3:269-289
4. El Tahry S H (1983) k–ε–Equation for Compressible Reciprocating Engine Flows. AIAA Journal of Energy, 7 (4):345-353

Investigation of Lattice Boltzmann Methods for LES

Rainhill K. Freitas, Wolfgang Schröder, and Matthias Meinke

Aerodynamic Institute Aachen, RWTH Aachen, Aachen, Germany
r.freitas@aia.rwth-aachen.de

1 Introduction

The Lattice Boltzmann Method (LBM) can be used to simulate continuum flows based on the gas kinetic assumptions of the Boltzmann equation. In the last decade the LBM has been proved to be a competitive alternative to conventional Navier-Stokes based CFD methods in certain fields of application, like e.g. multiphase flows and flows through porous media. One of the main advantages of the LBM is its computational efficiency and parallel scalability being a consequence of the locality of its collision operator, which does not require the computation of differentials.

In the present work the application of the Lattice Boltzmann Method for large eddy simulations has been investigated. A well known standard LB model, the Bhatnagar-Gross-Krook (BGK) method [1] has been applied.

2 The Boltzmann Equation

The Boltzmann equation is a nonlinear integro-differential equation, describing the evolution of a molecular distribution function

$$\frac{\partial f}{\partial t} + \sum_{i=1}^{3}(\xi_i \cdot \frac{\partial f}{\partial x_i}) + \sum_{i=1}^{3}(\frac{\partial f}{\partial \xi_i} \cdot \frac{F_i}{m}) = \int_{\xi_1'} \int_{A_C} (f' f_1' - f f_1)\ \mathbf{g}\ dA_c d\boldsymbol{\xi} \quad (1)$$

Here the quantity f is the particle distribution function, ξ_i is the i-th component of the molecular velocity vector, F_i is the i-th component of an external force and m is the particle mass. f' and f_1' are the pre- and f and f_1 are the post-collision distributions. For a derivation of the Boltzmann equation and a detailed description of its assumptions and limitations, see, e.g. [2].

A simplified collision operator for the Boltzmann equation has been proposed by Bhatnagar, Gross and Krook [1]. In the BGK model the collision term reduces to $\omega(F - f)$ resulting in the BGK equation

$$\frac{\partial f}{\partial t} + \xi_i \cdot \frac{\partial f}{\partial x_i} = \omega(F - f), \tag{2}$$

in which ω represents the collision frequency and F is the Maxwell equilibrium distribution function.

Discretization

In the present work the D3Q19 model has been applied. The DnQm notation, in which n stands for the number of dimensions and m stands for the number of discrete velocities, has been introduced by Qian et al. [3], where also a detailed discussion on other valid phase space discretizations can be found. The numbering of the discrete velocity set for this model is shown in Fig. 1. It reads:

$$\xi_i = \xi_0 \begin{cases} (0,0,\pm 1); (0,\pm 1,0); (\pm 1,0,0) & \alpha = 1..6 \\ (\pm 1,\pm 1,0); (\pm 1,0,\pm 1); (0,\pm 1,\pm 1) & \alpha = 7..18 \\ (0,0,0) & \alpha = 19 \end{cases} \tag{3}$$

Discrete BGK Equation

The discrete BGK (2) and the discrete multi-dimensional equilibrium distribution possess the form:

$$f_i(\mathbf{r} + \xi\delta t, t + \delta t) = f_i(\mathbf{r}, t) + \omega\delta t \cdot (F_i(\mathbf{r}, t) - f_i(\mathbf{r}, t)), \tag{4}$$

$$F_i(\mathbf{r}, t) = \rho\, t_p \big[1 + \frac{\xi_{i,\alpha} u_\alpha}{c_S^2} + \frac{u_\alpha u_\beta}{2c_S^2}(\frac{\xi_{i,\alpha}\xi_{i,\beta}}{c_S^2} - \delta_{\alpha,\beta})\big] \tag{5}$$

where $\alpha = 1,2,3$ and $\beta = 1,2,3$ represent the space dimensions and $\delta_{\alpha\beta}$ is the Kronecker delta. The coefficients t_P are weighting factors to be chosen such that macroscopic symmetry, conservation of mass, momentum and energy are satisfied. For the D3Q19 model one obtains $t_0 = 1/3,\ t_1 = 1/18, t_2 =$

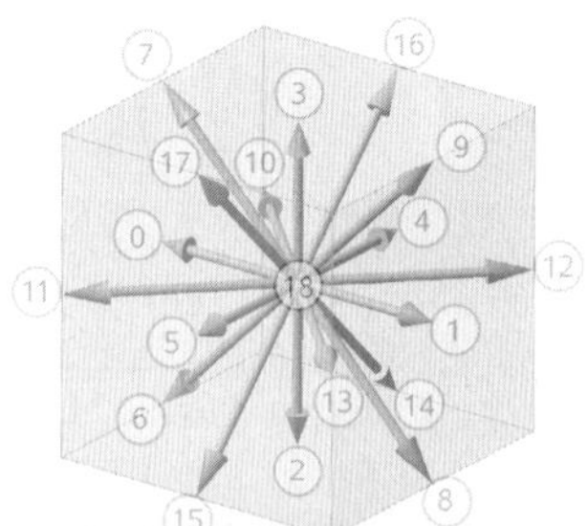

Fig. 1. The D3Q19 model

1/36 , and $c_S = \xi_0^2$, where the index describes the square modulus of the discrete velocities ξ_i. To derive the weighting factors, see, e.g. [4]. The macroscopic flow values are determined by summation over the base moments

$$\rho = \sum_{i=1}^{i=i_{max}} f_i = \sum_{i=1}^{i=i_{max}} F_i \tag{6}$$

$$\rho u_\alpha = \sum_{i=1}^{i=i_{max}} \xi_\alpha f_i = \sum_{i=1}^{i=i_{max}} \xi_\alpha F_i \tag{7}$$

$$\rho(e + u_\alpha^2) = \sum_{i=1}^{i=i_{max}} \xi_\alpha^2 f_i = \sum_{i=1}^{i=i_{max}} \xi_\alpha^2 F_i \tag{8}$$

3 Results

The computational domain of the channel flow possess a length (L) , height (H), and width (W), $\pi H \times 2H \times 0.289\pi H$, according to the minimum specifications for a periodic channel, given by JIMÉNEZ & MOIN [5]. The turbulent channel flow [6] applies a Reynolds number based on the shear stress velocity and the channel half-width (H) Re_τ of 180. The resolution of the uniform grid is $256 \times 128 \times 64$ grid cells, resulting in a y^+ of 2.8. Figure 2 shows that the velocity distribution according to the log law is obtained. However, deviations from the theoretical and other LES [7] and DNS [6] distributions occur in and slightly above the viscous sublayer, i.e. up to a y^+ value of around 15.

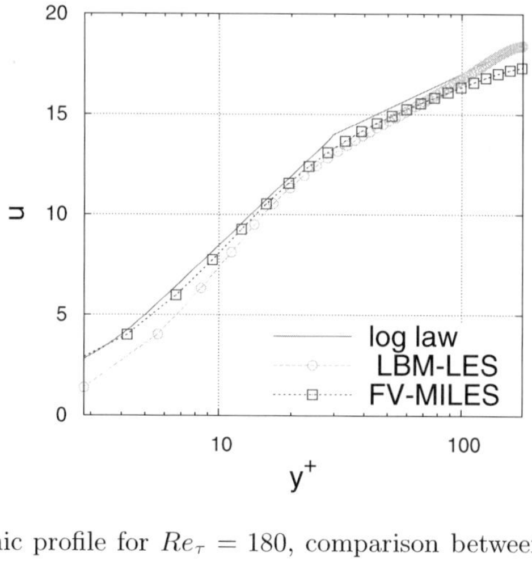

Fig. 2. Logarithmic profile for $Re_\tau = 180$, comparison between LBM-LES, FVM-MILES and DNS

Conditional Averaging of the Fully Developed Stationary Ribbed Duct Flow Using Q Criteria

Máté Márton Lohász[12], Patrick Rambaud[1], and Carlo Benocci[1]

[1] Environmental and Applied Fluid Dynamics Department, Von Karman Institute for Fluid Dynamics, Rhode-Saint-Genèse, Belgium lohasz@vki.ac.be, rambaud@vki.ac.be, benocci@vki.ac.be

[2] Department of Fluid Mechanics, Budapest University of Technology and Economics, Budapest, Hungary lohasz@ara.bme.hu

1 The Investigated Flow

The fully developed flow in a square section ribbed duct was simulated (see Fig. 1 (left)). The computation was carried on with a Reynolds number of 40000, defined by the bulk velocity (U_b) and the hydraulic diameter in the undisturbed region (D_h). The present configuration was characterized by a blockage ratio (rib height versus hydraulic diameter) of 0.3 and a pitch ratio (rib height versus pitch distance) of 10. This configuration is representative of the internal cooling channels in turbine blades and had been investigated by experimental means [1]. The calculation was performed applying periodic boundary conditions along the streamwise direction approximating fully developed flow. The flow was investigated by Large-Eddy Simulation (LES). The commercial finite volume code Fluent 6.1.22 was used applying the monotonic integrated approach to model the sub grid scale stresses with the use of a second order upwind scheme. The wall condition is imposed applying the lin-log law. For the present grid the first point is located, on the average, at a distance of $y^+ = 5$ from the wall. This value is insufficient to resolve the wall layer,

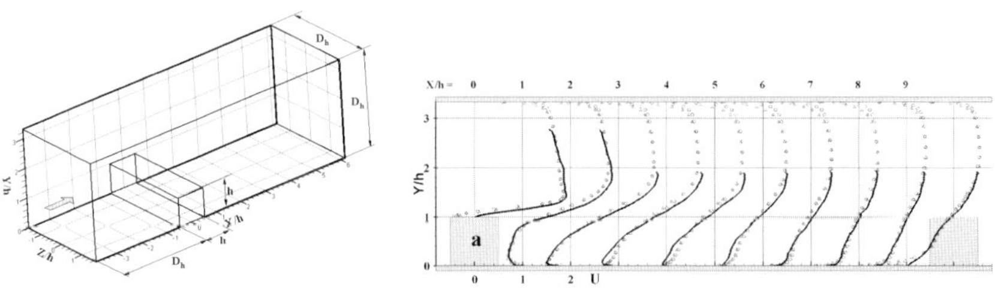

Fig. 1. Left) The computational domain. Right) Validation of the result against PIV measurement (solid line PIV, symbols LES)

but still ensures an acceptable estimation of the wall shear stresses. A time step of $\Delta t = 0.005 D_h/U_b$ was used to ensure an average CFL value of 0.3. To ensure sufficient convergence of the conditional average, 360 flowtroughs (p/U_b) were simulated on a grid made of 468900 hexahedral cells. More details can be found in [2]. The flowfield was validated against the extensive PIV measurements of [1] in six planes comparing in-plane velocities and Reynolds stresses. An example of the streamwise velocity in the symmetry plane of the duct is shown in Fig. 1 (right).

2 Condition Averaging

The present paper proposes new results using the concept of conditional averaging. The aim is to obtain a detailed understanding of the effect of coherent structures (CS) on the turbulent field. The Q criterion [3] was chosen for the detection of the vortices. Q is the second invariant of the velocity gradient tensor, and can be written in the following form: $Q \doteq \frac{1}{2}\left(\Omega_{ij}\Omega_{ji} - S_{ij}S_{ji}\right)$, where $S_{ij} \doteq \frac{1}{2}\left(\partial_i u_j + \partial_j u_i\right)$ is the rate of strain tensor and $\Omega_{ij} \doteq \frac{1}{2}\left(\partial_i u_j - \partial_j u_i\right)$ is the vorticity tensor. When positive, this invariant selects only the vorticity where rotation dominates the shear. Conditional averaging [4] of all first and second order quantities was determined relying on four Q classes. These classes were defined by the following three ad-hoc Q thresholds: $0, 200, 1500 U_b^2/h^2$. Figure 2 shows three different classes visualised in an instantaneous flow realisation. Considering these classes (Q_α) an indicator function was defined:

$$I_\alpha \doteq \begin{cases} 1 & Q(\mathbf{x},t) \in Q_\alpha \\ 0 & Q(\mathbf{x},t) \notin Q_\alpha \end{cases} \tag{1}$$

The conditional average of a scalar variable φ is then defined as ($\langle\rangle$ denotes Reynolds averaging):

$$\langle\varphi\rangle^\alpha \doteq \frac{\langle\varphi I_\alpha\rangle}{\langle I_\alpha\rangle} \tag{2}$$

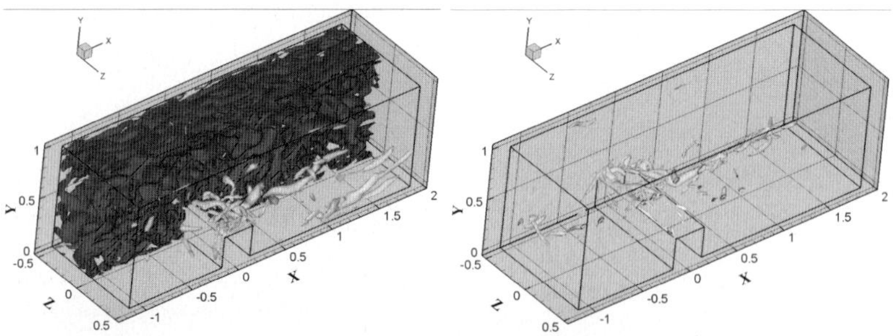

Fig. 2. Instantaneous Q isosurfaces, dark gray $Q = 0$, grey $Q = 200U_b^2/D_h^2$, white $Q = 1500U_b^2/D_h^2$

3 Results

A first important result consists in the probability density function of the vortices from the domain. It is usually believed that the instantaneous turbulent motions can be efficiently described by typical large energy containing vortices [5], which justifies the interest in the investigation of their more probable location. Figure 3 (left) shows the symmetry plane of the duct with the probability of fluid packets not belonging to any vortices. The high value associated to the class 1 ($\langle I_1 \rangle$) in the impinging region of the upstream side of the rib points out the destruction of vortices at this location. On the other hand Fig. 3 (right) highlights regions whith a high population of vortices. They are very probable on the top of the rib, directly downstream of the leading edge (the most important source of the vortices [2]). A high probability of presence of vortices is also found at the upstream corner of the rib, associated with the high vorticity separated region of the average field. The vortices are then convected downstream in the free shear layer of the wake of the rib. The probability first decreases and then increases along the streamlines till the reattachment. Reattachment seems to enhance the formation of the vortices. Figure 4 (left) shows the spanwise extension of high vortex probability region. Vortices are important also in the region close to the lateral wall, which corresponds to a region of high vorticity in the averaged flow presenting structures arching over the rib (described in [2]). In the Fig. 4 (right) streamlines in the symmetry plane of the duct are shown. The streamlines for the entire averaged flow are compared to the streamlines of field identified by the conditional average. The latter field is integrated from the vector field of the $\langle U \rangle^3 \langle I_3 \rangle + \langle U \rangle^4 \langle I_4 \rangle$ and $\langle V \rangle^3 \langle I_3 \rangle + \langle V \rangle^4 \langle I_4 \rangle$ conditional averaged velocity fields. It is clear from this picture that vortices ($Q > 200U_b^2/D_h^2$) behave different than the average fluid. A different behaviour of high energy vortices with respect to the average flow is noticed above the rib: the uplift is stronger for these structures. In the wake of the rib the recirculation region of the vortices is smaller and its centre is displaced upstream with respect to the overall average flow. The streamlines of the vortices (with $Q > 200U_b^2/D_h^2$) in the recirculation region do not correspond to a reattachment, but the corresponding positive bifurcation line is lifted up from the bottom surface to create a free stagnation "point" around $X/D_h = 1$. In other words, the main tendency of the vortices is to impinge the

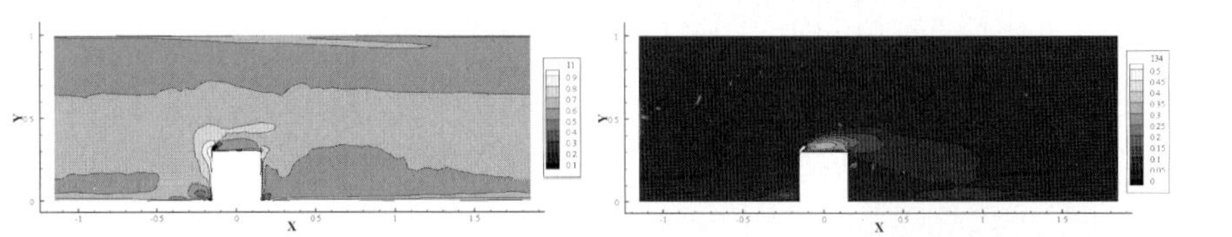

Fig. 3. Left) Probability of the $Q < 0U_b^2/D_h^2$ class ($\langle I_1 \rangle$). Right) Probability of the $Q > 200U_b^2/D_h^2$ class ($\langle I_3 \rangle + \langle I_4 \rangle$)

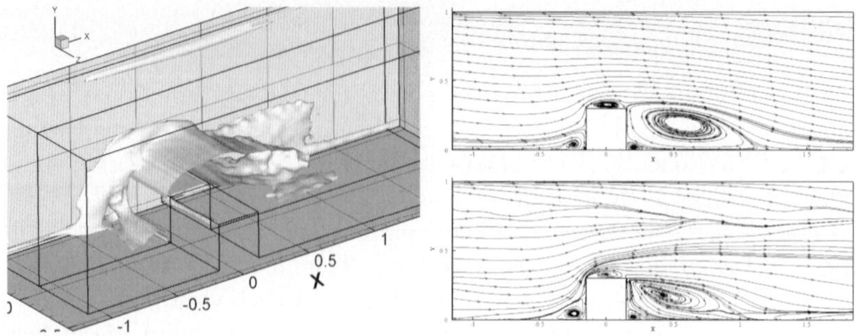

Fig. 4. Left) Isosurface of $\langle I_3 \rangle + \langle I_4 \rangle = 0.055$ Right) Streamlines in the symmetry plane of the duct ($Z = 0$) Upper) Averaged for all classes. Lower) Averaged for $Q > 200U_b^2/D_h^2$ classes

bottom wall boundary layer before being entrained by the main stream. This may complete the interpretation of the misalignment of the reattachment line with the maximum heat-transfer position ([6]).

References

1. Casarsa L, Arts T, (2005), Experimental Investigation of the Aerothermal Performance of a High Blockage Rib-Roughened Cooling Channel, J Turbomach, 127:580–588
2. Lohász M M, Rambaud P, Benocci C, (2005) Flow Features in a fully developed ribbed Duct Flow as a Result of LES, Proceedings of ERCOFTAC Int. Symposium on Engineering Turbulence Modelling and Measurements, ETMM6, May 23-25, 2005, Sardinia, Italy, 267–276
3. Hunt J C R, Wray A A, Moin P, (1988) Eddies, Stream, and Convergence Zones in Turbulent Flows, Center of Turbulence Research, Proceedings of the Summer Program, 1988:195–208
4. Antonia R A, (1981) Conditional sampling in turbulence measurement, Annu Rev Fluid Mech, 13:131–156
5. Hussain A K M F, (1986) Coherent Structures and Turbulence, J Fluid Mech, 173:303–356
6. Keating A, Piomelli U, Bremhorst K, Nešić S, (2004), Large-eddy simulation of heat transfer downstream of a backward-facing step, J Turbul, 5:020

Prediction of Flow and Heat Transfer in a Czochralski Crucible using LES with Interface Tracking

A. Raufeisen[1], T. Botsch[1], V. Kumar[2], M. Breuer[2], and F. Durst[2]

[1] Process Engineering Department (VT), University of Applied Sciences Nuremberg, Wassertorstr. 10, 90489 Nuremberg, Germany
alexander.raufeisen@fh-nuernberg.de

[2] Institute of Fluid Mechanics (LSTM), University of Erlangen-Nuremberg, Cauerstr. 4, 91058 Erlangen, Germany breuer@lstm.uni-erlangen.de

1 Introduction

The Czochralski (Cz) method is the preferred process for growing large silicon single crystals for wafers which are used in electronic and photonic devices. The liquid silicon ($Pr = 0.013$) is contained in an open crucible which is rotating, while the counterrotating crystal is pulled from the melt. Due to this setup, centrifugal and Coriolis forces, buoyancy, and Marangoni convection occur in the fluid as well as thermal radiation from the surface and the phase change due to crystallization. The shape of the interface between melt and crystal is crucial for the quality of the resulting crystal. Therefore, the influences of all effects on the crystallization front need to be investigated.

Unfortunately, the flow inside the melt is fully turbulent ($Re \approx 10^4$, $Ra \approx 10^7$), which makes numerical predictions difficult. It can be shown that the turbulent structures in this case are highly anisotropic and thus classical RANS models are not applicable. Highly accurate Direct Numerical Simulations (DNS) require high resolution and therefore use massive computational resources. Large Eddy Simulation (LES) combines the advantages of both: The large turbulent scales are computed directly, whereas the small scales are modeled. Thus a relatively high accuracy is achieved with moderate computational effort, so that parametric studies can be conducted easily.

2 Numerical Details

To validate the LES method for Cz simulations, computations of an idealized cylindrical case are compared to DNS reference data [2]. The crucible and crystal diameter are 340 mm and 100 mm, respectively, and the crucible is rotating with 5 rpm, while the crystal rotation rate is -20 rpm. A fixed temperature distribution obtained from an experiment and interpolated to the new geometry is prescribed at the crucible walls. The crystal/melt-interface is not moved and kept at the melting temperature of silicon. Furthermore, a fixed heat flux from the free surface accounts for radiation.

The computational grid for the LES consists of 627,200 control volumes, and a time step size of $\Delta t = 10^{-2}s$ is applied to achieve a CFL number of about 1. The simulations are run 60,000 steps; after 6000 steps time-averaging is conducted. For comparison with the DNS data set, the field is also averaged in circumferential direction. For the LES, implicit filtering is used. The subgrid scales are approximated with the Smagorinsky model with a constant $C_s = 0.065$. Near the walls, a van Driest damping function is applied. The temperature field is modeled with a constant turbulent Prandtl number $Pr_t = 0.9$.

The results are in good agreement with the reference data (Fig. 1). In the flow field, the characteristic vortices are resovled. The temperature distribution shows very little deviations, see also Fig. 2. Even simulations with a coarser resolution (192,000 CVs, $\Delta t = 5 \times 10^{-2}s$) lead to satisfactory results (not shown here). The computational effort for the LES is approximately 100 times less than for the DNS (LES in the order of $10^0 - 10^1$ CPU-hours on a NEC SX-8 using 5 processors).

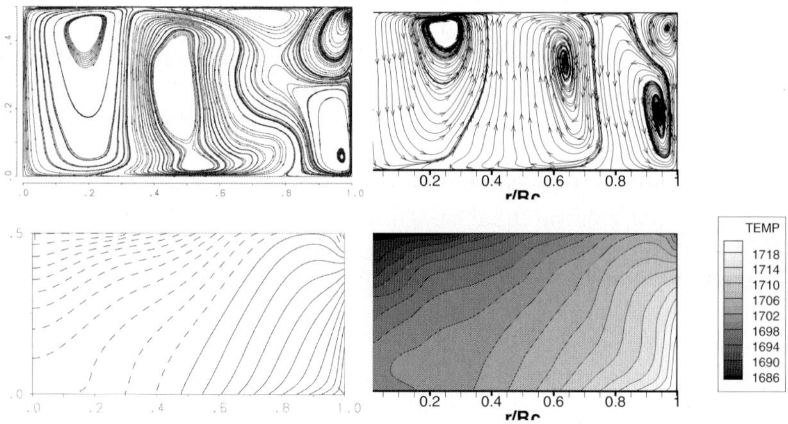

Fig. 1. Streamtraces and mean temperature by DNS [2] (left) and present LES computation in a vertical cut through the crucible.

3 Interface Tracking

The crystal/melt-interface is simulated using a moving grid approach, which allows the exact determination of the interface position [5]. To maintain continuity, an additional Space Conservation Law (SCL) has to be fulfilled. Moreover, to avoid grid distortions, a smoothing algorithm is applied. Test computations of a melting process of Gallium show very good agreement of the interface shape with experiments (Fig. 3).

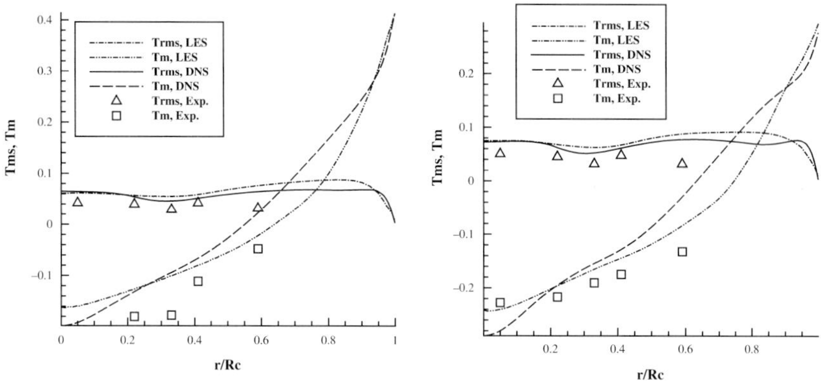

Fig. 2. Mean and rms temperature distribution along horizontal planes at $z/R_c = 0.21$ and $z/R_c = 0.32$ obtained by DNS [2], present LES prediction, and experimental measurements by Wacker [2].

With LES it is now possible to simulate industrial cases taking all occuring effects into account. A realistic Cz test case by Togawa [4] was computed to investigate different mechanisms used to control the interface shape. Due to the small size of the crucible (6 cm diameter), the flow is only weakly turbulent, so that a 2D axisymmetric approximation is possible, as was also shown by comparison with 3D simulations. As can be seen in Fig. 4, increasing the rotation rate or enhancing the crystal cooling can deflect the crystallization front upwards. This is achieved by breaking up the big convection roll, which is normally pulling the interface downwards, into small vortices that direct the flow under the crystal upwards.

However, to obtain realistic (thermal) boundary conditions, one has to consider the surrounding equipment as well. As the turbulent scales of the thermal field are much larger than those of the flow, this can be done by a

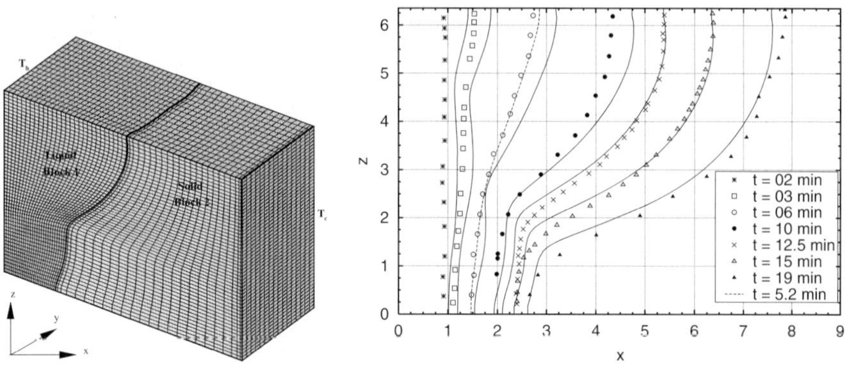

Fig. 3. Gallium melting in a square cavity heated from the left: Experimental measurements of the interface movement (symbols) by Gau and Viskanta [3] compared to present 3D simulation results (lines)

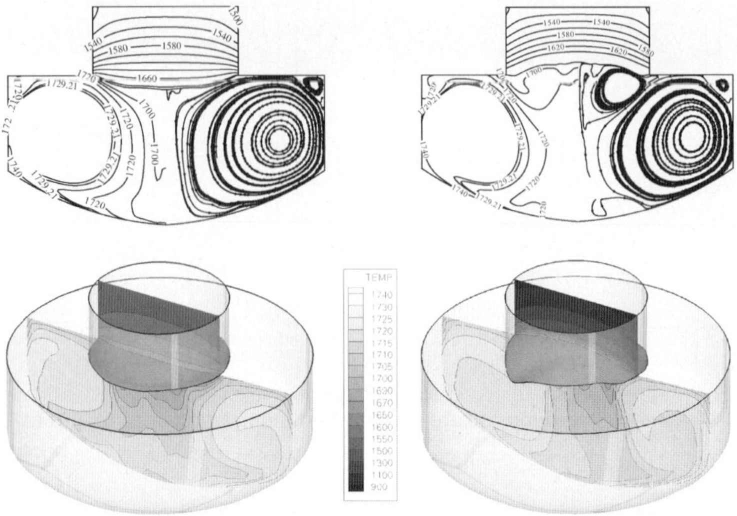

Fig. 4. Simulations of a Cz model crucible. Temperature and flow fields in 2D (top), temperature distribution in 3D case (bottom). An increase of the crystal rotation rate causes the melt/crystal-interface to change from a downward shape (upper left) to an upward shape (upper right). The same effect can be observed for an increase of the crystal cooling (lower left to lower right)

global 2D heat transfer computation coupled with the local 3D LES of the crucible and crystal. This is the subject of ongoing research.

4 Summary and Conclusions

LES computations of flow and heat transfer in an idealized Cz case were validated against reference DNS data. There is a good overall agreement, however, a detailed analysis especially of turbulent quantities has to be conducted. The melt/crystal-interface can be computed exactly by the moving grid approach. Therefore, it is possible to investigate the numerous influences on the interface. Parametric simulations of a model Cz crucible show the effects of changing the crystal rotation rate and cooling of the crystal. To obtain realistic boundary conditions, it is necessary to use global 2D simulations coupled with the local 3D simulation.

References

1. Kumar V et al. (2003) J Cryst Growth 255:27–39
2. Wagner C, Friedrich R (2004) Int J Heat Fluid Fl 25:431–443
3. Gau C, Viskanta R (1984) Int J Heat Mass Trans 27:113–123
4. Togawa S et al., (1996) J Cryst Growth 165:362–371
5. Jana S, Ray S, Durst F (2006) Appl Math Model, in press

Numerical Study of Turbulent Flow Around an Obstacle at the Wall of a Rectangular Duct

A. Teruzzi[1], S. Salon[2], F. Ballio[1], V. Armenio[3]

[1] Politecnico di Milano, p.zza Leonardo da Vinci 32, 20133 Milano, Italy
[2] Istituto Nazionale di Oceanografia e di Geofisica Sperimentale - OGS, B.go Grotta Gigante 42/c, 34010 Sgonico (TS), Italy
[3] Università di Trieste, p.zzale Europa 1, 34127 Trieste, Italy
armenio@dic.units.it

The turbulent flow field around a trapezoidal obstacle located at a vertical wall of a rectangular duct is investigated by resolved Large Eddy Simulation; this problem is archetypal of the flow field around bridge abutments; the study has shown that the distribution of the pressure and tangential stresses around the structure are intimately connected with local scour phenomena. In particular, the stresses at the bottom wall exhibit high mean values near the obstacle edge, as well as significant fluctuations in the upstream region. Our results also show the typical mean vortex systems developing around the obstacle.

1 Introduction and Problem Definition

The knowledge of the turbulent flow field at a bridge abutment is relevant for the comprehension of sediment-flow interaction phenomena. Over the last decades, a large number of literature works has been devoted to the analysis of local scours around bridge piers and abutments, see for example [3] and literature cited therein. In spite of this, there is a limited knowledge of the flow characteristics around abutments, see [2], and a general poor understanding of the interaction between the fluctuating turbulent flow field and the scour process at the rived bed. With aim to improve knowledge on the characteristics of the turbulent field around abutments, we carried out *resolved* Large Eddy Simulations (LES) (thus directly resolving the wall layer) in a rectangular cross-sectional *closed* duct with a trapezoidal body placed on the wall, representing an abutment. Figure 1 shows a sketch of the geometry of the duct and the coordinate system.

The simulation discussed herein uses the generalized coordinate Lagrangian dynamic mixed sub-grid scale model presented in [1]. Since we consider a closed rectangular duct no-slip boundary conditions are set on the duct walls; the inlet condition is a series of cross-section velocity fields previously obtained

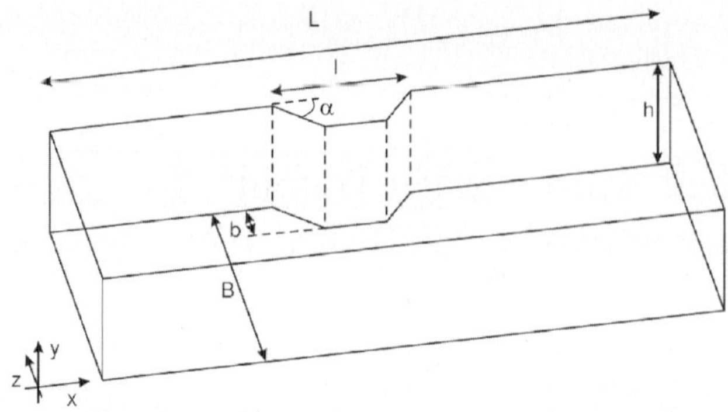

Fig. 1. Sketch of duct geometry and coordinate system. Duct and abutment dimensions: $B = 2h$, $L = 13.5h$, $b = 0.5h$, $l = 1.5h$, $\alpha = 45°$

from a developed and constant mass-rate flow in a straight rectangular duct, with same dimensions and flow-rate of the actual simulations; the Reynolds based on the duct height and the bulk velocity number is set equal to 7000. Numerical simulations were carried out using $256 \times 64 \times 96$ grid cells, respectively in the x-, y- and z-direction. The grid points are clustered near the walls, in the y-z plane and in the obstacle region along the streamwise direction x. The minimum grid spacings are: $\Delta x^+_{min} = x\, u_\tau/\nu = 10$ and $\Delta y^+_{min} = \Delta z^+_{min} = 2$, ensuring the direct resolution of the viscous sublayer. The quantity u_τ is the mean friction velocity in the corresponding straight duct. The time step in numerical simulations is constant giving a value of Courant number about 0.4; after a statistical steady state was reached we carried out 4500 time steps, thus computing about 10 large eddy turn over time ($LETOT = h/U$ where U is the bulk velocity in the straight duct). The mean quantities presented in Sect. 2 are obtained by averaging the flow field in time and over the two *symmetric* parts of the duct separated by the $(x, h/2, z)$ plane.

2 Results and Discussion

2.1 Vortex Structures

The mean flow field presents typical mean vortex structures, detected by the use of the well known Q-criterion: in particular the iso-Q surfaces plotted in Fig. 2 identify vortices in the three-dimensional space. Due to symmetry along the y direction, the bottom half duct is shown Fig. 2. A main vortex structure (denoted by I) is detected, similar to the horse-shoe vortex developing in junction flows. The width of such structure, on average is approximately equal to $0.09h$, corresponding to about 35 wall units. Obviously this structure cannot develop symmetrically around the obstacle due to the presence of the vertical solid wall $z = B$. The presence of the obstacle also causes a clockwise

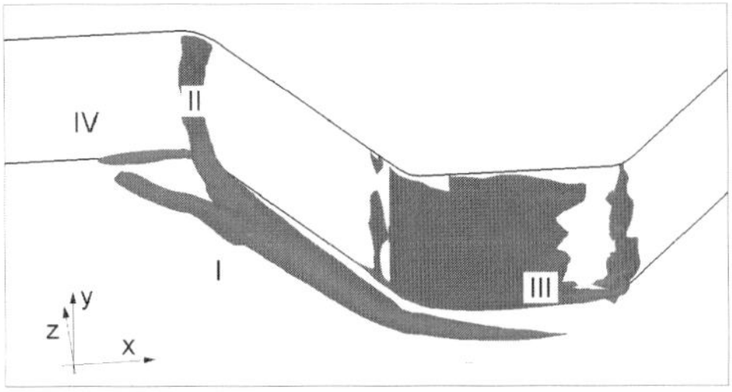

Fig. 2. Iso surface of the scalar Q in the obstacle region. The iso-surfaces (dark gray) $Q = 3U^2/h^2$ are reported

recirculation identified by the structure (II) in Fig . 2; due to the interaction with the bottom wall this structure tends to be curved and to merge into the main structure (I). The vortex structure II has a width approximately equal to 25 wall units, it is located at the wall-obstacle junction near the symmetry plane $y = h/2$ and progressively detaches from the junction line going down toward the bottom wall. The vortex system (III) identifies vortex sheet detaching from the obstacle edge. Finally (IV) identifies a structure that rises due to the mutual interaction between the vertical and the horizontal walls upstream the obstacle. This structure has a width approximately equal to 12 wall units and is placed at the junction between the vertical and the horizontal walls.

2.2 Wall Stresses

Figure 3(a) shows the module of the shear stress $|\tau| = \sqrt{\tau_x^2 + \tau_z^2}$ (where τ_x and τ_z are the stream wise and the cross-stream components of the shear stress at the bottom wall respectively) made non dimensional with $\bar{\tau}$, the averaged shear stress on the walls of a straight rectangular duct with same dimension of the present one. Large value of shear stress, up to $10\bar{\tau}$, are located in the region near the obstacle edge, but a wide area upstream the abutment exhibits

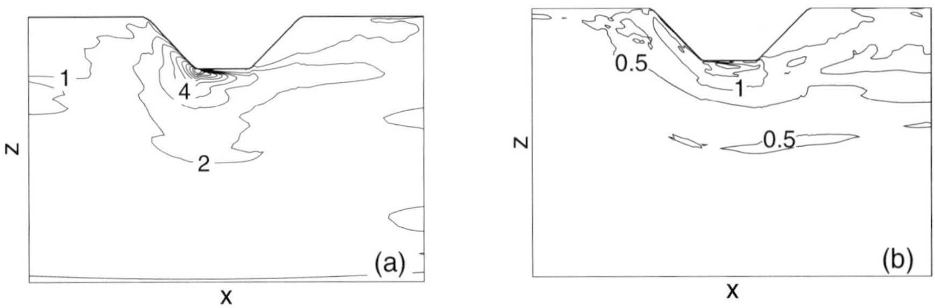

Fig. 3. Non-dimensional shear stress contours on the (x, z)-wall: **(a)** mean values; **(b)** *rms* values

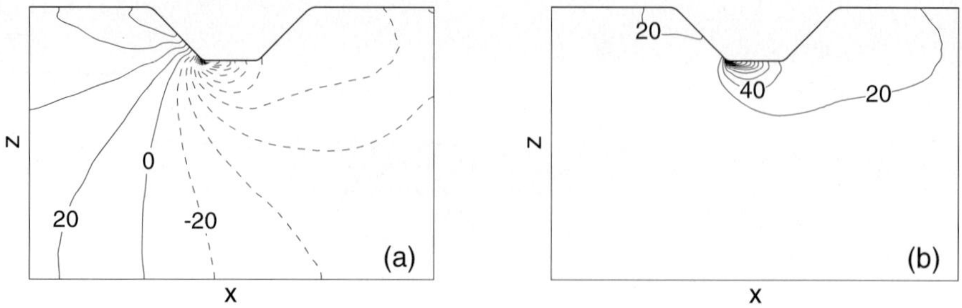

Fig. 4. Non-dimensional pressure contours on the (x, z)-wall: **(a)** mean values; **(b)** *rms* values

values of the shear stress as large as twice $\bar{\tau}$, thus detecting a region with high erosion capability.

Shear stress fluctuations are associated to the intermittency of the large scale structures present in the flow field. Specifically, large values of the shear stress fluctuations on the (x, z)-wall are detected near the obstacle edge as well as at the corner region upstream the obstacle, Fig. 3(b). The maximum value of τ_{rms} is of the order of $\bar{\tau}$.

The distribution on the bottom wall of the non-dimensional pressure $p/\rho u_\tau^2 = p/\bar{\tau}$ is plotted in Fig. 4. The mean values present a minimum at the obstacle edge related to the local acceleration, revealing an uplifting sediment capability of the flow in this region; the pressure fluctuations show maximum values at the obstacle edge but also significant value in the region in front of the obstacle, related to intermittent characteristics of the flow.

The combined effects of wall stress mean values and fluctuations obtained from our simulations result in a region with erosion capability at the edge and in front of the obstacle. In spite of our simulation restrictions (closed duct and fixed bed are not the real conditions of erosion around abutment) these results are consistent with literature experiments, see for example [4], since the scour hole shape nearly corresponds to the highlighted region with high erosion capability.

References

1. Armenio V, Piomelli U (2001) Flow Turbul Combust 65:51–81
2. Chrisohoides A, Sotiropulos F, Sturm TW (2003) J Hydr Engrg 129:177–186
3. Ettema R, Muste M (2004) J Hydr Engrg 130:635–646
4. Kwan RTF, Melville BW (1994) J Hydraul Res 32:661–673

Energy Content of Large-scale Turbulence in Wide Open Channel Flows

Wim S.J. Uijttewaal[1], Wim van Balen[1], and Bram C. van Prooijen[2]

[1] Delft University of Technology, P.O.Box 5048, 2600 GA Delft, The Netherlands
w.uijttewaal@citg.tudelft.nl
[2] Svasek Hydraulics, P.O.Box 91, 3000 AB Rotterdam, The Netherlands
Prooijen@Svasek.com

1 Introduction

Most environmental flows as found in rivers and coastal areas can be considered as shallow. The shallowness gives rise to a strong anisotropy in the turbulence properties since the vertical length scales are much more confined than the horizontal length scales. This is particularly clear when an obstacle that is placed in the flow introduces and extra length scale that can be much larger than the water depth. Under those conditions quasi two-dimensional flow structures can be formed that predominantly move in the horizontal plane. The dynamics of those flow structures is governed by the horizontal shear in a virtually 2D-flow, whereas the bottom boundary layer causes a strong three-dimensional mixing. Previous studies have shown that the interaction of the quasi-2D turbulence with 3D turbulence has two important aspects. First the small scale 3D turbulence causes an enhancement of the effective (eddy) viscosity. Second, the large-scale component of the bottom turbulence has a strong influence on the emergence and intensity of the quasi-2D eddy structures as they are formed in horizontal shear layers [1, 2]. In order to obtain a better view on the role played by the large flow structures we first need to know the distribution of the turbulent kinetic energy over the very large-scale turbulence in a bottom boundary layer [3, 4].

2 Approach

In order to obtain an impression of the spatial distribution of the turbulent kinetic energy in the horizontal plane, a very large domain was used, providing a sufficiently large bandwidth to cover structures that are much larger than the water depth. The properties of a very-wide open channel are calculated using Large Eddy Simulation with a simple Smagorinsky sub-grid model and the Schumann-Grötzbach wall-function [5] for the bottom and a free-slip boundary condition for the free surface. We adopt x as the streamwise, y as the transverse and z as the wall-normal coordinates with associated velocity vector (u, v, w). The horizontal dimensions of the periodic domain comprise 80×80 water depths D covered with a grid of $1024 \times 1024 \times 16$ nodes. In order to check the proper representation of the small scale turbulence in the boundary layer the simulations are validated with DNS-data [6]. Within the range of resolved scales a very good agreement was obtained for the mean velocity, turbulence intensities and spectral distributions of turbulent kinetic energy. This ensures the three-dimensional dynamics to be properly represented. The simulations are performed for $Re_\tau = 395$, $Re_\tau = 590$ and $Re_\tau = 1000$. Here, emphasis is put on the results for $Re_\tau = 590$.

3 Results

In Fig. 1 the energy-density distribution is shown for the depth-averaged velocity fluctuations in the two-dimensional horizontal wave number space for $Re_\tau = 590$. The velocities, and consequently the energy-densities, are made dimensionless with the friction velocity u_τ being equal to unity, balancing the forcing by the pressure gradient. This figure clearly shows the strong anisotropy for the streamwise flow component and isotropy for the transverse component, in accordance with the streamwise elongated flow structures as found in a turbulent boundary layer. The maximum values appear at wave numbers that are of the order of the water depth ($k = \pi$), whereas a significant part of the energy is present on length-scales larger than the water depth ($k \leq \pi$). It should be noted that only through the large horizontal dimensions, such details for small wave numbers can be obtained.

The depth averaging procedure not only provides a smooth spectrum but also removes the information of the sub-depth scales. In order to obtain more information on the vertical distribution of the energy and the relative

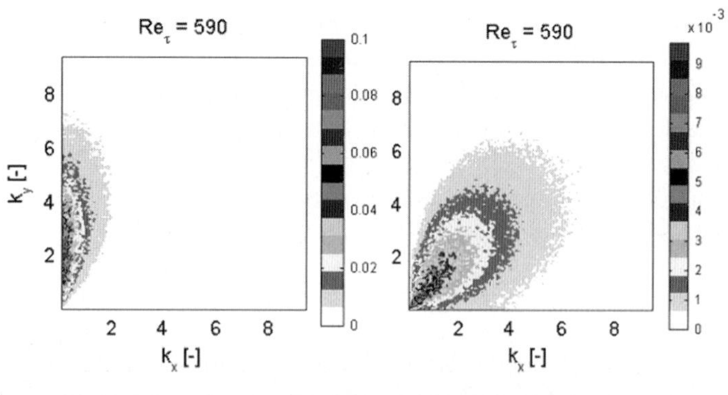

Fig. 1. Two-dimensional energy density spectra of the streamwise velocity fluctuations E_{uu} (left plot) and the transverse velocity fluctuations E_{vv} (right plot)

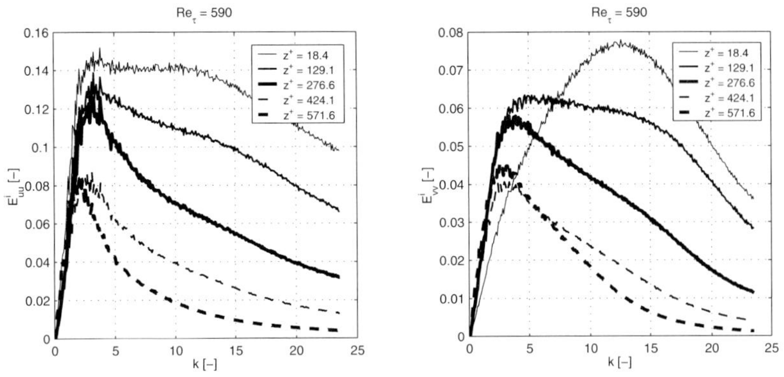

Fig. 2. Two-dimensional energy density spectra of the streamwise velocity fluctuations E_{uu} (left) and the transverse velocity fluctuations E_{vv} (right) for different z^+, for $Re_\tau = 590$

importance of the small scales, the two dimensional energy density spectra were integrated over circle segments of constant k for different vertical positions.

Figure 2, shows (for $Re_\tau = 590$) that the maximum energy densities are again found around the $k = \pi$, associated with the water depth. In the range of the spectral wave numbers $0 < k < \pi$, the energy density increases from zero to the maximum value almost linearly. The curve closest to the bed is most likely affected by the wall function. Higher in the water column the contribution of the small scales to the turbulent kinetic energy becomes smaller, in absolute and in relative sense. Similar results, in particular the linearity on 'supra-depth' length-scales, are found for $Re_\tau = 395$ and $Re_\tau = 1000$.

4 Relevance

An analysis of a shallow mixing layer has shown that shear instabilities attain a high amplitude more rapidly when they grow from disturbances of a comparable length scale containing significant energy [1]. The dominant role of large-scale disturbances in the merging process of quasi two-dimensional vortices was also demonstrated in a recent numerical study by Van Balen [7]. A decaying field of eddies initially disturbed on a large length scale, of the order of the domain size, reaches its equilibrium state much faster and with 50% less energy loss when compared with the same eddy field subjected to small-scale disturbances, see Fig 3. Similar effects were found with forced mixing layers [8]. This indicates that the evolution of the large eddies in a shallow shear flow very much depends on a proper representation of the large-scale bottom boundary layer turbulence. The present study has revealed how the energy content is distributed on the largest scales.

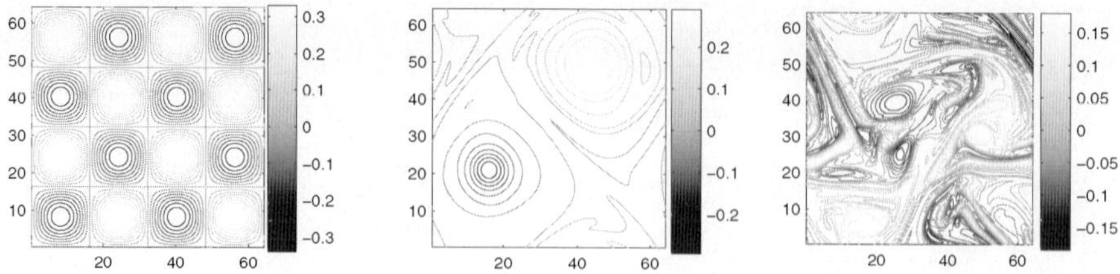

Fig. 3. Evolution of an eddy field ($t = 0$, left), initially disturbed on a large length scale of the order of the domain size ($t = 990$, middle), disturbed on a scale smaller than the initial eddy size ($t = 990$, right). All quantities are made non-dimensional using the amplitude $\hat{u}$ of the initial vortices, the depth δ and the viscosity ν

5 Conclusions

- A large domain resolved on a fine mesh is necessary to obtain a clear view on the large-scale ('supra depth') structure of the turbulent motions in a shallow open channel.
- For $0 < k < \pi$ the energy density increases linearly towards its maximum value located at $k = \pi$, indicating substantial energy being present on length-scales larger than the water depth.
- Instabilities in a horizontal shear flow can be highly affected by these large-scale motions that act as seedlings for the horizontal flow structures.

References

1. Prooijen, B.C. van, W.S.J. Uijttewaal (2002), A linear approach for the evolution of coherent structures in shallow mixing layers, *Physics of Fluids*, 14(12), pp. 4105-4114.
2. Uijttewaal, W.S.J. and G.H. Jirka (2003), Gridturbulence in shallow flows, *J. Fluid Mech.* 489, pp. 325-444.
3. Kim, K.C. and R.J. Adrian (1999), Very large-scale motion in the outer layer, *Physics of Fluids*, 12(2), pp. 417-422.
4. Piomelli, U. and E. Balaras (2003), The inner-outer layer interface in large-eddy simulations with wall-layer models, *Int. J. Heat and Fl. Flow* 24, pp. 538-550.
5. Schumann, U. (1975), Subgrid scale model for finite difference simulations of turbulent flows in plane channels and annuli, *J. Comp. Phys.*, 18, pp. 376-404.
6. Moser, R.D., J. Kim and N.N. Mansour (1999), Direct numerical simulation of turbulent channel flow up to $Re_\tau = 590$, *Physics of Fluids*, 10, pp. 1417-1423.
7. Balen, W. van (2005), *Large-scale coherent structures in turbulent shallow flows*, MSc-thesis, Delft University of Technology.
8. Oster, D., I. Wygnanski (1982), the forced mixing layer between two parallel streams *J. Fluid Mech.* 123, pp. 91-130.

Subgrid Modeling in Large-eddy Simulation of Complex Flows

A.W. Vreman

Vreman Research, Godfried Bomansstraat 46, 7552 NT Hengelo, The Netherlands
bert@vremanresearch.nl

1 Introduction

Two recent papers on large-eddy simulation of complex flows will be revisited and extended: a paper presenting a simply implementable subgrid model that is able to deal with transition and near-wall behavior [1], and a paper introducing the notion of adjoint filtering, which is required to ensure that the basic conservation properties of the Navier-Stokes equations are not violated after general nonuniform filtering [2]. Section 2 extends the model proposed in [1] to account for compressibility effects and presents advanced simulation results for turbulent supersonic mixing layers. Section 3 shows new adjoint modeling predictions for subgrid backscatter in turbulent channel flow.

2 Engineering Subgrid Model Applied to Supersonic Flow

In [1] an eddy-viscosity model has been presented, which is essentially not more complicated than the Smagorinsky subgrid model. Unlike the Smagorinsky model, the proposed model is applicable to transitional flow and vanishes near walls. The model is expressed in first-order derivatives only and, in contrast to the well-known and successful dynamic model [3], it does not involve explicit filtering, ensemble averaging or clipping procedures. The foundation of the model relies on an algebraic classification of all three-dimensional flows. The construction of the model is such that the model and the theoretical subgrid dissipation vanish for the same classes of incompressible flows. In [1] the model was tested for a subsonic transitional mixing layer at high Reynolds number and for plane channel flow. In both cases the model outperformed the Smagorinsky model and was found to be as accurate as the computationally more demanding dynamic model. The range of flows tested for the new model is extended in this section.

into a subgrid model, while the globally dissipative behavior remains ensured, analytically. We illustrate this for the Smagorinsky model, but we could have used another eddy-viscosity equally well, for example the one presented in the previous section.

The adjoint-filtered Smagorinsky model reads [2]

$$\tau_{ij} = -F^a(2C_S^2\Delta^2|s|s_{ij}), \tag{5}$$

where $s_{ij} = FS_{ij}$, $|s|^2 = 2s_{ij}s_{ij}$ and $F = I - G$, $F^a = I - G^a$. G is an arbitrary explicit filter operator and G^a its adjoint. The definition of the adjoint operator partial integration proves that the total dissipation of the latter model is positive for arbitrary G. Nevertheless, the model is able to predict backscatter (locally negative regions of $-\tau_{ij}S_{ij}$). Model (5) is related to the variational multiscale approach [8] with two main differences. Firstly, the latter approach defines the large scales with use of a projection operator, which can always be written as a kernel filter [2]. Another difference with [8] is that in equation (5) the 'small-scale extraction' operator F is applied to the strain-rate and not to the velocity. Only in this way nonuniform filtering does not prohibit a tensorial form of the model. The tensorial form is necessary to use the standard expression for the definition of backscatter [9].

The turbulent channel LES configuration described in [1] was used to test model (5) for a three-points self-adjoint filter G ((52) in [2] was implemented for $\gamma = \frac{1}{2}$) with satisfactory results. Fig. 1 (right) clearly shows that the model predicts backscatter near the wall. The total amount of backscatter is about -13% of the entire subgrid dissipation.

References

1. Vreman AW (2004) An eddy-viscosity model for turbulent shear-flow: algebraic theory and applications. Phys Fluids 16, 3670
2. Vreman AW (2004) The adjoint filter operator in large-eddy simulation of turbulent flow. Phys Fluids 16, 2012
3. Germano M, Piomelli U, Moin P, Cabot WH (1991) A dynamic subgrid-scale eddy viscosity model. Phys Fluids A 3, 1760
4. Nicoud F, Ducros F (1999) Subgrid-scale stress modelling based on the square of the velocity gradient tensor. Flow Turb Comb 62, 183
5. Vreman AW, Sandham ND, Luo KH (1996) Compressible mixing layer growth rate and turbulent characteristics. J Fluid Mech 320, 235
6. Papamoschou D and Roshko A (1988) The compressible turbulent shear-layer - an experimental study. J Fluid Mech 197, 453
7. Vreman AW, Geurts BJ, Deen NG, Kuipers JAM (2004) Large-eddy simulation of a particle-laden turbulent channel flow. In: Friedrich R, Geurts BJ, Metais O (eds) Direct and Large-Eddy Simulation V. Kluwer, Dordrecht, pp271-278
8. Hughes TJR, Mazzei L, Jansen KE (2000) Large eddy simulation and the variational multiscale method. Comp Visualiz Sci 3, 47
9. Piomelli U, Cabot WH, Moin P, Lee SS (1991) Subgrid-scale backscatter in transitional and turbulent flows. Phys. Fluids A 3, 1766

SPRINGER PROCEEDINGS IN PHYSICS

66 **The Structure and Conformation of Amphiphilic Membranes**
Editors: R. Lipowsky, D. Richter, and K. Kremer

67 **Nonlinearity with Disorder**
Editors: F. Abdullaev, A.R. Bishop, and S. Pnevmatikos

68 **Time-Resolved Vibrational Spectroscopy V**
Editor: H. Takahashi

69 **Evolution of Dynamical Structures in Complex Systems**
Editors: R. Friedrich and A. Wunderlin

70 **Computational Approaches in Condensed-Matter Physics**
Editors: S. Miyashita, M. Imada, and H. Takayama

71 **Amorphous and Crystalline Silicon Carbide IV**
Editors: C.Y. Yang, M.M. Rahman, and G.L. Harris

72 **Computer Simulation Studies in Condensed-Matter Physics IV**
Editors: D.P. Landau, K.K. Mon, and H.-B. Schüttler

73 **Surface Science**
Principles and Applications
Editors: R.F. Howe, R.N: Lamb, and K. Wandelt

74 **Time-Resolved Vibrational Spectroscopy VI**
Editors: A. Lau, F. Siebert, and W. Werncke

75 **Computer Simulation Studies in Condensed-Matter Physics V**
Editors: D.P. Landau, K.K. Mon, and H.-B. Schüttler

76 **Computer Simulation Studies in Condensed-Matter Physics VI**
Editors: D.P. Landau, K.K. Mon, and H.-B. Schüttler

77 **Quantum Optics VI**
Editors: D.F. Walls and J.D. Harvey

78 **Computer Simulation Studies in Condensed-Matter Physics VII**
Editors: D.P. Landau, K.K. Mon, and H.-B. Schüttler

79 **Nonlinear Dynamics and Pattern Formation in Semiconductors and Devices**
Editor: F.-J. Niedernostheide

80 **Computer Simulation Studies in Condensed-Matter Physics VIII**
Editors: D.P. Landau, K.K. Mon, and H.-B. Schüttler

81 **Materials and Measurements in Molecular Electronics**
Editors: K. Kajimura and S. Kuroda

82 **Computer Simulation Studies in Condensed-Matter Physics IX**
Editors: D.P. Landau, K.K. Mon, and H.-B. Schüttler

83 **Computer Simulation Studies in Condensed-Matter Physics X**
Editors: D.P. Landau, K.K. Mon, and H.-B. Schüttler

84 **Computer Simulation Studies in Condensed-Matter Physics XI**
Editors: D.P. Landau and H.-B. Schüttler

85 **Computer Simulation Studies in Condensed-Matter Physics XII**
Editors: D.P. Landau, S.P. Lewis, and H.-B. Schüttler

86 **Computer Simulation Studies in Condensed-Matter Physics XIII**
Editors: D.P. Landau, S.P. Lewis, and H.-B. Schüttler

87 **Proceedings of the 25th International Conference on the Physics of Semiconductors**
Editors: N. Miura and T. Ando

88 **Starburst Galaxies**
Near and Far
Editors: L. Tacconi and D. Lutz

89 **Computer Simulation Studies in Condensed-Matter Physics XIV**
Editors: D.P. Landau, S.P. Lewis, and H.-B. Schüttler

90 **Computer Simulation Studies in Condensed-Matter Physics XV**
Editors: D.P. Landau, S.P. Lewis, and H.-B. Schüttler

Printing: Krips bv, Meppel
Binding: Stürtz, Würzburg